UNDERSTANDING BALLISTICS

Basic to advanced ballistics;
simplified, illustrated and explained.

Written and illustrated by
Robert A. Rinker

Mulberry House
Publishing Company
P. O. Box 575
Corydon, IN 47112 U.S.A.

Understanding Ballistics

First Printing Copyright © 1995. This Revised Second Edition Copyright © 1996, By Robert A. Rinker.
All rights reserved. No part of this book may be reproduced or transmitted in any form, including photocopying and recording, except for the inclusion of brief quotations in a review, without written permission from the author or publisher.

Published by
Mulberry House *Publishing Company*
P.O. Box 575
Corydon, Indiana 47112 U.S.A.

ISBN: 0-9645598-1-1

Library of Congress Catalog Card Number: 96-94109

Printed and Bound in the United States of America

Revised Second Edition
Third printing

Printed in the USA by

Morris *PUBLISHING*
3212 E. Hwy 30
Kearney, NE 68847
800-650-7888

ACKNOWLEDGMENTS

In preparing this book, I have received invaluable help from several sources.

Mr. William C. Davis, Jr. of Tioga Engineering Co., Wellsboro, Pa., not only gave his generous permission to reprint some of his work from *The American Rifleman* magazine, he even corrected and revised one of the tables. Mr. Davis is highly qualified and his help is greatly appreciated.

I also received permission to reprint some material from an excellent magazine, *American Rifleman*, a publication of The National Rifle Association. Many thanks to Ron Keysor, the Managing Editor at the time.

With their kind permission, I have also copied some facts published by Winchester/Olin Corp. Winchester is certainly one of the best and it is a pleasure to give credit where it is due. Also, a chart each is credited to the Federal Cartridge Co. and the Sporting Arms and Manufacturer's Institute. It is impossible to report on a subject as complicated as firearm ballistics without the research of the manufacturers.

I have also made frequent mention of various government agencies and military ordinance groups. Their vast resources are invaluable in these areas of research. For an additional benefit, the results are generally available to the public.

I wish to thank Betty E. Rinker for reading the manuscript and pointing out errors in spelling and poor wording.

Last, but by no means least, I want to thank one of my grown children, Teresa. She not only spent many hours reading the manuscript in the pursuit of errors, she has offered invaluable advice and moral support.

To all of these people and organizations, I wish to express my deep appreciation and thanks.

TABLE OF CONTENTS

Acknowledgments	Page	III
Disclaimer	Page	V
Forward	Page	VI
Chapter 1 A Basic Start	Page	1
Chapter 2 Mathematics	Page	7
Chapter 3 Basic Laws of Physics, Newton & Galileo	Page	12
Chapter 4 Primers, Powder & Gas Expansion	Page	18
Chapter 5 Interior--Basics & Pressure	Page	40
Chapter 6 Recoil	Page	59
Chapter 7 Basic Steel Properties & Applications	Page	73
Chapter 8 Gyroscopic Stability	Page	81
Chapter 9 Spin--Gyroscopic Drift	Page	96
Chapter 10 Ricochets & Brush Deflection	Page	105
Chapter 11 Barrels	Page	109
Chapter 12 Rifling--Twist	Page	119
Chapter 13 Velocity & Drag	Page	139
Chapter 14 Ballistic Coefficients	Page	159
Chapter 15 Cartridge Details	Page	174
Chapter 16 Bullet Design & Performance	Page	182
Chapter 17 Trajectory	Page	196
Chapter 18 Minute of Angle & Evaluation of Target Groups	Page	223
Chapter 19 Maximum Range	Page	231
Chapter 20 Wind Deflection	Page	237
Chapter 21 Shotgun--Pellet & Buckshot Ballistics	Page	250
Chapter 22 Handguns	Page	280
Chapter 23 Air Guns to Muzzle-loaders to Cannons	Page	290
Chapter 24 Terminal Ballistics, General	Page	296
Chapter 25 Terminal Ballistics, Game & Hunting	Page	312
Chapter 26 Terminal Ballistics, Self Defense	Page	317
Chapter 27 Miscellaneous Ballistic Information	Page	335
Chapter 28 Glossary of Ballistic Terms	Page	347
Index	Page	365
About the author	Page	373

PLEASE READ THIS DISCLAIMER.

Neither the author nor the publishing company accepts any responsibility for accidents, injuries or any other problem arising from the use of data reported in this book or any part thereof. Although every effort was made to be accurate, much of the information is a result of experience and tests that go back numerous years by many people. Apply this data with common sense and caution. Use of firearms involve many variables, any one of which could influence the ultimate result and possibly cause injury or death. Therefore, we must disclaim all possible liability, consequential, incidental, actual or otherwise, for any damages or injuries caused by the use of this information which is supplied without representation or warranty of any kind.

The technical data in this book is of value to police, hunters, target shooters, reloaders, trap shooters, experimenters and research people. Nothing printed here is intended to encourage criminal activity or the unlawful killing of animals.

FORWARD

"A knowledge of ballistics will increase our proficiency with firearms." It is with that idea in mind that this book was written. A desire to increase the skill and ability of the sportsman, the police officer, and the honest citizen who owns a firearm for self-defense.

Personal opinion, as opposed to proven facts known to the scientific community, have no place in a book of this type. Many writers sprinkle them liberally throughout the text with "I" this and "I" that. This book will not have "I" used in that context except in this *forward*. The reader is not interested in my personal opinion; only the facts as determined by the basic laws of mechanical physics, testing and research.

As an engineer, I have found ballistics to be a very frustrating area of science. There are so many variables that it is almost impossible to use formulas that are based on known laws of physics without also including information based on actual tests. It is difficult to make positive statements that are not backed up by some form of testing. The expression, *if all else is equal,* or something similar, will be used frequently in regard to tests. And getting *all other things equal* is not as easy as it sounds because it has to be done in a literal sense, not a figurative one.

First, let us forget the idea that ballistics, even advanced ballistics, is too complicated for the average person. If explained properly (and this book will try hard to do that) it is within the ability of almost anyone.

Second is the fear that ballistics can not be understood without mathematics. Wrong again. The subject can be understood with no math involved what-so-ever. Math is included in this book if the reader wants to calculate an answer to an exact amount. Also the math helps in comprehension and understanding, but it can be skipped if the reader is not interested. Chapter 2 provides guidance in this area. And remember, a cheap calculator can make a math whiz out of almost anybody. Even if the reader does not understand the math and has no interest in learning it, casually read through it anyway. Knowledge and more understanding will be gained, even if the reader doesn't realize it at the time.

To fully understand any section of this book, a knowledge of other areas may be required. The subjects are all intertwined and meshed together. Like a jig-saw puzzle, each piece depends on several other pieces and many have to be in place before the picture can be seen.

Most of the information presented is not in doubt or dispute. People may quarrel or bicker over technical areas they do not understand, but it is, nevertheless, fact not fiction. A few areas, notably terminal ballistics and brush deflection, are hotly argued. This book will always try to give the correct answer to each problem even if public opinion is counter to that answer. A democratic system does not work in scientific technology. A poll where 80% believe one answer is correct and 20% another does not make the 80% answer correct.

Probably 98% of firearm ballistics is not controversial. It is based on scientific facts that have been proven by tests conducted on the firing range and in the laboratory. A few areas are subject to different beliefs, depending on which authority is giving the opinion. Luckily, there are very few points with inexact and undetermined answers. All subjects that fall into this category will be identified. In other words, if it is controversial, that will be pointed out. A lot of text, sketches and mathematics is in the form of proof. Some readers will take what is said on faith and others will need evidence and confirmation.

Books, articles, and bar room conversations are full of errors that will be corrected. For the most part, people will not intentionally make false statements, but few in the general public have the engineering and technical background to realize their mistakes. If one person says or writes something that sounds reasonable, most will believe it and pass it on as the truth.

Wording that explains everything to everyone's satisfaction is a concern. The reader with a Ph. D. will desire a different style of writing than a gun enthusiast with an eighth grade education. Yet both have an equal status. All readers may not have equal knowledge about firearms. Some, in fact, may have never fired a gun and others may be experts. Nevertheless, it is important to use words that properly describe things. Most readers can follow a text even if a few of the words go over their head.

Not all words are used in science and technical writing in the same way as in common usage. Also, all special subjects have their own slang and ballistics is not an exception. Notice **there is a glossary of ballistic words and terminology near the end,** just before the index.

We all have strong opinions. That is our right. It is also our responsibility to change our position if it is overcome by intelligent facts to the contrary. Many people are not up to this challenge. These controversial areas will be covered with objectivity and without my personal opinion.

Fortunately, as Cyrus Gordon reminded his readers in *Before Columbus,* "Ignorance is a curable disease."

Our knowledge increases each year. Yet what we know is as trivial as a whisper in a vast desert. How much wisdom is ignored because it does not fit into our pattern of accepted thought? How much is looked at but not seen? Read with an open mind. When the door to knowledge is opened even the smallest crack, the light of curiosity will tempt us to open it wider.

Robert A. Rinker

CHAPTER 1

BASIC HISTORY & FIREARM OPERATION

Some people believe that it is just as well to ignore ballistic details and concentrate on shooting. Intelligent sportsman and police officers believe it is best to help the odds by learning as much as possible and also concentrate on shooting. Both are important.

BASICS

The phenomena that is ballistics is divided into three independent areas. All three are intertwined and meshed together by their physics but they are divided by their location. Some students of ballistics believe that it is filled with inexact and uncertain details, but the laws of nature cover all three very well. The basic laws of momentum, deceleration, gravity, etc. cover all the actions completely. That isn't to say it is simple. Far from it.

(1) Interior ballistics covers what takes place inside the firearm. This concerns powders, ignition, bore friction, pressure, etc. While the facts are just that, facts, they can not easily be seen. Therefore, they have to be taken by the average reader, on faith. **(2) Exterior ballistics** takes over as soon as the projectile leaves the barrel. It covers the flight of the bullet from muzzle exit to impact. **(3) Terminal ballistics** is the final section. This term, although incorrect in some respects, is used for the science of how the projectile and target (paper, game or human in self defense) react at impact and in the following seconds or minutes.

Some of these subjects can be confusing because not everything in the world is as we see it or believe it to be. Multi-colored rainbows and our own mirror-reflections are clearly visible and can be photographed, yet from a substance standpoint, neither one exists.

The total number of physical factors that are involved in the problems of interior and exterior ballistics is almost infinite. The elimination of most of them is warranted only by careful testing and computations.

A BRIEF HISTORY OF BALLISTIC STUDY

Much can be learned from the history of the science of ballistics. Its different eras are marked by many ingenious and elaborate experiments and by investigations of the highest value. Like most other sciences, the theories and methods of one period have been found inadequate for the problems of later times. Occasionally this has led to entirely new mathematical techniques for handling the problems.

The practical handling of artillery and small-arms trajectories extends back not only to the first use of gunpowder as a propellant, but to the crossbow and the catapult.

Prior to the early 1500's, references to the shape of the trajectory speak of the first part as a straight line, the middle part as the arc of a circle, and the last part as straight again.

The science of ballistics is usually considered to have been started by the physicist Nicholas Tartaglia, who published a treatise on the flight of projectiles in 1537. He demonstrated that a trajectory is curved throughout, attaining a maximum range at an angle of departure of 45 degrees. (He was wrong.)

Galileo in his fourth *Dialogue on Mechanics* demonstrated that the path described by a projectile, being the result of a joining of a uniform transverse motion with a uniformly accelerated vertical motion, must be a parabola with a vertical axis. (Don't give up if the big words throw you. They will all be explained later and the subject will become clearer. Besides, Galileo was also wrong.)

Newton was the first to take into consideration the resistance of the air. As a deduction from basic physical laws, he showed that when the velocity is so little in comparison with the mobility of the air as to make it possible to ignore the diminution of pressure in the rear of a projectile, and the stream line motion of the air generally, then the resistance must vary as the square of the velocity. To test these basic principles and to determine the constants involved, he dropped from the dome of St. Paul's Cathedral, various weight spheres filled individually with air, water, or mercury, thus repeating Galileo's research only with new refinements. Newton's results, while in themselves satisfactory, were soon found not to be directly relevant because they dealt with ideal conditions not found in the real world. (As we shall see later, he wasn't completely right either. But he was getting closer; especially at low velocity.)

Instruments in the form of rotating vanes, similar to the anemometers employed by the Weather Bureau, have been used as the basis of computations by numerous investigators, such as Borda, Didion, Hutton, Morin, Piobert, Thibault, etc.

Benjamin Robins published his *New Principles of Gunnery* in 1742. It was based upon results obtained from using his *ballistic pendulum*. This was the first device to record, with any degree of accuracy, the velocities of projectiles. While Newton's results were confirmed at velocities under about 900 f.p.s., Robins was able to measure velocities up to 1,700 f.p.s. The discrepancies between observation and Newton's *square law* became enormous. Further extensive measurements by the ballistic completed by Hutton in England in 1791 and by the Commission de Metz in France (Piobert, Morin, Didion) in 1839-1840.

In 1840, the practical use of the ballistic pendulum became obsolete. That was when Wheatstone suggested measuring the velocity of the projectile through time intervals. This was recorded by means of electric contacts affected as the projectile passed through successive screens at carefully measured distances.

Mathematical solutions for air resistance have gone through various historical periods. For a long time, air resistance was ignored. No one knew of its existence. Then the scientists of their time started using complete mathematical solutions obtained by Bernoulli's method. This was incorrect. Next were methods that reached approximate solutions through numerical integration and other ways that are explained in Chapter 14. A few other methods were tried with little success until today, we use variations of what is actually an old turn of the century method, updated by the use of modern computers.

FIREARM OPERATION -- All explained in detail in later chapters.

Only a few hunters or target shooters realize the thousands of complex chemical and physical actions that occur in the small fraction of a second between the firing pin striking the primer and the projectiles impact. It almost seems as if it is instantaneous.

There are a series of actions or operations, that every firearm must go through. A .22 caliber revolver or a .50 caliber machine gun. The order may be different and some of the steps may be performed by hand and some by an automatic action of the mechanism, but the end result is always the same. The ejection of a projectile on an arc known as a ballistic curve toward the target. (No, it is not a parabolic curve, but we will discuss that later in Chapter 17.)

The sequence of events usually begins with pulling the trigger. Through mechanical means, the firing pin strikes the primer. This creates a tiny explosion. Not shock but heat flashes through the hole and ignites the powder. The powder does not explode but burns very rapidly. The surface of the grains begin to change from a solid to a gas. This process rapidly spreads inward. The nitrocellulose molecules separate from the composition and create smaller molecules that move away at high speed. These rapidly moving molecules exert a pressure against each other and the cartridge walls, chamber walls, bore, the projectile's base and everything else they come in contact with. The projectile will start to move because the pressure is greater than the holding force of the case neck. This may push the primer backwards in the pocket while the case is held forward. As the powder burns, the pressure and temperature rise rapidly. The pressure increases and the neck and body walls of the case expand to meet and grasp the inside chamber walls. The case will stretch some lengthwise with the thinnest side wall, if one exists, stretching the most both lengthwise and radially. The bullet is pushed out of the case by the pressure and into the rifling in the bore. The molecules continue to move rapidly and this continues to push against the base and to accelerate the projectile. The hot gas extends evenly through the bore behind the projectile as it expands to over a 1,000 (thousand) times the volume of the original powder charge. The average velocity of the molecules begins to decrease because of momentum lost to the projectile. The frequency of molecular collisions and their temperature and pressure are reduced by the ever increasing volume they must occupy. Meanwhile, the projectile continues to accelerate as it rotates it's way toward the opening at the muzzle.

If the peak pressure is low, the cartridge brass will not have taken a set and will return to almost its original state as the pressure drops. If the peak pressure is more than the elastic limit of the brass, the case will hold its new size. The elastic limit is the greatest stress that can be applied without leaving a permanent deformation upon release of the pressure.

The case head will move back and hit the bolt face or breech face. If the head is not square with the case walls, it may or may not be permanently squared, depending on the head-space, temperature in the chamber, peak pressure, and the modulus of elasticity and resilience of the brass case. This impact of the case head may contribute to a barrel whip.

That is a lot to happen in such a short time, perhaps about .003 second. (3 thousandths of a second.) More or less. **Later, for an interior ballistics study, we will break these items down and study each on its own and how they relate to each other.**

Throughout this rapid sequence of events, the breech would have been locked to keep the pressure from escaping in any direction except out the muzzle. The breech must be unlocked by a pump handle, bolt or some other mechanical operation. It may be either by hand or automatic. The case will have to be extracted, and the case will have to be ejected. These seem like one operation but they are really two separate actions that may be performed with one as the continuation of the other. Somewhere in this sequence, the firing mechanism should be cocked and made ready by storing energy in the firing pin for the next shot. Another cartridge could be fed into the chamber to replace the ejected case. The breech-block must again be locked to retain the hot expanding gases that will be developed when the gun is again fired.

The shooter feels a kick to his shoulder or hand as the gun recoils from the reaction of the projectile. This recoil is a precise measurement that can be stated in foot pounds of energy, which may or may not be the same as the kick the shooter feels.

The projectile leaves the muzzle for the exterior flight toward the target. Its curve is acted upon by gravity and air resistance and affected and controlled by gyroscopic action imparted by the spin from the rifling in the bore. It is influenced by nutation, yaw, spin drift, the earth's rotation, and dozens of other phenomena that most sportsmen, hunters and other gunners have never heard of. Wind tries to either blow it to one side or speed it up or slow it down, depending on the direction it is blowing from. Momentum is both its savior and its enemy.

If an object such as a twig is struck by the projectile as it flies toward the target, it may ricochet off in a new direction or shatter into pieces. It also could break off the twig and continue on as if nothing had happened.

The laws of nature that affect and exert control over a bullet from a high-powered rifle also have the same influence over a handgun bullet, a shotgun slug, a shotgun load of pellets, or about anything else that is launched or fired on a trajectory.

At the target, the flight may be over, but the work is not finished. The main purpose of the trip is yet to be done. The bullet has to penetrate and expand as the job requires. For a paper target, it may be easy and the job about finished. If the object is to stop a large grizzly bear that is attacking someone, there is a lot of serious work yet to do. And it has to be done fast and efficiently with no room for errors.

Only a few of the hundred or more actions on a projectile are mention here because they all will be discussed in the following chapters. Frankly, to list any more here is a waste of time and space. The main thing to remember at this point is that not only is there much to study, but it is not all as it first appears to be. That is not to say that any of the laws of physics or chemistry are strange or mysterious. Not at all. But nature does not always do what we want or expect.

VARIABLES

No two guns are exactly alike. Nor are loads of powder or bullets or primers or anything else that is connected with ballistics. The variables are why ballistics is not as precise or accurate a science as we wish it were. The mathematics may be precise, for example, but the numbers fed into the equations are based on variable amounts. That is why the figures for your gun or situation may be different from someone else's, and yet both can be correct.

A BRIEF FABLE

As a method of slowing easing into the scientific study of ballistics, let's review a short scene that will demonstrate how a knowledge of ballistics can be helpful.

It was a cold morning in the high hill country with a light drizzle falling. Steve was chilled to the bone. If it was a few degrees colder it would freeze or snow. The top of the mountain was hidden by mist like a veil conceals the face of a grieving widow.

Slowly the hunter looked over the cliffs. After what seemed like a long time, he spotted his game standing on a bluff high above him but still well below the wind blown mist. Steve knew the game hadn't seen him. He also knew it wouldn't smell him because there was a strong wind blowing down the valley from his left which caused the rain to wet his left side. A high mountain blocked the wind from the game and the last three fourths of the range, but Steve could feel the wind was strong where he was. He shivered from the wind chill as he raised his bolt action .30-'06 rifle. A quick guess told him the range was 300 yards across with another 300 yards up. He thought the elevation angle looked like about 45 degrees.

Steve's heart started to beat faster from excitement. He forgot that he had a piece of plastic electric tape stuck over the muzzle to keep out moisture and dirt. As he placed the rifle butt to his shoulder and looked through the scope, he wondered if he was correct in trusting ammo that had been stored for ten years and then he remembered that he had not cleaned the barrel after sighting in the gun the previous day. Strange what a person will think of during stressful moments.

Steve had sighted the gun at 300 yards so he figured the 45° angle was farther and raised the cross-hairs about 4" for the extra drop. Then he aimed 5" higher because he wanted to offset the gravity of the steep up-hill angle for a total allowance of 9". At the last minute he remembered the left cross wind but he decided it was only blowing strong near him and not on the rest of the range, so he made no correction for what he thought was a 10 m.p.h. wind. As the cross hairs lined on his aim-point he inhaled a slow breath of cold air into his lungs. He held it for a second and then, as his finger tightend on the trigger, the rifle butt kicked his shoulder.

That is enough. Think about it and decide if he would score a direct hit on his trophy or perhaps a wild shot that misses. Then check to see how you would do after reading the book. No cheating. Read the book first, not the answer. In Chapter 27, after these subjects and many more have been discused, we will determine how Steve would have made out.

CHAPTER 2

MATHEMATICS

The casual reader may disregard the mathematics that is included in this book and still gain a great amount of knowledge. So why is the math included? There are two main reasons. **First,** readers that can follow the math will gain a deeper understanding of ballistics. **Second,** there may be occasions when the reader will want information that can be obtained by working out a problem. Wind drift and rifling twist are two common problems that are easy to solve.

In every case, the mathematics may be omitted if the reader so wishes. The text itself will give a good understanding of each subject area. When there is math presented, as there will be in most sections, it will usually be explained in easy terms. An attempt will be made to show the reasons behind each step instead of just listing the formula with no background or details. **Every effort will be made to give the reader an understanding of the material without the math. But, it the reader studies and understands the math, not only can he or she obtain answers to questions, it will also give a deeper grasp of the subject.** In many instances, the math confirms that a statement or comment is a proven truth rather than a theory.

With a few exceptions, all the mathematics in this book is easy if a scientific calculator is used. Easy enough that even if you flunked math in school, you can still work the problems. If you don't have a scientific calculator, it would be advisable to purchase one. It is not necessary to have a top of the line expensive model. A few dollars spent at a discount store will do. A regular calculator is a great help, but the scientific models will do things with a key stroke that with the others would require tables and tedious calculations. A scientific model will enable someone who doesn't understand trigonometry to do trig. problems. The book with the calculator is valuable. Refer to it as needed.

The math formulas supplied are usually the easiest that will do the job. Sometimes great accuracy is required and if several formulas are available the easiest will not always be as accurate. In this case, two or more formulas will be given as roads to the same answer. Rules of thumb and their simplistic math are included where appropriate.

The mathematician and astronomer Carl Friedrich Gauss, the German genius who dominated the field in the 19th. century, said, "Nothing shows the lack of mathematical education more clearly than excessive accuracy in calculating." There are practical limits that vary from problem to problem, but in each case, there is no need to exceed them. (Although most of our examples exceed what is necessary.)

BULLET WEIGHT

To convert from pounds to grains or from grains to pounds is common in calculations involving ballistics. The formulas will not work if the wrong numbers are used and yet this is a frequent mistake. If you have pounds, multiply by 7,000 to obtain grains. If you have grains, divide by 7,000 to obtain pounds. With even the cheapest calculator it is very simple. The hard part is remembering to do it. **Example:** 200 grains divided by 7,000 = .0285 pounds.

CONSISTENT UNITS

All units used in a problem must be consistent. This is one of the major causes of wrong answers. If the person is lucky, it will be by such a huge amount that an error is obvious. If the formula calls for grains, pounds will not work. If yards are needed, feet will not give the correct answer.

FEET PER SECOND AND MILES PER HOUR.

A fast way to convert feet per second (f.p.s.) to miles per hour (m.p.h.) is to divide the f.p.s. by 3 and multiply the answer by 2. The result will be within 3%. Need the answer dead on? Multiply the f.p.s. by 0.6818 to obtain m.p.h. or by 0.0114 for miles per minute (m.p.m.). **Example:** 900 f.p.s. divided by 3 = 300 times 2 = 600 m.p.h. **Example:** 900 f.p.s times 0.6818 = 613.6 m.p.h. **Example:** 900 f.p.s. times .0114 = 10.26 m.p.m.

For readers who prefer a chart:

f.p.s.	m.p.h.	f.p.s.	m.p.h.
800	545	2000	1364
900	614	2100	1432
1000	682	2200	1500
1100	750	2400	1636
1200	818	2600	1773
1300	886	2800	1909
1400	955	3000	2045
1500	1023	3200	2182
1600	1091	3400	2318
1700	1159	3600	2454
1800	1227	3800	2591
1900	1295	4000	2727

CONVERSION TABLES FOR BALLISTICS

Multiply	ounces	by	437.5	to	obtain	grains
"	pounds	"	7000	"	"	grains
"	grams	"	15.43	"	"	grains
"	grains	"	0.0648	"	"	grams
"	degrees	"	60	"	"	minutes
"	degrees	"	3600	"	"	seconds
"	feet	"	0.3048	"	"	meters
"	feet	"	0.3333	"	"	yards
"	feet/sec.	"	0.6818	"	"	m.p.h.
"	meters	"	3.281	"	"	feet
"	meters	"	1.094	"	"	yards
"	miles/hr.	"	1.467	"	"	feet/sec.
"	yards	"	0.9144	"	"	meters

SYMBOLS

The use of computers has changed the symbols used to alphanumeric characters. An asterisk (*) is used instead of an (x) to indicate multiplication. A slash mark (/) indicates division. The number to the left is the numerator (top number) and to the right is the denominator (bottom number). At first, this may look strange. Just remember that when you see a * , it means to multiply the same as if the symbol had been an x.

While we are discussing fractions, an inverse proportion or to vary inversely are just fancy words that mean as one quantity increases, the other decreases. As one goes up, the other goes down or the other way around.

A later section has a few formulas that use the Greek sigma. (σ) If you are not familiar with it, don't be concerned. It was established by Swiss mathematician Leonard Euler as the symbol for the sum of a *finite* number. Sum is the key word as it is used to indicate the process of adding. When working on problems involving average or central tendency, it is used for cumulation (successive additions).

SCIENTIFIC NOTATION FOR HUGE NUMBERS

Frequently the numbers involved are long with a lot of zeros. Many calculators cannot accept and deal with over 8 numbers and that is not enough for some of these problems.

There is a method to handle extra long numbers called *scientific notation*. The huge number is expressed as a *mantissa* or base number times 10 and raised to an exponent or power. Confused? It means to take the base number times 10 and raise to a power by multiplying it by itself the amount of times the power indicates. The decimal point is placed the number of spaces as the exponent. A + or - can be used for positive or negative numbers. Positive shifts the decimal to the right and negative to the left.

Example: (Mantissa * 10 power.) $3.8895 * 10^8 = 388{,}950{,}000$.
The decimal is moved 8 places to the right (positive) adding zeros as required.

Example: $3.7982 * 10^{-16} = .000\,000\,000\,000\,000\,37982$
The decimal is moved 16 places to the left (negative) adding zeros as required.

Most modern scientific calculators have keys for scientific notation. Check your manual or instruction book.

The following are a few formulas using scientific notation.

Energy $= 1/2 \ mv^2 =$ **(bullet weight)** $v^2 / 4.51 \ (10^5)$

momentum $= mV =$ **(bullet weight)** $v / 2.25 \ (10^5)$

Don't let that scare you. First, it is not as hard as it looks and second, both of those formulas have much easier versions in their own chapters. And to repeat what has already been said; even if you cannot add 2 plus 2 and get 4, **most ballistic science can be understood without the math, but it must be included for the readers that want it. Sort of like pickles on a sandwich. Most places give them to you, but if you don't want them, you may skip them.**

Formulas may be expressed in different ways. If the proper numbers are put into the formula and it is worked correctly, then the answer will be what is requested. But there may be different ways of expressing the same thing. Formulas can use different information or equations can be combined together. Therefore, we could arrive at the same place by a different road. For example; we can say the sine of angle x is y. We can also say $y =$ sine x. Both look and sound different but they say the same thing.

TRIGONOMETRY

Many students panic at the thought of studying trigonometry. They hear stories of how difficult it is and that it is impossible to understand. This is not true for most people. Especially as we work the problems within this book. With a scientific calculator, most trig. problems can be worked without understanding the theory by inserting the proper numbers in the equation and pressing the correct keys. Except for introducing a few new words and using some strange looking letters from the Greek alphabet, trigonometry should create no problems. The computations are, for the most part, just simple algebra or arithmetic. The formulae are important and must be followed carefully in the proper order and with the proper units.

While it would be nice if a few of the readers who know nothing of trig. would become interested and want to learn more, the main idea is to be able to work the problems in this book. With a calculator and a little thought, this can be done.

Trigonometry is simply the study of triangles and the relationship or ratios between the sides and the angles. It can also be applied to other problems in which angles or directions are involved. For example, if a gun is pointed in a direction 15 degrees east of north and the barrel is at an angle of 22 degrees with the horizon, trigonometry would be involved in any calculations we would do. These ratios are represented by the cosine (cos), sine (sin) and tangent (tan). These change in numerical value as the size of the angle changes.

Angles are kept separated by using Greek letters to denote them. Lower case *theta* θ and *beta* β are examples. Each time these are required, it will be explained and easy to understand and use.

CALCULUS

As the bullet arcs toward the target, at any given moment the bullet is moving in a definite direction. In respect to the ground, this will be an instantaneous slope and a change in the bullets altitude. This can be regarded as changes with respect to the elapsed time since the bullet left the muzzle. At any specific point, we can discuss the relationship between the distance and altitude and time. In calculus, this would be a derivative (differential coefficient) and written as simply dy/dt, where t = time.

A derivative, or limiting value of a ratio, is the fundamental process with calculus. The logic and mathematics of calculus can be used on problems of time, points on a curve and many other ballistic problems. It can be described as the developing difference between the situation at one moment and the situation at the next moment. This gives clues or evidence of how a situation is shaping up. If the ratio of the net changes that take place is judged as a limit neared as the space between the moments approaches zero, then the limit shows how rapid the situation is developing.

This same rule is applied to the classic equation for gravity and acceleration of a falling object which is discussed later. Calculus permits us to stop motion and break down movement into points that can be traced through both space and time. We can calculate the velocity and acceleration at a specific moment.

Students that have not studied calculus think it is the ultimate math. Something to be feared and approached with caution. A few of the problems in this book involve basic calculus but they can be solved by just following the directions. If the reader is interested in learning more, the libraries and book stores have excellent material on mathematics.

CHAPTER 3

BASIC LAWS OF PHYSICS - NEWTON & GALILEO

Every aspect of ballistics is governed by the laws of physics and mechanics. Later, we will be discussing trajectory and drag in terms of physics. Rifling deals with spin and moment's of inertia. Almost every single area of discussion is involved in one or many fields of basic physics. This brief section will touch on a few of them now, while other areas will be covered in the section where they are involved.

Like mathematics and gyroscopic stability, which are elsewhere in this book, this must be understood to really grasp ballistics. **If you believe you know it, feel free to skip this section.** If you don't understand it and don't read it now, then you may have to come back and read it later.

This is not intended to be a course in physics. This is basic, simple and easy to understand. As with the chapter on mathematics, readers who have a good background in physics may skip this section. Most people, if they are honest with themselves, will need it. Relax. It isn't that hard.

THREE BASIC LAWS

Much of the theory behind firearm ballistics depends on the discoveries and physical formulations of English mathematician, physicist and astronomer, Sir Isaac Newton (1642-1727)

This is the basis for much of our study. Sir Isaac Newton's three basic laws of motion which were listed in the last half of the 17th. century.

1st. law: "A body at rest tends to remain at rest, and a body in motion tends to remain in motion at the same speed and in the same direction unless acted upon by a force."

This means that nothing stops or starts moving unless some force makes it do so. For example, the bullet won't start in motion until the expanding gas pushes it and once moving it will do so until something overcomes its inertia.

2nd. law: "When a body is acted upon by a constant force, the resulting acceleration is inversely proportional to the mass of the body and is directly proportional to the applied force."

This is not as hard as it sounds. It deals with overcoming the 1st. law. Acceleration, in this case, can be both positive as in starting up and negative as in deceleration and stopping. In effect, it says that **force = mass * acceleration.**

3rd. law: "Whenever one body exerts a force on another, the second body always exerts on the first a force which is equal in measure but opposite in direction."

This is stated more commonly and briefly as "for every action there is an equal and opposite reaction." The fundamental principle behind the movement of a jet airplane or how a gun may recoil and give you a *kick*.

KINETIC ENERGY and GRAVITY

Energy is the capability to do work. A body may possess this capability through its position or condition. When a body is held so that it can do work if released, it is said to possess energy of position or potential energy. When a body is moving with some velocity, it is said to possess energy of motion or kinetic energy. The fundamental law of *Conservation of Energy* states that energy can be neither created nor destroyed. There are different types of energy and energy can, however, be transferred from one type to another and from one body to another.

Briefly, a foot pound (ft. lb.) is a unit of kinetic energy but that information helps little at this point. Let's look at it from the basic facts.

Originated by Italian physicist and astronomer, Galileo Galilei in 1585 with a simple equation, $y = 16t^2$. This shows how gravity acts on a free falling object; a rock, a bullet or a cannon ball. Y is the distance fallen in feet and t is the elapsed time in seconds from the start of the fall.

Newton, by differentiating the equation once, determined that the speed an object is falling at any moment equals about 32 times the number of seconds which it has been falling. Differentiating the equation a second time, he determined that the objects acceleration is always 32 feet per second, every second. This does not change and represents a law of nature.

When an object falls, starting from a dead stop, it accelerates, that is gathers speed, at the rate of 32.17 feet a second for each second. Therefore, after one second, the falling object (what it is doesn't matter) is dropping at 32.17 feet per second. After 2 seconds, 64.34 f.p.s. After t seconds it would be t times 32.17 or 32.17t. Of course the object will not fall 32.17 feet the first second as it starts at 0 and has to accelerate, but it will increase in speed at 32.17 feet for each second.

◆ Velocities are given in feet per second (f.p.s.) and accelerations are given in feet per second per second. (2) A small but important difference.

◆ The sum of the kinetic and potential energies of a body acted on by gravity alone is constant. If a body is at a certain height above the ground and is motionless, its potential energy is equal to the work done in raising it to that height. If that body falls to the ground, its velocity v will be the **square root of $2gh$,** and its kinetic energy will be $wv^2/2g$. The potential energy before starting to drop is, therefore, the same as its kinetic energy when it reaches the earth. At any point during its fall, its total energy, obtained by adding the kinetic and potential energies, will be the same

$wv^2/2g$, where v is the velocity that the body would have if it fell freely to the earth from the original height.

◆ The final velocity of a falling body is proportional only to the vertical distance through which it falls, and is completely independent of the path it follows. In other words, it does not matter if a body (or projectile) falls straight to earth or falls in a different path because of an outside force. The time may be more, but the final velocity will be the same.

◆ If a body is thrown downward, (or a projectile is fired downward) to the constant force already found must be added the impulsive force given the body. This is proportional to the velocity imparted and the time of its action. In other words, the velocity at impact would be the total of the two.

◆ If a body, (or a projectile) is launched upward, the direction of the body is opposite to gravity and its velocity will be diminished each second by the quantity $g = 32.17$. Therefore, the time of rise will be found by dividing its original velocity by g.

Remember, in all of these cases the actual figures are never realized in practice because of air resistance and other variables.

◆ If we raised a 50 lb. weight to 10 ft., we would expend 500 ft. lbs. or a 10 lb. weight to 30 ft. we would expend 300 ft. lbs. Weight times the number of feet equal foot pounds. If the 50 lb. object fell the 10 feet, the kinetic energy is the energy of the object in motion during the fall. All objects in motion have kinetic energy.

◆ **kinetic energy = $MV_2 / 2$**

Where: M = mass
V = velocity

◆ **For bullets use: energy in ft. lbs. = $WV_2 / 450400$**

Where: W = weight of bullet in grains.
V = Velocity in feet per second.

EXPLANATION OF NUMBERS USED

For readers who have followed ballistics in other publications or studied kinetic energy in school, a brief explanation is required at this point.

The formula for energy can be found with both 450,240 and 450,400 as the denominator. Also the gravitational constant is seen as 32.2 and 32.16 and 32.17 and perhaps a few others as well. The key is the gravity constant because the long number in the denominator is related to it.

The pull of gravity is not the same at different locations. It is the strongest on the earth's surface and less above it and below it. Even on the surface it will vary. Example: 32.258 ft./sec. at the poles and 32.144 at the equator. The variation is slight, as is the result of calculations using the different figures.

Gravity is the term for the attraction between material bodies. The Newtonian law of universal gravitation declares that every mass attracts every other mass with a force which varies directly as the product of the attracting masses and inversely as the square of the distance between them.

In actuality, the earth is not a sphere. Its shape is closer to a spheroid whose equatorial radius is 21.6 kilometers longer than its polar radius. Also it rotates and the centrifugal force due to rotation reduces gravity, especially near the equator. The earth's surface is also irregular in outline and the density variable, at least near the surface. Thus the formulas for the variation in gravity at different parts of the earth's surface are complicated.

For years, 32.2 was used for basic calculations and 32.16 or 32.17 for advanced calculations. Lately, the standard value of 32.1741 ft./sec. has been set by international agreement. This makes the more simplistic 32.17 the number to use.

The denominator of 450,240 was correct for a gravity value of 32.16. Using the gravity value of 32.1741 the denominator works out to 450,437.4 and with 32.17 it is 450,380.0. For ballistic purposes, the modern trend is to use **32.17 ft./sec. squared** for the gravitational constant and **450,400** for the denominator in energy formulas. In any case, the difference in the final result is slight.

This long number is the *energy factor* or the 1 / 2 M in some energy formulas. To obtain, use the formula: **1 / 2 * g * 7000.**

The digit 2, because we must either multiply by 1/2 or divide by 2, and the 7000 is the number we divide the grains of the bullet's weight to obtain pounds. Therefore: **2 * 32.17 * 7000 = 450,380.**

WEIGHT AND MASS

Weight and mass are not the same, although many Americans without a background in engineering or physics may believe they are. People that use the metric system have an easier road to understanding. They use the kilogram for mass and the newton for force or weight. In the English system of physics, mass is weight in pounds divided by the acceleration of gravity, 32.17.

To be technical, in engineering terms the pound is not a unit of weight but of force. The slug is the term used for mass. For some, slug conveys a mental picture of a bullet. Other readers may think of a piece of metal used in place of a coin in a vending machine or of a small snail like creature with no shell. To printers of an earlier generation, it is a strip of type from a Linotype machine. The slug is also called the *geepound* or *the engineer's unit of mass*. (The English language is strange at times.)

In physics, a slug is the mass that an unbalanced force of 1 lb. will create an acceleration of 1 ft./sec. squared. Let us word it slightly different *one slug one foot per second per second.* In other words, an object with a mass of 1 slug is accelerated on the surface of the earth at 32.17 ft./sec. squared by gravity. The pull on the object would be 32.17 lbs. which would be its weight. This brings us back to the formula just given. Mass in slugs is weight in pounds divided by 32.17 (for gravity).

When we weigh a body, we do not find out how much mass or matter there is in the body. If we use a spring scale to weigh an object and then take it to a high altitude, it would become lighter. At return to earth it would be back to its original heavier weight. The object itself did not change during its trip to a higher altitude, only the pull of the earth's gravity changed. Material or mass is related to weight, but different.

For another explanation; while the object was at altitude, if we measured its acceleration as we dropped it from the cabin roof to the floor, and performed the same test close to earth, the result would be a slower drop at altitude. **The ratio of the acceleration between the 2 altitudes would be the same as the ratio of the different weights measured at the 2 different altitudes. Mass tells us the quantity of matter, while weight tells us the gravitational force on the body or object.**

The **kinetic energy** is directly proportional to the mass of the moving object. For example, if their speed is the same, a 2,000 lb. car will impact at half the force as a 4000 lb. car.

Kinetic energy increases as the square of the velocity. For example, double the speed of a car and the impact force will be increased 4 times. If the speed is increased by 3 times, the impact force will be increased by 9 times. Bullets respond the same way.

If two cars are moving at the same speed, the heavy car will do more damage to an object it hits than would the lighter car. If the cars are of equal weight, the faster car will do more damage than the slower and at a higher rate than expected. Remember, **the energy increases as the square of the velocity.**

POTENTIAL ENERGY

Kinetic energy is a result of mass and motion. **Potential energy** is static and is possessed by a body when it is in a position where it may descend or fall with the force of gravity. (Old movie fans may be reminded of a piano being lifted to a second story by a frayed rope.)

As the total energy is unchanged, an increase in velocity (kinetic energy) will be accompanied by a decrease in potential energy. For an example, it is similar to a ball rolling on a smooth surface. If the ball rolls downhill, the potential energy due to position is exchanged for the kinetic energy of motion. If there were no friction, the change of potential energy would equal the change in kinetic energy.

ACCELERATION

We all know what acceleration is as it relates to our automobile. What is good for a worn out jalopy would be poor for a late model sports car. With projectiles the basic idea of acceleration is the same. It is the change in velocity per unit time. (**acceleration = velocity / time**) This equation can be switched to obtain velocity. (**velocity = acceleration * time**) Acceleration can be either uniform or varying. If it is varied, it can be plotted on a graph as a *space-time curve*. The slope of the curve at any point will represent the acceleration at that time. *Uniform motion* is when the velocity is constant and the acceleration is zero. There are mathematical means to calculate relations between space, time, velocity and acceleration.

MOMENTUM

The momentum is nothing more than the product of its mass and its velocity. **mass * velocity = momentum** The velocity in the equation indicates that the object has to be moving, which is true. Time is also involved. A force is required to obtain the movement and the momentum is the result of the force acting for a length of time. This is called *impulse*. From this we can see that **time * force = impulse.** This is, in turn, equal to momentum. In other words, we could say that **momentum = mass * velocity = force * time.**

It was just shown in Newton's 3rd. law that $F = M * a$. In acceleration it was shown that $a = V / t$. From this we can see that $F = M * a = M * V / t$ or $F * t = M * V$. This may be just a bunch of confusing letters and symbols to readers who are not interested in math. Readers who are math inclined will see that this shows why the two equations, one momentum and the other impulse, are actually the same.

TORQUE

Torque, often called moment, is the twisting or rotating tendency created by a force or a combination of forces applied to an article. (Called torsion.) Torque is calculated by multiplication of the force and the shortest perpendicular distance of the force. Energy from rotary motion is equal to torque times angular displacement. Torque is different at various points on the object and it is zero for all points exactly on the line of action of the force.

Evaluating a torque requires only that we measure a distance and a force and multiply them together. Engineers put the force before the distance to avoid confusion with energy units which is also force times distance. In other words, torque is lb.-ft. and energy is ft.-lb.

Torque has a relationship to the twisting of both the firearm and the projectile created by the rifling. Chapters 6 and 12 cover these subjects in detail.

CHAPTER 4

PRIMERS, POWDER & GAS EXPANSION

PRIMERS

Primers date back to an invention, actually more of a discovery, by a Scotsman named Alexander Forsyth in 1807. He found that fulminate of mercury could be used to ignite the powder in firearms. Seven years later the idea was formed to put a small amount in a cap placed over a tube leading to the chamber. This could be hit with a hammer on the side of the gun and send the flame to the main charge. This was the percussion cap and was used for many years. The cap that is used today in the base of a center fire cartridge is similar to the original percussion cap.

Older primers were corrosive and were the main source of bore corrosion in the early history of cartridge use. They used fulminate of mercury and some potassium chlorate. The main source of trouble was the salt deposited by the decomposition of the potassium chlorate. (The mercury caused problems, but primarily with case cracking.)

The chemical ingredients have been modernized. Several chemical mixtures were tried through the years, both in the U. S. and Europe. Remington lead the way in the U. S. with patents purchased from two German inventors. Now lead styphnate is the main ingredient. They are nonmercurial and noncorrosive which means their chemicals will not damage the chamber, bore or case.

The modern percussion primer consists of a brass or gilding metal cup that contains a pellet of sensitive explosive material secured by a paper disk and a brass anvil. A blow from the firing pin on the center of the cup base compresses the primer composition between the cup and the anvil. This causes the composition to explode. Holes or vents in the anvil or closure cup allow the flame to pass through the primer vent in the cartridge case and ignite the propellant (powder).

Rimfire ammunition, such as the caliber .22 cartridge, does not contain a primer cup. Instead, the primer composition is spun into the rim of the cartridge case. The propellant is in intimate contact with the composition. The firing pin strikes the rim of the cartridge case, compressing the primer composition and starts its explosion.

There are several attributes that a suitable primer must have. A good primer must <u>always</u> fire when struck by the firing pin and yet not fire by ordinary bumps and shocks that may happen in normal use. It also must always have a uniform flash that is hot enough without being too violent. In other words, it must always consistently produce the proper amount of heat.

The primer case must not rupture and permit the gas to move rearward. Finally, it must be chemically stable and not affected by changes in climate.

As we will soon explain, gun powder does not explode, it burns rapidly. Primers are a high explosive called *igniters*. The primer is composed of chemicals which have a fragile bond. The molecules are held together very weakly and come apart in an *explosive wave*. The white-hot flame is extended into the case through the primer vent hole where it ignites the powder in a microsecond. The primer ignites the powder by heat, not by shock. On the other hand, high explosives in military and demolition are activated by detonators that produce high shock as well as heat.

There are two basic primer types in use today for center fire cartridges. The Boxer system developed by the British is in use in the U.S., and the Berden which was developed in the U.S., is used more in Britain and Europe. Strange and backwards, perhaps, but true. Both types are named for their inventors. The main difference is the Boxer type has a built in anvil and uses a central flash hole. Cases using this type can be reloaded. The Berden type has the anvil built into the case and no central flash hole. For them, reloading is not usually practical. (For readers who are interested, Boxer's full name was Edward Boxer, a British Army officer. The other man was Hiram Berdan, a U. S. Army Ordnance officer.)

The anvil is necessary for ignition. The explosive priming compound is driven into the anvil by a blow from the firing pin and detonated. The outer part of the primer cup is thinner for pistols, which have a light firing pin pressure, and thicker for rifles which strike harder.

Primers come in different strengths and two basic sizes; .175" and .210" diameters. Handgun primers use less priming compound and have soft, thin cups, as compared to rifle primers. Handgun cartridge cases require less flash and the firing pins usually produce less of a blow. Magnum primers produce a bigger and longer flash to ignite larger charges. They are useful for slow-burning and slow igniting powders.

Bench-rest primers are highly consistent and uniform. Other primers can fall victim to the normal problems of mass production. In other words, primers should not be mixed or changed with out expecting mixed or changed results. Not as much of a change, perhaps, as with powders or bullets, but a difference nevertheless.

Hot primers are not the great wonder for hand-loaders that some people believe. Data published by Winchester shows a classic case where a hot primer produces a lower pressure than a standard primer by 1,000 p.s.i. There are other cases where the same is true. In shot-shell use, some people believe the hotter primer compresses the wad and pushes it and the shot forward a slight amount. This creates more space for gas expansion which holds down the pressure. This is not a proven fact at this time, but the end result is a fact and the explanation is logical and based on sound principles.

GUNPOWDER

Gunpowder is a greatly misunderstood product. Most people, including hand-loaders who use it on a regular basis, probably do not understand much about it because they don't believe they need to. As long as their cartridges work fairly well and they have had no horrible accidents at their reloading bench, so what? For the millions who have an interest in learning and a thirst for knowledge, here are some basic and advanced facts about gun powder.

Gunpowder, as a propellant, has a chemical energy which is converted to the energy that moves the projectile down the barrel's interior and toward the target. This transformation involves 3 steps. (1) Chemical as the propellant converts or decomposes almost completely into a gas. (2) Thermodynamic as the stored chemical energy is changed into heat which in turn creates motive-power. [Thermodynamics is the science that deals with the relationship of heat and mechanical energy and the conversion of one into the other.] (3) Physical because as the projectile is pushed by the hot gas, it reacts to the friction and creates torque, recoil, barrel whip, etc.

All gunpowder produces the force to move a projectile as the result of 3 things. (1) When it burns, it produces a huge quantity of gas. (2) As it burns, it produces a huge amount of heat. (3) After ignition, it creates its own oxygen and needs no outside air. All 3 are required. At first, the need for heat may not be as obvious as the other 2, but hot gas expands and requires more space than cold gas. This adds to the pressure increase in the chamber. Heat is also necessary for rapid burning.

BLACK POWDER

The invention of gunpowder is usually credited to the Chinese, although they supposedly used it in fireworks for many years without realizing its full potential. In the mid 1200's, Rodger Bacon used a form of black powder that is still used today in a modified form. A finely ground mixture of charcoal (carbon) at 14%, sulphur at 10%, and saltpeter, (or *petre* if you prefer) at 76%. (Potassium nitrate, a colorless crystalline compound [KNO_3]) Those are the percents used by the U.S. Military in the 1860's and they are by weight, not volume.

The saltpeter can be increased and increase energy, but only to a point. It is very important in black powder and as its quality varied, so did the power and efficiency of the powder and hence the firearm itself.

The charcoal is nothing more than wood or other organic matter partially burned or oxidized in kilns from which air is sometimes excluded.

It is hard to control and its purity and quality influence the powder's performance. It is the fuel in the mixture.

The potassium nitrate is the oxidizer. The 3 atoms of oxygen from the KNO_3 combine with the carbon to form carbon dioxide and carbon monoxide. By combustion, this creates high heat energy and the heat expands the gas. The sulfur burns and leaves the rotten smell of hydrogen sulfide. Its main purpose is as a catalyst. The sulphur will vaporize at 444.6 degrees C. (832.28 degrees F.). This is higher than the ignition temperature of gunpowder. About half of the sulphur is expelled un-burned. About one third of the gas created is nitrogen. Between 54% to 59% of the results are solids that must be expelled with the projectile. This takes the appearance of a grayish white smoke.

The 3 ingredients were ground into powder separately and then mixed together and ground again. As you can imagine, this was very dangerous. After they were mixed, a static spark could, and frequently did, cause a violent explosion.

In the early history of black powder, it was just that. A powder packed so close that no air space was left between the particles. Among other problems, sometimes the sulfur would settle out of the mixture. Frequently it would either not ignite or just smolder.

In the 1500's, it was discovered that with the addition of moisture in a controlled amount, the powder would solidify into a mass that could be pressed into cakes. These could be broken into uniform grains by passing it through a screen. It would pass through the larger mesh and be collected on a lower screen that it would not penetrate. The size of the holes in the screen controlled the grain size and this in turn was discovered to control how fast the powder would burn (the burn rate). Experience and experimentation taught early gunners that grain size controlled the speed of combustion and they began to seek a size that suited their needs. The moisture also held the mixture together. The moisture content was, and is, important. Too much will slow the burning rate and a point is reached on the high side that will prevent it from igniting.

Modern smokeless powder (a misnomer) was invented to overcome many of the problems with black powder. The large cloud of gray smoke emitted from black powder use is one reason modern powder is called smokeless. Black powder was weaker, so larger amounts were required. This increased the kick as this weight is added to the bullets in recoil computations. (See recoil chapter.) Longer barrels were required for proper burning and gas expansion. This steers people today to sometimes believe they need a barrel much longer than is required. Shotgun pressures of 5,000 p.s.i. and rifle pressures of 25,000 p.s.i. were normal. This is roughly half of todays pressures.

Black powder must be handled carefully because it is easy to ignite and will burn violently with high pressure. Years ago, when black powder was in common use by the military, it was tested to almost 100,000 p.s.i. In the late 1800's, Nobel and Abel gauged black powder at 96,000 p.s.i. One pound of black powder will have close to 600,000 ft. lbs of stored chemical energy. Nevertheless, guns rarely blew up because of too much black powder. Modern powder is not so forgiving. By comparison, one pound of modern powder has stored energy from 1,200,000 to 1,500,000 ft. lbs.

FFFFg was the smallest grain with the fastest burn rate and is used mainly in handguns. FFFg and FFg were larger for small bore rifles and shotguns. Fg was the largest with the slowest burn rate for big-bore rifles.

MODERN POWDERS (PROPELLENTS)

The invention that brought on modern propellants was originally discovered by Christian Schönbein in Basel, Germany in May, 1846. He mixed cotton and nitric and sulfuric acid. His invention, nitrocellulose, burned completely with a little over 3 times the energy of black powder. It was an effective substitute for black powder, but it was dangerously unstable and burned too rapidly. In Europe in the mid 1800's, a lot of workers were killed in accidents when gunpowder plants exploded. It was not until 1884 that the manufacturing process became practical. That was when Paul Vieille, a Frenchman, discovered how to gelatinize it and make it workable by dissolving it in a solution of ether and alcohol. It removed the fibrous content and formed a congealed substance that could be cut and shaped as desired. The ether and alcohol are then evaporated away leaving a hard substance. This became *Poudre B* which was the first practical smokeless powder. Propellent is really a better word as it was no longer a powder. It is an interesting sidelight that modern synthetic fibers, celluloid, plastic, and many other substances we now take for granted, were a direct result of Vieille's work with nitrocellulose.

Other major contributions were the discovery of nitroglycerine by an Italian, Ascanio Sobrero in 1846. Nitroglycerin is a chemical compound which has a huge amount of oxygen. It is unstable but has the unique ability to change into a stable gas. A sudden jolt will cause this change to start and proceed so rapidly it will be an explosion. Nitroglycerin soaked into a porous material is the basis for dynamite.

This lead to Alfred Nobel, a Swede, with the invention of ballistite in 1888. It was 60% nitrocellulose and 40% nitroglycerine.

Modern smokeless powders (a misnomer) are manufactured to perform in a predictable manor by adjusting the chemistry, coating and grain size and shape. The speed of burning rate is adjusted to fit the different requirements.

There are over 100 different powders available to the hand-loader. A suitable powder will be satisfactory for the full temperature scale, from both high to low. It will also be uniform with regard to velocity and accuracy. The ambient temperature performance, muzzle flash and overall ballistic uniformity will be as required and expected. With the bullet seated correctly for chamber and freebore length, magazine length and what experience has shown to be best for accuracy, the powder should be either not compressed or at the most, lightly so. Varying seating depth is one way to improve accuracy. (See the index for loading density & freebore.)

BURN RATE

Modern powder burn rates are controlled by the size of the grains and the chemical composition. The smaller grains burn faster because they have more surface area exposed. Invented by U.S. Army Col. Rodman, perforations were added to powder grains to expose the inside. This increases the gas released and the thrust. Therefore, the exposed area has a lot to do with how fast the fire travels through each grain and through the load. It could be said that the rate of burning is proportional to the propellant free to burn.

The burn rate can be controlled so it produces pressure for a longer time by coating it with something which will retard the initial combustion. As this coating burns off, the rate of combustion increases. (See progressive powders.) Powder grains are glazed with graphite which gives it a strange dark look. This has a retardant effect to control burning rate. It also is done to prevent the hazards of static electricity. An extra side benefit is a better flow when loading cartridges by machine. Temperature has an influence on burning rate. A slow moving bullet will let the temperature rise as the gas attempts to push it out of the cartridge and the bore. The cartridge shape also is a factor. (See index for cartridge subjects.)

Degressive burning. As strips and cords burn, the burning surface decreases continuously until the grain is consumed. Such burning is characterized as degressive.

Neutral burning. A single-perforated grain burns in opposite directions. By controlling the initial diameter of the perforation, the total burning surface hardly changes during burning. Such burning is characterized as neutral.

Progressive burning. A triperforated grain can be so designed that the burning surface actually increases until burning is nearly completed and slivers are formed. Such a grain is said to burn progressively. This characteristic can be made more pronounced if the grain is multiperforated. The slivers may not burn at all and will be blown out of the bore with the escaping gases. Note that triperforated means a grain with 3 perforations or holes and multiperforated usually means 7 perforations.

Mention of a fast powder denotes the burning rate and not to the velocity it can push a bullet. The faster powder will develop a higher peak pressure. While powders are listed as being faster or slower, no actual value is placed on them because the expression *burning rate* has no precise definition. No list could be drawn up that would be safe and accurate for hand-loaders.

Relative quickness is the term used in interior ballistics. It is the time that a propellant uses to burn up completely. This is obtained by burning powder charges in a *ballistic bomb* (also known as a sealed vessel or closed vessel or closed bomb) and measuring the pressure increase. The vessel is filled 5% to 20% with the powder and then ignition is provided by electrical means. Pressure gauges record peak pressures. The ballistician can figure burn rate, pressure rise, energy content, etc. The information that is obtained will only be an estimate of the way the powder will perform in a particular cartridge and gun. The experts do not use this as a final charge weight but as a basis for loads that will be tested in pressure and velocity barrels under controlled conditions.

Detonation is described by Webster's dictionary as a violent explosion and that is what many people believe happens inside the cartridge when it is fired. In reality, the powder does not explode or detonate, it burns; or more properly, it *deflagrates*. Another look at Webster and we see that *deflagrate* means to burn very rapidly with intense heat, and that is just what powder does.

An explosion would destroy the firearm, while the rapid burning builds the pressure and accelerates the gas and the projectile down the barrel. The pressure must build slowly, as compared to an explosion, to supply the thrust required. This is a hard concept for some people to understand because it appears to be as rapid as an explosion. In a modern shotgun, time from propellant ignition to the shot leaving the muzzle is only .003 second. (3 thousandths of a second.) The powder burns inward and may be capable of producing pressures in excess of 100,000 p.s.i. at a rate up to 60 inches per second. (Bullseye brand burns at 3.6 inches per second at about 10,000 p.s.i. At 100,000 p.s.i., it would burn at 30 inches per second.) By comparison, an explosion or detonation can be in the neighborhood of 2,000,000 p.s.i. (yes, million) and the speed up to 300,000 inches per second. This includes a shock wave. Sudden pressure peaks are not detonations or explosions, although a person who has had a prized gun destroyed in his hand will certainly think so.

In the open air, powder burns quickly but with nowhere near the rapidity required to accelerate a bullet. When it is confined, as in a cartridge case, the first heat that is created cannot escape. This causes the burning to move faster and faster through the load. This faster burning, in turn, generates more heat and pressure. It could be described as feeding on itself.

Pressure is required for proper powder burn rates. A small mound of powder in the open air will burn like a piece of celluloid in perhaps 3 to 10 seconds. Put the same amount in a suitable size cartridge case, confine it in a gun chamber, and it will burn in a small fraction of a second and create much heat and pressure. It burns rapidly only under high pressure.

Pistols, rifles and shotguns will generally use different powders. Pistols, because of their short barrels, require a fast burning powder. Shotgun shell powder is made in dense form. It is important for handloaders to follow manufacturers' suggested loads for safety.

Powder is so specialized that almost every conceivable use has a powder that is best for that purpose; not in brand or manufacturer but in class or type. The rate of burn is a good indication of use, but not always.

Older *bulk* powder could be loaded in the same amount as black powder. Other powder is more dense. Some is nonprogressive and others are progressive. Bulk powder is used by the dram (16 drams = 1 oz. avoirdupois) and dense powder is used by the grain (437.5 gr. = 1 oz.). Factory ammunition may use dense powder by the grain but the box may mention "drams equivalent". This indicates the muzzle velocity will be the same as if the stated amount of bulk powder had been used.

Handgun cartridges and other small volume cases such as the .22 Hornet and light bullet combinations are typically filled with a fast burning powder. Rifle cartridges frequently use slower burning types because they have a large case, but not always. The slower-powder will require a longer barrel to push the bullet through in order to reach full velocity.

Surprisingly, smokeless powder is not as dangerous to store and handle as many other every-day items. If your pick-up truck has 10 gallons of gasoline (about 60 lbs.) in the tank and you have 1000 lbs of smokeless powder in the cargo bed, which has the most latent explosive energy? You are right, it is a trick question. **They are equal and the same.**

POWDER CHEMISTRY
COMPOSITION

Of the over 100 elements know in the universe and the atoms they are composed of, only 4 are necessary for ballistics and firearm propellants.

SYMBOL	ELEMENT	ATOMIC WEIGHT	USEFUL WT
C	Carbon	12.011	12
H	Hydrogen	1.008	1
N	Nitrogen	14.008	14
O	Oxygen	16.000	16

It is sufficient to use simplified numbers for the weights because the actual numbers are all close to them. There are none like 1.7 or 3.4 that will not round-off without inducing an error.

Molecules are the smallest particle of an element or compound that can exist in a free state and still retain the characteristics of the element or compound. Atoms, of course, are extremely small bits that combine to form molecules or compounds. The molecules of elements consist of 1 atom or 2 or more similar atoms. The molecules of compounds consist of 2 or more different atoms. The atomic weight is a number representing the weight of one atom of an element as compared with a number representing the weight of one atom of another element which is used as a standard. (oxygen at 16.0000)

A radical is a group of 2 or more atoms that acts as a single atom. It goes through a reaction unchanged or is replaced by a single atom. A radical will not exist by itself and is always a part of a larger and more complex molecule.

CELLULOSE

We have already discussed cellulose, but it is important to remember that it is a lot more than simply cotton or wood pulp. It is the chief substance composing the cell walls of plant life. A carbohydrate may be of unknown molecular structure but having the composition represented by $(C_6H_{10}O_5)_x$. It is also an organic compound, that is to say it is from a living thing with carbon molecules primary. It can and does contain other molecules. Cellulose is also a carbohydrate. Its molecules contain carbon, hydrogen and oxygen and there is double the hydrogen atoms than oxygen atoms. (As in water.) It can also be described as polymeric because it is composed of the same chemical elements repeatedly joined together. As a cellulose, cotton is important because it is almost pure cellulose and easily available.

We have already explained that cellulose is nitrated and to this we should add that the percentage is, for propellants, between 12.6 and 13.8. This nitrocellulose percentage is a measure of the average degree of cellulose nitration. This involves replacing the hydroxyl groups by the nitrate radical and it is never perfect. (OH groups are changed to ONO_2.)

Cellulose, by itself, is a very stable material. The addition of nitric acid (HNO_3) dramatically changes the situation. The nitric acid molecules react with the hydroxyl radical (OH) and generate molecules of water (H_2O). This creates nitrate radicals (ONO_2) as replacements. Each hydroxyl group creates more water. **This is the basic reaction that creates an explosion.** ($HNO_3 + OH \rightarrow ONO_2 + H_2O$) The sulfuric acid molecules become attached to the water. This prevents the water from becoming so prolific that it slows down the process. Absorption of the water is not quite correct, but it is a term that makes the process easy to understand.

Before ignition, the nitrocellulose is full of a large amount of stored chemical energy which it can unleash on demand; or sooner, as the next section on stability will explain.

Nitroglycerin (glycerin trinitrate) is created in the same general way. Molecules of nitric acid acting on the hydroxyl radicals in a glycerin molecule are changed to nitrate radicals.

STABILITY

The nitrocellulose that is created by the above process is basically unstable. That means that it can explode unexpectedly from its own heat build-up. The heat is created from the nitrate radicals (ONO_2) that are in the material coming in contact with nitric acid and creating nitrogen dioxide, NO_2. A chain reaction slowly builds heat as the nitrate radicals are attacked and become nitric acid molecules. When the air is humid, this action is accelerated by the moisture (H_2O) in the atmosphere combining with the nitrogen dioxide (N_2O) to create nitric acid (HNO_3). These reactions create heat which can easily reach a temperature in the 300 to 400 degrees F range. This is the propellents ignition temperature. Most people have heard of *spontaneous combustion*. An excellent example is rags soaked in linseed oil which will ignite from the rapid oxidation of the oil. As we have just seen, this early powder had a similar problem. It would ignite from the heat generated from internal chemical changes.

As we previously mentioned, Paul Vieille, a Frenchman, discovered how to gelatinize it and make it workable by dissolving it in a solution of ether and alcohol. This removed the fibrous content and formed a congealed substance that was safer. But, unplanned explosions still occurred. It was not until the early 20th. century that diphenylamine was used as a stabilizer. With the addition of as little as 1%, it will absorb the excess nitrogen dioxide (NO_2) molecules and release a hydrogen (H) atom.

COMBUSTION--DECOMPOSITION

Heat is the ingredient added to cause the chemical reaction that releases energy. The composition is released and the atoms are no longer united. They are free to move about or pair up again; divorced, we might say. The percent of nitration influences the temperature of the reaction and in turn, the temperature influences the way the atoms, primarily oxygen, pair up.

These are the gaseous products of combustion from smokeless powder. There are no solids to produce smoke. They are all gases.

SYMBOL	ELEMENT	WEIGHT
CO	Carbon Monoxide	28
N_2	Nitrogen	28
H_2	Hydrogen	2
CO_2	Carbon Dioxide	44
O_2	Oxygen	32
H_2O	Water (as a gas)	18

To the list we can also add Nitric acid (HNO_3) with a molecular weight of 63. Also, note the water is a gas, because of the extremely high temperature involved. (This should not be confused with atomic weights which are different. Oxygen, for example, has an atomic weight of 16.0000 and a molecular weight of 32.)

SECONDARY MUZZLE FLASH

A careful study of the above list also explains the reason for the bright muzzle flash that is extra noticeable at dusk, night time or early dawn. There is a lot of CO and H_2 which are starved for oxygen. When they hit the atmosphere, they combine with the oxygen and are still hot enough to flash over.

All shooters are used to the muzzle flash caused by the initial cartridge firing. Normally about the same with each shot, it is the result of the hot gas and a few grains of powder that are still burning. And of course, some guns and cartridges create more of a show than others.

There is a secondary muzzle flash that was common in earlier times but modern powder development has made it rare. Also a problem in artillery and cannon, this secondary flash is much bigger and brighter and can be described as a ball of fire.

As previously pointed out, the oxygen required to burn the powder is included in the powder itself. When a secondary flash occurs, there has been more flammable hydrogen produced than burned in the barrel. This phenomenon is usually caused by a lack of oxygen which is present in quantity after exiting the muzzle.

At times, the expanding gas is still hot enough to ignite on its own with the addition of the needed oxygen in the atmosphere. At other times, a burning grain of powder is expelled and becomes the perfect *lighter*. Friction between the molecules of the expelled gas and the gaseous atmosphere can also increase the temperature enough for ignition. After all, the gas is very hot and needs only a small increase to reach the flash point.

An extra note; if friends see a secondary flash from a muzzle and the shooter doesn't, perhaps the shooter is shutting his eyes when firing.

TYPES OF POWDER
SINGLE BASE POWDER

A single base powder (nitrocellulose) is composed of a cellulose which may be cotton or wood pulp (guncotton), treated with nitric and sulfuric acid. (Much more on this later.) This mixture is too loose so it is dissolved with alcohol and ether or acetone. This forms a sticky paste which is forced through holes of a specific diameter. The result is a spaghetti like string which is cut to required length. This idea is credited to Vieille, a French chemist, in 1885. Another manufacturing method involves forming balls in water where their size is governed by temperature, speed and timing.

The nitrocellulose is the chief ingredient. In addition to a stabilizer, they may contain inorganic nitrates, nitrocompounds and such nonexplosive materials as metallic salts, metals, carbohydrates, and dyes. There is no nitroglycerin in a single base powder.

DOUBLE-BASE POWDER

This idea is from the genius of Alfred Nobel. (The same Nobel of the *Nobel Prizes*.) Ballistite invented in 1888 is also called Nobel powder and was a former trademark. It was 60% nitrocellulose and 40% nitroglycerine. Nitroglycerine is a thickening agent for nitrocellulose making a gelatinous substance.

Today, nitrocellulose, a double-base powder has nitroglycerin added at from 10% to 40%. This is a liquid formed by sulfuric and nitric acid on glycerin. Double-base powders burn very hot and have a reputation of being rough on barrels. On the plus side, they produce high velocity at low pressure.

A double-base composition contains nitrocellulose and a liquid organic nitrate, such as nitroglycerine, which is capable of gelatinizing nitrocellulose. Like single-base powders, double-base powders frequently contain additives in addition to a stabilizer.

Ballistrate was the type of double-base powder used for many years. Today's powders contain less nitroglycerine. The single-perforated grains having these compositions are coated with dinitrotoluene or centralite and glazed with graphite. Although they have somewhat less ballistic potential than the ballistite type of powders, they are more stable, cause less erosion of barrels, and have less tendency to flash.

Cordite is one, and there are others. Some use less nitroglycerin to reach for a balance between velocity and barrel damage. Hercules powders and many of the Winchester-Western Ball Powders are double base.

PROGRESSIVE POWDER

Progressive powder is a modern powder that will push heavy loads to higher velocity. It has a burning rate that increases as the pressure in the shell or case increases. The burning rate is related to the pressure and it will maintain thrust and push as the bullet moves down the bore. It is believed to be easier on the firearm because the velocity is reached without the peak pressure and temperature going as high as with other powders. These powders will have a uniform pressure distribution which is held longer. The peak pressure is not as high because it is not attained as quickly and is more spread out. The grain size, perforations and coating are factors in burn rate.

CORDITE

Cordite is a type of double-base powder, named because its stringy appearance resembles cords. This is a British development which is not in grains, as U.S. hand-loaders are familiar with. The spaghetti like strands are loaded into the case before it is necked. All other loading operations will have been finished, so after the Cordite is installed, a cardboard divider is installed and the case is necked and the bullet seated. It has a reputation for damaging gun barrels. It is made of 58% nitroglycerin, 37% guncotton and 5% Vaseline.

BALLISTITE POWDER

Ballistite powder was the first smokeless powder. Invented in 1888, it is sometimes called Nobel powder. It was 60% nitrocellulose and 40% nitroglycerine. It was a double-base powder.

PYRODEX

Pyrodex is a modern substitute for black powder. It is slightly harder to ignite but burns faster and produces more pressure. Pyrodex is substituted on a volume for volume basis and not by weight. It is safer than black powder with less bore fouling. It has a higher flash point than black powder, which adds to safety, but it can not be used in flintlocks as a priming charge

IMR POWDERS

IMR stands for Improved Military Rifle and the letters are usually followed by a number designation. This is a group of powders that have a deterrent coating which controls the rate of burning. They are noted for the ability to push a projectile to high velocity without excessive pressure (by comparison). IMR powders are from DuPont and they are single-base.

BALL POWDER

Ball powder was invented by Fred Olsen, an American with Western Cartridge Co., in 1933. It is a nitrocellulose, double-based powder, which uses a manufacturing process which lowers the production costs. The grains are normally spherical and firm and 0.02 or 0.03 inch in diameter. The graphite coated grains flow easily through powder measures and are well suited for hand-loaders.

During manufacturing, the powder is in a watery slurry which can be pumped from place to place during the process. This makes it a much safer and easier operation. Basically, it is wet nitrocellulose in a solvent (e.g., ethyl acetate). Chalk is added to counteract acids that are present. Diphenylamine is added to make the powder stable. Other chemicals are added to control various reactions and the nitroglycerine is added near the end of the process along with a solvent to aid in stabilization.

SHOTGUN POWDER

Powder used in shotshells must burn faster than rifle powders. The pressure must drop off more, by comparison, by the time the shot reaches the muzzle. Modern powder is no longer loaded bulk for bulk as compared to black powder, as it once was. Nevertheless, it is still loaded to *dram equivalents* or *equivalent drams*.

THERMODYNAMICS / PUSHING the PROJECTILE

Thermodynamics is the science that deals with the relationship of heat and mechanical energy and the conversion of one to the other; the availability of energy for work and with the stability of chemical substances. In other words, changes in which energy is involved. This describes the way the chemical energy in the powder is changed into heat which in turn creates motive-power. Heat is energy in transit from one mass to another because of a temperature difference between the two. Whenever a force acts through a distance, *work* is involved. Like heat, work is energy in transit.

The **first law of thermodynamics** is a statement that energy can be neither created nor destroyed. When a transformation takes place, the amount of one will always be equal to the other. (Formed by Carnot in 1832 and Mayer in 1832 but first stated unambiguously by Helmholtz in 1847)

The **second law of thermodynamics** is a statement that conversion of heat to work is limited by the temperature at which conversion occurs. **It is impossible to bring about any change or series of changes which will result in the transfer of energy as heat from a low to a high temperature.** (Formed by Carnot-1832, Clausius-1850 and Kelvin-1851.)

TERMS DEFINED

Let's pause here for an explanation of some confusing terms that will be needed very soon in this discussion.

MOL or MOLE

Mol, also correctly spelled *mole*, is the molecular weight of a substance. A pound mol is the weight in pounds equal to the molecular weight. A 1 pound mol of oxygen weighs 32 lb. At the same pressure and temperature, the volume of one mol is the same for all perfect gases. An Italian, Amedeo Avogadro, discovered in 1811 **that equal volumes of gas at the same pressure and temperature would contain equal numbers of molecules.** No matter how many molecules the gas contained, the number in a container of equal size would be the same. The weight would be different according to the weight of gas inside. This is the *molecular weight* or mol (mole) from the word molecular. If expressed in pounds, it is called the *pound-mol*; if in grams, the *gram-mol*. If the substance is an element the term *atom* is used instead of mol.

If this is all confusing, remember that the number of molecules remains the same in equal containers. The weight does not. For example: hydrogen (H_2) has a molecule weight of 2, oxygen (O_2) has a molecular of 32 and sulphur dioxide (SO_2) has a molecular weight of 64. If each sealed container held the same number of molecules, the hydrogen container would be light in weight while the oxygen container would be 16 times heavier. The sulphur dioxide container would weight twice as much as the oxygen container and 32 times as much as the hydrogen container. All this while each holds the same number of molecules. The mol is a measure of how many gas molecules are in a container. (The actual weight in pounds of 1 cu. ft. at standard atmosphere and 68 degrees F. would be .005234 for hydrogen, .08305 for oxygen and 0.1663 for sulphur dioxide.)

(While the molal volume of any gas under standard temperature and pressure is the same, the molal volume of liquids and solids varies. It depends upon their specific chemical nature and their density. The bulk density of solids depends upon their physical form and state of subdivision.)

RANKINE TEMPERATURE

Rankine is a temperature scale used in engineering, chemistry and thermodynamics. Temperature is the condition of matter that determines the flow of heat between bodies, their physical state from solid to liquid to gas, etc. It cannot be measured by absolute standards. It must be determined in relation to standard bodies under conditions known to be constant and reproducible. In other words, reference points like the boiling point or freezing point of water. The Rankine scale is 460 degrees higher than the Fahrenheit scale. To be extra accurate;

degrees Rankine (R) = degrees Fahrenheit (F) + 459.69.

This gives it a zero the same as absolute zero which is -460 degrees F. (-273 degrees C.).

PRESSURE & TEMPERATURE

The extreme heat from gun powder is created by the chemical reaction when oxygen combines with hydrogen and forms water or combines with carbon and forms either carbon dioxide or carbon monoxide. The energy to push against the projectile is created by the sudden change of the solid to a gas. The temperature will increase from the flash temperature of the mid 300 degrees F. to around 5,550 degrees F., average, in a tiny fraction of a second.

Pressure and *temperature* in thermodynamics are used slightly different than in everyday use. The two are knit closely together. The higher the temperature, the more the molecules in a gas move about and collide with one another and the walls of their container. The velocity and frequency of collisions increase and in turn, increase the pressure. This is based on average velocity of movement because some will move faster than others due to collisions with each other and the walls. Therefore, pressure can be considered a determination of how frequent are the collisions and temperature is a determination of the average velocity of the molecules. (This molecular movement has been scientifically checked and for most molecules, the velocity, number of collisions per second and distance between collisions at various temperatures are known.)

From the preceding paragraphs, it can be seen that the volume of the container also influences pressure. If the volume is increased, that is the size of the container is made larger, the molecules will still move at the same velocity as long as the temperature stays the same. But, they will now have more room to move about in and will not collide with each other as often. Therefore, the pressure will be less. A decrease in volume with no temperature change, will increase pressure because of the decrease in distance between molecules and the increase in collisions.

Pressures of 55,000 p.s.i. are common at the peak but may drop to 6,000 p.s.i. by the time the bullet reaches the muzzle. With most powder, the flame temperature will reach about 5,550 degrees F at the peak pressure. Most barrel steel will melt at half that temperature or about 2,500 degrees F. The high temperature peak will only last for a small fraction of a second and, with the exception of machine guns, will cool a lot between each shot.

MATHEMATICS

An important equation for interior ballistics, called an *equation of state* or *perfect gas law* is based on gases *P-V-T* properties being expressed by the simple relation;

$$p * v = R * T$$

Where: R = a gas constant.
v = volume per unit weight of the gas in cu. in.
p = gas pressure in p.s.i.
T = absolute temperature of the gas in degrees Rankine

If v is expressed as volume per unit weight, the value of the constant R will be different for different gases. If v is expressed as the volume of one molecular weight of gas, then R is the same for all gases in any chosen system of units. In general, for any amount of gas, the perfect gas equation then becomes:

◆
$$P * V = N * R * T$$

N = quantity of gas in pound mols (the no. of mols)

(Note: chemistry and engineering students may recognize this as a combination of Boyle's and Charles' laws.)

The volume that is filled by the expanding gas expands with it. It has to. The projectile starts to move from the case immediately at the split second the powder starts to burn. It has to move and provide an ever increasing area for the gas to expand into. If it did not, the gun would be destroyed in the blast and possibly the shooter as well. This creates a problem with the volume part of the equation. It can not be based on the cartridge volume or the answer for pressure would be too high. The volume must be an estimate using a figure of from 4 to 10 times the original.

There are several equations and variations that use an empirical constant to take into account high pressure. Others use a volume corrective term, such as α.

◆ A modified equation that takes into account the higher pressure is:
$$P(V - Cb) = C * R * T$$
Where in addition to the earlier notation, we add:

C = weight of the propellant in pounds

b = the actual volume involved with the gas molecules. For our purposes in ballistics, use 26.3 in.^{3} / lb.

This formula is basic to several important questions. How much will the pressure increase when the temperature is raised by a known number of degrees? How much will the temperature increase by a certain increase in pressure? How much will the pressure decrease by an increase in the volume of the container?

If v is expressed as the volume of one molecular weight of gas, then R is the same for all gases in any chosen unit system. If v is expressed as volume per unit weight, the value of the constant R will be different for different gases. R can be determined from the equation if we know 4 of the 5 quantities. Also, advanced books on chemistry and engineering have charts on properties of gases that list the constant R for each gas. For our purpose, the value of the constant R for degrees Rankine using pressure units of pounds per sq. ft. and volume units of cu. feet is 1,545.0. Other values are used for different measuring systems and all are listed in chemical engineering books. Another is 18,510 for pounds, inches with degrees still in Rankine. The average molecular weight of the products of combustion from standard gun-powder is 24.1. If this is divided into 18,5810 we have 768 as a constant.

5,550 degrees F. is a good average temperature for burning propellent under the conditions of pressure and other involved items.

The reader should be familiar with the laws of motion that are discussed in the previous chapter. The knowledge will be needed to properly understand gas expansion and the projectiles movement.

♦ **Velocity** can be determined by another important equation. It will determine the velocity of a projectile accelerated down the bore from the pressure of the expanding gas.

$$V = P * A * t (g/m)$$

Where: V = velocity in ft./sec.

A = bore cross sectional area in inches ^{2}.

P = average pressure in lbs. / inches ^{2}.

g = gravitational constant

t = time of pressure's action in seconds

m = mass of the projectile in pounds

♦ **Energy** of a projectile at the muzzle can be obtained from:

$$E = P * A * X$$

where: E = energy (energy = force * distance)

P = average pressure in lbs. / inches ^{2}.

A = bore cross sectional area in inches ^{2}.

X = distance

♦ **Momentum** of a projectile at the muzzle.

$$I = P * A * T$$

Where: I = Impulse or momentum (momentum = force * time)

P = average pressure in lbs./ inches ^{2}.

A = bore cross sectional area in inches ^{2}.

T = time of pressure's action in seconds.

IRON NITRIDES

The nitrogen (N_2), under the heat and pressure, chemically combine with the outer layer of steel in the bore to create iron nitrides. This is visible as a white covering which is heaviest where the pressure is highest. This covering not only changes the size and form of the area it covers (the chamber throat), it damages the steel beneath it.

This gas is heavy with nitrogen and nitrogen derivatives. It is also heavy with a molecular weight of about 28. This requires some of its energy to push the gas down the barrel, as well as the projectile.

OXYGEN REQUIREMENTS

Cartridges can be fired in a vacuum or underwater because no outside air (oxygen) is needed for combustion. **Any air left in the cartridge at loading is unneeded and a full cartridge case performs better.**

Three items are needed for combustion and they are always the same, whether in a house fire or the fuel in an internal combustion engine. (1) Kindling temperature. (2) Combustible material (3) Oxygen. To ignite and burn, any fire needs all three to start and all three to sustain combustion (keep burning). Any fire can be extinguished by removing any one of the three; by cooling it, cutting off the air, or eliminating the burning material.

As previously explained in detail, smokeless powder is nitrocellulose (cellulose nitrate). Nitroglycerine is added to some powder. When the powder burns and changes to a gas, the complex molecules of carbon, hydrogen, nitrogen and oxygen are released and changed. Yes, oxygen. The oxygen is supplied internally in the powder's chemicals. In black powder, the oxygen is supplied by the saltpeter (nitrate) and the sulphur and carbon are the fuel.

MUZZLE LOADER

Remember that we just said that high pressure is required for proper powder burn-rates? The bullet or ball that is seated by inserting from the muzzle end will not hold enough pressure to offer proper burning in a muzzle loader. If enough smokeless powder is used to push the projectile to the same velocity it was enjoying with black powder, the gun may explode.

Black powder is said to be very corrosive. This comes mainly from the old caps that contained chlorates. Black powder is only mildly corrosive, and this arises primarily from the sulphur. Cleaning a firearm after use will prevent this problem.

REDUCED LOADS

A lot of internal ballistics has to be taken on faith. We simply cannot properly see what is going on inside a closed breech or a cartridge.

One unproven theory was used in a Kentucky court case. (Schuster v. Steedley, Oct. 1966, 406 S. W. 2d 387.)

Related by an expert witness from H. P. White Laboratory, the theory stated that when a cartridge is loaded so that the powder charge leaves a lot of space in the cavity between the primer and the base of the bullet, the primer flash can cross the open space while the powder is laying on the bottom. (The cartridge being horizontal at the time.) This flash may move the bullet out of the case but it will have insufficient energy to properly engage the rifling. This happens a short instant before the powder is ignited and creates the main gas expanding force. More force is then required to move the bullet than is normal and excessive pressure is exerted on the cartridge's base and the breech mechanism of the gun. Also, the bullet may become lodged in the barrel and create an obstruction for a successive shot.

Another theory that is about the same and maintains that the powder may partially ignite and create a "slow burn". This forces the powder to the front where it presses against the bullet and can create a pressure wave caused by runaway burning.

Norma, the Swedish ammunition manufacturer, cautions about the hazards of light loads that fill the case to only about one third of its volume. They warn that the primer may flash along the surface of the powder and only ignite part of it. This will press the bullet into the rifling where it may stop. A microsecond or two later when the balance of the powder ignites, it will not be able to move the lodged bullet and will create extra high pressure. Remember, the bullet plugs the barrel so the gas does not have its normal means of escape. Also, if the gun survives this shot without the shooter being aware of the problem, the barrel will be plugged when the next shot is fired.

Powder is supposed to burn at a controlled rate and the pressure is a controlling factor. This produces the expanding gas in a hurried yet orderly fashion. A small amount of fast burning powder in a much larger case can explode and do extensive damage. Most people believe that explode is what they want the powder to do, but as this book explains, explode is too sudden and violent. There may be only a small time difference involved, but that small variation is important and necessary.

It is interesting that tests by the Frankfort Arsenal show changing the powder position in an unfilled case has a big effect on performance. With .30 caliber M72 Match ammo, the shift from forward to rearward powder position can alter the strike point 2 or 3 inches at 300 yards and at 1,000 yards from 16 to 30 inches. These are significant amounts for such a simple cause. It is a good idea to hold the powder in position by the use of a filler. Use only approved fillers such as kapok or Dacron and for safety, start with a reduced load.

Whether the powder is positioned forward, rearward or level is not as imperative as for it to be the same for each shot. Not much can be done about the situation when hunting, but many target and bench-rest shooters raise the barrel to position the charge to the rear or a side to side shake to level it before each shot This is an issue that is very easy for an amateur to prove for himself. Hand-loaders will notice some powders and cartridges are more inclined toward this problem than others. The same holds true for factory ammunition.

Reduced powder charges, sometimes called squib-loads, are useful with a big bore gun for practice and target shooting. They are effective at reducing barrel wear. Although problems are rare and not completely understood, squib loads are not without risk

"COOKING-OFF"

Cooking-off is the term used to describe a cartridge that is fired or exploded because of the application of heat. A blow on the primer is not applied. It can happen in many ways and the results are as varied as the means.

First, we need to say that testing by the National Rifle Association and Army Ordinance and other independent groups has shown that the average rifle and pistol cartridge will cook-off at high 200 to low 300 degrees Fahrenheit. The average shot-shell a little higher at high 300 degrees F. Average being an important word in that statement. Some may be lower or higher.

As we can easily understand, this is too high to cook-off in boiling water which is 212 degrees F. Not to mention the water can leak into the case and wet down the powder. Still, the temperatures are low enough that weapons fired very rapidly, such as machine guns, can develop enough heat to fire without pulling the trigger.

Cook-offs are frequently prevented in machine guns by the use of an *open breech*. This does not, of course, mean the breech is open during firing. The bolt is held to the back against a compressed spring which brings the bolt forward and picks up and loads a cartridge. It is fired upon chamber seating and the bolt is again opened. The open breech allows cooling air to circulate and no cartridge is held in a hot chamber. They are only chambered when fired and not before.

The high temperature that steel can obtain in direct sunlight, especially in the south and the tropics, can also cause a cook-off in standard firearms. If a few rounds are fired, the temperature can raise the few extra degrees needed. A cartridge that is confined in the chamber will expel the bullet and be as deadly as any unaimed shot. Military crash crews are taught how to disarm aircraft weapons at a fire scene and not to be in front of the gun until it is safe.

Ammo not in a chamber, during a fire for example, will explode quickly. With the average temperature of a cook-off being below 400 degrees F. and a house fire temperature reaching a couple thousand or more, they will react quickly. The case will explode outward and while the bullet may be thrust forward some, as the case may be thrust back, with out the confinement of the chamber the force will be in all directions and not just toward the base of the projectile. Remember some of our earlier discussion? The chamber not only directs the pressure toward the base of the projectile, it also adds pressure which is vital for rapid burning.

After fires in homes where ammo has been stored, and also in gun-shops and hardware stores that sell ammo, there are usually stories of all the explosions and sounds of cartridges firing. For the most part, the talk is much worse than the event.

In early April of 1970, a store that sold ammunition in Scotch Plains, N J. was heavily damaged by fire. The fire was reported in the news reports and by *The American Rifleman*.

Two firemen were hit by either bullets or some other unknown objects. In both instances, the velocity was so low that only a slight sore was made. Many cartridges did not explode even though they were near the fire. There was no explosion of smokeless powder even though some was on hand for sale. Rifle cartridges that exploded had a split case with the bullet staying nearby. Generally, other fires have a similar effect and don't cause near the damage from the ammunition that is expected.

This is not intended to indicate that it is safe. No one but a fool would stay in a room with a fire that was exploding stored ammo. Also, other confined places, such as tubular magazines, can create a bullet propulsion that can be deadly.

A good summation would be that a cooked off cartridge is not as dangerous as it would at first appear, but as in all things with firearms, variables can change the situation so extreme caution should be used.

EXTRA NOTES

◆ Hand-loaders should never use burn-rate tables to determine charge weight. It is not a safe practice.

◆ Muzzle loaders should use caution to avoid excessive pressure. They do not have a case which may extract with difficulty or to examine for signs of excessive pressure.

◆ Barrel life is strongly influenced by peak pressure and gas volume. Experiments have shown that a double charge will reduce barrel life by 75% while a charge lowered by 10% will increase barrel life by about 20%. These are impressive numbers, especially when it is considered that the benefits of a heavy charge are very small. (Usually a slightly reduced wind drift and a little flatter trajectory.)

◆ Rules of thumb can get people in trouble. Therefore, there is a concern that the readers will **use proper caution**. Nevertheless, here is one that is interesting for powder and case volume. It says that as a general rule, we can say that one grain of powder for each 3 grains of bullet weight will not be too much for the proper volume between case and bore. **Hand-loaders should follow their manuals. This rule is listed for academic purposes only. Literature should be checked for the proper powder and amount.**

Hand-loaders who like to experiment can sometimes hit the perfect combination. For each gun, case and bullet, there will be one powder, which if loaded in the correct amount, will perform better than all the others. Knowledge of the subject will make the testing easier, safer, faster, and more fun.

CHAPTER 5

INTERIOR--BASICS & PRESSURE

PRESSURE

Many intelligent people believe that all of the danger with a firearm is in being on the wrong end of the barrel. Sadly, this is only a part of the picture. The high pressure that hand-loaders can obtain by using the wrong type of powder or an excessive amount can do more than give a flatter trajectory and a higher velocity. It can also split the case or blow the gun apart and injure the shooter or a bystander.

The same danger exists in using a cartridge that will fit a chamber but is not what the manufacture intended. Some of the older guns, made for weaker ammo than we use today, are frequently a victim of this problem. While some firearms may handle 60,000 p.s.i. with ease, others will explode at perhaps 10,000 p.s.i. If we try to get into the range of 80,000 p.s.i., few guns today are safe.

Ballistic efficiency is figured by the ratio of pressure to velocity. Always be aware of pressure if experimenting to obtain higher velocity.

FACTORS AFFECTING PRESSURE

There are many factors that affect the pressure inside a firearm. Some are more obvious than others. **(1)** The primer type, composition and size. **(2)** The powder type, composition, grain size, grain form, density, and amount. **(3)** The bullet type, weight, shape, diameter. **(4)** The crimp between the bullet and case. **(5)** The condition of the throat and the bore. **(6)** The loading density. (This is a special subject that is covered in detail in Chapter 15.) **(7)** The cartridge type. **(8)** chamber type & condition. **(9)** The length of freebore. **(10)** The temperature of the cartridge and gun at the time of use. **(11)** The temperature and humidity of the location where the ammunition was kept prior to use.

Any change in any of these factors will change the pressure in the chamber. The reason behind 1 and 2 is explained in chapter 4. Many of the others are based on the fact that anything that slows down the acceleration of the projectile as it moves down the bore, will permit the expanding gas to build-up and increase pressure. A tight crimp makes it harder to move the bullet. The same for a heavy bullet or one that is tight in the barrel. The bullet's shape controls how much area (length) is in contact with the interior walls and the rifling grooves. A longer length or tighter fit will increase pressure because the projectile will be harder to get moving.

Erosion, abrasion and wear in the chamber, throat and bore will lower pressure. The areas affected are enlarged and permit the expanding gas to escape around the projectile. The bullet simply does not seal the bore. A small bullet can have the same effect.

Temperature and humidity is discussed in several places in this book, so to save space, it will be omitted here. Item no. 9, freebore, is also covered in detail elsewhere. The index will help locate these areas.

LOSS OF EFFICIENCY

HEAT LOSS: Not all of the pressure that is developed by the propellent is used to push the bullet out of the bore. Heat that is transferred to the gun is a loss of energy. Usually, a high percent of energy is lost in this fashion. Rifles have a loss of about 10%. With shotguns, it is much higher. The projectile is in the bore for a longer time period and the bore area is larger. This permits a heat loss of about 25% of the propellent's energy.

PRESSURE GRADIENT: Another loss is due to the propellent's expanding hot gas not exerting a full push against the projectile as it moves down the bore. The gas that is being added to the expanding interior just cannot keep up or catch up. The pressure in the chamber will be higher than the pressure pushing against the bullet's base. This is called a *pressure gradient*. (In this use, the word *gradient* means a rate of change.) The pressure gradient will be very low as the bullet starts to move. It will increase and it can be very high by the time the bullet reaches the end of the barrel.

FRICTION loss is a variable that depends on the fit of the bullet in the bore and the barrel length. There is also a friction loss from the gas molecules on the barrel's interior. While the loss from gas contacting the bore is small, it has to be considered.

EXPANSION RATIO is covered in detail in Chapter 11. For now, it is important to understand that it is the ratio of the volume of the bore from the cartridge base to the muzzle, to the volume of the powder portion of the cartridge. It is easy to see why it is sometimes called a volume ratio. This influences the amount of propellent energy that can be converted into projectile energy. The higher volume ratios are better and more efficient. A longer barrel will increase the ratio and a large volume cartridge case will lower it. In even the most efficient guns with an excellent expansion ratio, only about 50% of the potential energy is converted to mechanical energy of the projectile.

MUZZLE ENERGY

Muzzle energy is discussed in the chapter on Newton's laws and a formula is included. This is a little more on the same subject only from a different angle.

Energy in use at the brief moment the projectile leaves the muzzle can be worked out from the energy that was available in the propellant load. We need to know the energy in foot pounds per pound that the powder has available. This will vary, but if the actual amount is unknown, an average figure of 1,400,000 can be used. We already know from the mathematics chapter that there are 7,000 grains in one pound so we can say there is 200 ft. lbs. of energy in a grain. (1,400,000 ÷ 7,000 = 200) It is then a simple matter of multiplying the grains in the load times 200. An example would be 47 grains times 200 for 9,400 ft. lbs. of energy. Sound high? It is. This is the total amount of energy produced by the powder. It varies, but the total energy imparted to the projectile will be only about 40% to 45% of the available energy.

There is much wasted energy. These values are all approximate, but the lowest would be the amount of energy used to create the sound. This would be so small that it can be discounted. About 1% to 2% is chemical energy and will be wasted in unburned fuel and simply blown out the muzzle into the atmosphere. It probably burned there where it was seen as a muzzle flash. (explained earlier) Another 2% to 3% is mechanical energy lost to friction. Thermal energy used to heat the barrel will take about 25%. The largest is more thermal energy in hot gas which was unused as the projectile departed. This will consume about 30%. All together, anywhere from 55% to 60% of the energy is unused for the purpose of propelling the projectile. As can easily be seen, a firearm is not an efficient machine.

The 1st. law of thermodynamics states that energy can be neither created nor destroyed and this applies here. The thermal energy that was absorbed into the barrel as heat will be changed into radiant energy as it cools. It is all accountable in one way or another.

EARLY RODMAN TEST

Early pressure testing was done by several methods, the most important being the Rodman pressure gage. Thomas Rodman was a Captain in the Ordnance Dept., U.S. Army, at the time of his invention in 1861. Later in his military career, he was promoted to Brig. General. His invention consisted of a small metal unit that mounted in a hole drilled into the top of a standard receiver. At peak pressure, a piston with a chisel shaped blade would be forced up into a copper plate. The plate and blade would be removed and an identical mark would be pressed into the plate by a press which had a pressure gage. The theory was simple; same mark, same pressure.

COPPER & LEAD METHODS

Many readers who are interested in ballistics have heard of the copper crusher method of taking pressure readings. The CUP figure is usually about 85% of the correct pressure and it is peak pressure only. It is also a *relative* pressure and not a *true* pressure. Still, it is an excellent method of comparison which is highly consistent and accurate. The copper crusher method is used for pressures over 15,000 p.s.i. and lead is used for pressure below this level. The principal is basic and simple. As the sketch shows, the pressure in the chamber pushes upward and compresses the copper cylinder. Actually, *push* is not the best word, because the rapidity of pressure on the gauge is the same as on the bullet. The piston slams the copper like a bullet.

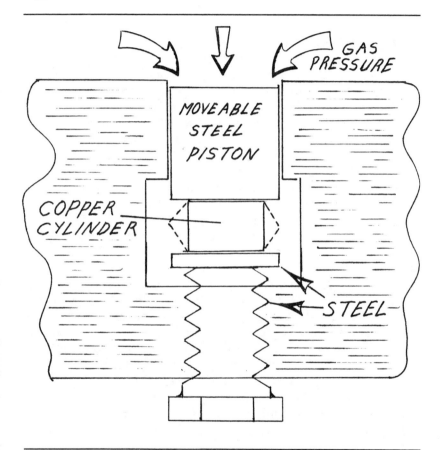

The copper is always made exactly the same in type and size so the amount of crush will tell, by past experience, the amount of pressure used to deform it. The past experience is recorded on a chart called a tarage table which is changed for each new lot of copper crushers. It is still the standard method used by ammo makers to check quality.

It must be noted that this is not installed in a standard gun. It is part of a special pressure testing unit designed and built for testing purposes only. A standard diameter hole is drilled in the case and paper is used to keep in the powder. The hole is placed under the piston. The chamber dimensions will be accurate and made to standard dimensions. The results will be compared to industry standard ammunition to obtain safety in firearms of different brands.

Occasionally the reader may see the initials L.U.P. for lead units of pressure or C.U.P. for copper units of pressure. These results indicate the method used in testing.

PIEZO-ELECTRIC GAUGE

There are other methods of measurement which gage and depict the pressure curve. Clark University's Dr. Arthur G. Webster (1863-1923) developed a method used at the Springfield Armory in 1921. Another early method was Thring's Pressure Recorder, described in the British *Textbook of Small Arms*, in 1929. There is no need to explain the operation of these early systems, but the piezo gauge is used at many factories and is worth a brief discussion.

The **piezo-electric** pressure gauge system gives a picture of the pressure as it increases to a peak and falls back to near zero in a time interval of perhaps .003 (three thousandths) of a second. Naturally, the time will vary as will the pressure and the curve.

The piezo-electric method of measuring pressure-time uses the fact that materials such as tourmaline and quartz generate a small electrostatic charge when compressed in a certain direction. The system works because the amount of charge is directly proportional to the amount of pressure used to compress it. Pieces of quartz crystal are mounted in Bakelite in alternate layers with thin metal plates and placed in special constructed units. A steel cylinder is pressed against the crystal by the expanding gas pressure. Each metal plate is wired and the total impulse is collected. The electric charge is small and has to be amplified because it will be less than one-ten-millionth of one coulomb. (A coulomb is the unit quantity of electricity and the ampere is the unit of electric current. One coulomb is the amount of electricity moved by one ampere in one second. One coulomb is equal to the charge of $6.24 * 10^{18}$ of electrons.)

The charge increases as the pressure increases and drops as the pressure drops. This creates a climbing and descending curve.

The amplified charge is sent to a cathode ray tube which is similar to a television screen. The rays, as in an oscilloscope, move across the screen horizontally to depict time and vertically to depict pressure. This graph-like picture can be recorded by a camera and converted to pressure and time. It is not fast and easy, but it is a faithful depiction of what we can not perceive in any other way.

HOME METHOD

Costly special equipment is used in laboratories for pressure testing and even an extra safe area is required. The high temperature and extreme speed make the tests difficult to do accurately and safely. If an inventor or hobbyist wants pressure testing, there are a few ballistic laboratories that are independent of the government and the ammo makers that can do the testing for a reasonable fee.

A new home method of measuring chamber pressure is sold by a chronograph manufacturer. A small strain gage is attached to the gun over the chamber area where it measures the steel expansion during firing. This information is digested and recorded by computer. This does not give a precise measurement in p.s.i. but will give an excellent *relative* pressure. *Relative* meaning in relation to another test or to a standard; a comparison.

The same unit will measure velocity by the standard screen method. With an acoustic target, it will accurately measure group size on bullets traveling at supersonic speed at the range involved. This is similar to the method used by industry. Three special microphones are used in a triangle position to pick up the sonic *crack* from the bullets passage. The computer then *triangulates* the position of the bullet as it passed through the invisible target.

What is most remarkable about this unit, is that it is available for home use at a reasonable cost. In these modern times, a hobbyist can own better equipment than a professional laboratory used just a few years ago.

There are methods to compute the pressure by math, but they usually arrive at a figure higher than the test results. Heat dissipation and heat loss make the math solution only fair in accuracy. There is tremendous heat generated quickly which is lost to the chamber walls, breech face, cartridge case, bullet, etc. It is possible to figure the temperature created but not the amount of loss. This makes it impossible to *accurately* calculate the chamber pressure, because the gas expansion is a direct result of heat.

CASE READING FOR PRESSURE SIGNS

Without expensive equipment, how can a shooter or hand-loader be safe and still experiment? One way is to read the signs on the fired cases. **Some people will not recommend this method because a dangerous situation can develop before it is noticed. While that is true, it can usually be done safely if common sense and caution are used. Be careful.**

First, we have to assume that all proper procedures are followed and case damage is correctly interpreted. Bad and overworked brass can have split case necks and other faults. Head separation can be caused by improper sizing. Primers can move out of their pockets from excessive head-space and/or low pressure. A bulged case may be caused by excessive pressure or it may be from a poorly reamed and oversized chamber. In other words, the pressure may be normal and the cases may still be flawed. But there are case problems that probably are caused by excessive and dangerous chamber pressure.

With the pressure levels used in modern rifle ammunition, a diameter check can be made on the case just above the extraction groove or if a rimmed case is checked, then close ahead of the rim. A belted case can be checked on the belt. The check is made for expansion and compared to factory ammo fired in the same gun. A good micrometer and a delicate touch is needed because a variation as small as .0002" (two ten thousandths) is important. Even better is .0001 if it can be checked with accuracy.

Inside the extractor groove is a good place to check, but special equipment is required. 0" to 1" blade micrometers can be used. These special instruments are a little larger than normal 0" to 1" micrometers. Instead of the standard round gauging surfaces, they have thin blades that will fit inside grooves. Sometimes, wires can be fitted and held inside the grooves and the check made over the wire with a standard micrometer. Not all cartridges are suitable for this and it requires some hand dexterity. It also requires 2 pieces of wire, one for each side, that are straight, accurate and just the right size. A set of *thread checking wires* is excellent, but the cost is not cheap.

Tip: Purchase a *standard*; something that is of a known size and very accurate, such as the 1" gage that comes with a set of 1" to 2" micrometers. The rounded disk type is better than the straight rod type. Use it to practice the delicate feel that is required for accurate micrometer readings. Running the micrometer down to zero gives the appearance of doing the same, but the two flat surfaces, to an experienced machinist, have a different feel than the flat micrometer surface against a round surface.

No matter where the check is made, make it at numerous places around the case because it will seldom be perfectly round. Check new brass and after firing it, check it again. Hand-loaded ammunition should not expand a case more than factory ammunition. Cartridge brass will all be different from manufacturer to manufacture and lot to lot. This will be important to keep in mind. Also remember this only measures expansion, not actual pressure.

Some guns, such as those with fluted chambers, can not be checked this way.

A blown primer is a certain sign of high pressure. There could be other reasons beyond the powder charge. For example, a too big diameter bullet, but it sure demands a close look.

Flattened primers can be an indication of high pressure, especially if the gun does not otherwise cause this problem. If factory ammo primers are flattened, the head-space is the first check to make.

Tight case extraction is frequently used as a check for excessive pressure. In some instances this will work, but the exceptions can be dangerous. Many times, the brass will not yield until the pressure is far beyond the danger point. There may be no warning until the damage is done. Another concern is the condition of the chamber itself. None are equal, and a rough chamber will grip a cartridge while one made properly, that is smooth and polished slick, may come out easily even if excessive pressure has expanded the case. Of course, case type and condition are also factors. If standard factory cartridges will extract properly and a hand-load doesn't, pressure is a probable cause. Repeating something said frequently in this book; use caution and think. Guns can be replaced. Fingers and eyes can not.

This problem with case expansion is not limited to high velocity rifle cartridges. They can occur with low pressure cartridges as used in some revolvers. 55,000 p.s.i. in a 45,000 p.s.i. case is not required. If a case is designed and manufactured for 18,000 p.s.i. and it is subjected to 23,000 p.s.i., it will expand to fill the chamber.

A primer pocket that is not defined as clearly as it was before firing is another caution sign. Also a change in primer can cause a change in pressure. So will a change in flash hole size. A larger than normal flash hole will create a slight pressure increase.

PEAK PRESSURE

Peak pressure is, as the name implies, the maximum pressure in the chamber. It has occurred and is measured before the projectile has moved over one forth the length down the barrel. Peak pressure is what really counts in avoiding damage to a gun or in reaching a velocity. A lot of elements are involved and all are important. Type of powder, amount of powder, bullet weight and design, seating depth of bullet, type of primer, installation of primer, cartridge design, cartridge internal capacity, firmness of bullet in case, temperature and humidity, chamber condition and dimensions, bore size in relation to bullet diameter, and bore condition. This is just a partial list, but as you can see, it is not easy or simple.

The pressure starts to increase from primer ignition and quickly reaches a peak and then drops back again to zero a small fraction of a second after the bullet leaves the muzzle. The peak pressure will be reached in from .0005 to .001 of a second. (five ten-thousandth to one thousandth). In most shotguns, the load will have moved in that short time from a half an inch to one and a half inches. In most center-fire rifles it will be about the same but it will be somewhat less for most handguns.

The peak pressure is what does most, although not all, of the work and can also do most of the damage. The Sporting Arms and Manufacturers Institute (SAAMI) has established safe maximum pressure standards for most U.S. made ammunition. Military ammunition is made to military specifications which will be very rigid and precise.

Velocity is the speed of the bullet in flight, usually given in feet per second (f.p.s.) and can be measured at any point from the muzzle to impact. More on this subject in exterior ballistics. Chamber pressure is a factor in velocity, as is the length of time the pressure is applied. *Progressive* powders which instead of burning at a uniform rate, burn at a rate that increases as the pressure in the case and chamber increase, will increase velocity. The pressure is held against the bullet for further down the barrel because the pressure lasts longer. Pressure and time (duration of pressure) are both important as are other factors we will discuss.

Chamber pressure is frequently measured in pounds per square inch (p.s.i.) and if everything else is equal (not an easy task), more pressure will increase velocity. A chamber pressure of 50,000 p.s.i. on a bullet with a base area of 1/5 of a square inch would give a pressure against the bullet of 10,000 p.s.i. (50,000 * 1/5). If the bullet were bigger and had an area of 1/4 square inch, then it can be easily seen that the pressure would be increased to 12,500 p.s.i. (50,000 * 1/4). If the pressure were held for the same period of time and the bullets weight were the same, the larger bullet would have the higher velocity. Even though the chamber pressure was the same at 50,000 p.s.i., the effective pressure on the base of the bullet would be more.

When we *neck down* a case to use a smaller caliber bullet, we obtain a higher velocity because even though the base of the bullet is less, the gas volume is the same (same case) and better fills the smaller diameter barrel. As a result, the pressure (push) continues for a longer time. Remember, both pressure and time are important. Now, we are not considering that the smaller bullet is lighter but that will also help increase velocity.

Factory ammunition is loaded moderately so it will never get near the maximum operating pressure. This is complicated for factories and hand-loaders alike because of the relationship between velocity and pressure. Fundamentally, the pressure percent will vary by about double the percent of the velocity varied. In other words, a 3% increase in velocity will be accompanied by a 6% increase in chamber pressure. If the cartridge is loaded close to the limit it could easily slip over the line of maximum allowable pressure. Factory ammo is held low so fluctuations in pressure will never become unsafe.

Hand-loaders can operate closer to the maximum safe allowable pressure by the use of better quality control than is possible in a factory environment. Keeping in mind, for safety, that a change in velocity will also cause a change in pressure at about double the percent. If operating close to maximum load, a small percent velocity increase can push pressure over the safe limit. (This is reminiscent of the old question, which came first, the chicken or the egg. To be precise, the pressure would come first and then the velocity as a consequence.)

Testing shows that an increase in pressure can, at times, actually lower velocity. Some things, such as bullet size, that increase pressure also increase the friction between the bore and the bullet. Others, like seating depth and tightness in the cartridge, will do the same. There are more, but the point is made that while a pressure increase will usually increase bullet velocity, it will not always do so.

AVERAGE PRESSURE

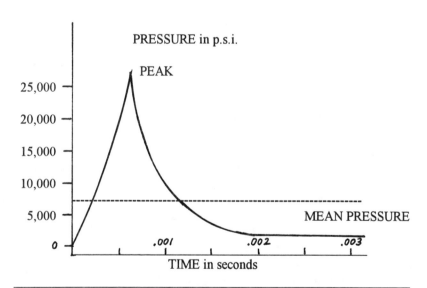

Many of the math equations use pressure. It is obvious that the lowest starting pressure would not be usable, but neither is the peak pressure. The pressure, if plotted on a graph as a curved line, will climb rapidly to a peak, drop rapidly and then drop slower as the projectile accelerates down the bore. What is required is average pressure.

◆ The average pressure is calculated from the energy and the volume.

Pressure = Energy / Volume

Where: Volume is the bore area based on cross-sectional area times length.

PRESSURE LIMITS

In 1971, the H. P. White Laboratory tested revolver pressure overloads for the Treasury Dept. With the .38 Special, they found that light-frame and cheap guns may experience a failure at 50,000 CUP. Medium frame guns of high-quality could stand up to 70,000 CUP. This does not mean that all cheap guns will be safe up to 50,000 CUP and better quality weapons at 70,000 CUP. One important lesson; there is a major difference between metallurgy of cheap guns and quality guns. The old expression that "you get what you pay for" applies to firearms.

◆ Actions and cartridges must be a suitable match for pressure as well as cartridge fit. For example, it is possible to change barrels and convert a .30-40 Krag to .30-'06. The .30-40 Krag produces a breech pressure of about 43,000 p.s.i. and the .30-'06 about 55,000 p.s.i. The Krag action and its single locking lug cannot handle the extra 12,000 p.s.i.

◆ The U.S. Springfield Trapdoor models of about 1873 to 1884 are safe at pressures to only about 25,000 p.s.i. maximum.

(Refer to *elastic properties*; later in this chapter.)

WHIP, FLEX, BULGE & OTHER BARREL PHENOMENA

Many of the phenomena discussed here are very slight, but they can be proven by testing and very much influence accuracy. They also are the reasons shooters wanting extreme accuracy use barrels and actions that are bedded so there is no contact with the wood. (Free floating) Also, heavy, stiff barrels are not practical to carry for hunting, but they are best for accuracy because of the reduced twisting, bending, flexing, etc. The lighter and thinner the barrel, the more it is affected by the forces.

That gas expands equally in all directions is a basic fact. Not just against the bullet's base and the rear of the chamber but outward in all directions. The pressure on the barrel diameter creates an expansion of the barrel. *Swelling* is another word that describes the action very well. The amount will depend on the type of barrel, thickness of wall, type of metal and other factors, but it **definitely will expand**. The expansion will be the largest at the breech end and will be less as the bullet moves down the bore and the pressure drops. As the projectile passes, the expansion diminishes.

A fair analogy would be the pictures seen on educational television of a giant snake swallowing a dog or small animal. The outside of the snake expands as the meal is swallowed. The location of the animal is easily seen by the slowly moving bulge. For all purposes, the same thing happens to the gun barrel. Faster and for different reasons, but still about the same thing.

This can be proven by firing a bullet into a substance such as water where it can be recovered without damage. Measure the diameter of the bullet and repeat the operation after cutting off an inch of the barrel at the muzzle. Continue to remove inches of barrel and measuring the bullet diameters. The bullets will be larger as the barrel is shortened. The bore diameter should have been the same for its entire length before firing. Of course, an extra heavy bull barrel will not expand as much as a thinner barrel because the cooler outside surrounds it and holds it in.

Upon firing, the fixed barrel is suddenly twisted by the bullet being forced to rotate in the rifling (helical cut grooves). The torque required to start the bullet rotating imparts a distortion on the barrel which increases as the bullet moves farther away from the action and other methods of holding it in place. This is a counter rotation to the bullet. An interesting point is that the barrel distortion created by torque, precedes the bullet down the barrel. (That is, it moves ahead of the bullet.) It stands to reason, then, that if the barrel is affected ahead of the bullet, that good accuracy cannot be obtained by laying or touching the barrel to any hard object at the time of firing. The barrel has to be able to move in a normal way.

There is tremendous tensile stress placed upon the steel barrel and chamber caused by rapid expansion from the intense heat and the subsequent cooling and contraction. The gas molecules colliding with the inside of the barrel impart thermal energy (heat) to the steel as they slow down from the collisions. Cracking and failure caused by stress fatigue is the normal consequence. The weak points in a steel's grain texture or faults in design or manufacture are the likely places.

Any object that is horizontal and supported only on one end and the middle will have a droop at the unsupported end (if only slightly) and a rifle barrel is no exception. We must also consider the barrel may not be perfectly straight from other reasons. In any case, the force of the bullet and the rapidly expanding hot gas will try to straighten the barrel. The molecular bombardment creates shock waves that influence the steel. Also, the combined result will be a *vibration,* as it is frequently called.

A vibration is described as a periodic reversal of motion. Some people believe the barrel movement is better portrayed by a different word. Probably vibration indicates more cycles up and down than actually happen. In many barrels, there will be a movement, but not in the form of cycles. But, call it whatever you wish, the movement is real.

A thin barrel, as used on the Winchester Model 70, will flex down as the bullet is fired and will most likely not be straight at the moment the bullet leaves the muzzle. If properly sighted in, it will present no problem so long as the ammunition remains the same. Any change in load or ammo will cause the bullet to leave the barrel at a different point in the arc and can cause a big difference in the impact point. A very thick diameter barrel will correct the problem, but is not suitable on a hunting weapon due to weight.

A fisherman's fly rod will demonstrate this downward flex. Hold the rod out horizontal and touching only the handle, give it a sudden snap upward. Notice that the tip started down for a split second before it followed the rest of the rod up. If it was hard to see, try it again. The similarity is close enough to envision the barrel doing the same only in a lesser amount.

Perhaps a good description would be "a deflection away from the direction of original movement, the magnitude of it building up logarithmically with respect to time."

For example, let us say we have a .22 rifle barrel that is .750 inch diameter and 24 inches long. The bullet will take about .004 (4 thousandths) of a second to exit the muzzle. If the barrel whips down 4 minutes below the zero line the upswing will be 0.19 minutes above the zero line. Of course the main concern is how the muzzle is curved from the main axis, not the barrel behind it. At the muzzle it will be about 0.5 minutes.

In the above example, a tapered barrel of the same weight would have less droop and less whip. The droop, if not supported at the receiver, would be 1.7 minutes and the more important muzzle figure would be 2.18 minutes from the zero line. Of course, in most firearms there would be support past the receiver, but this example shows that droop and whip are not fantasy. The barrels are made of thick steel that people can not bend, but other forces can.

The whip may be extreme enough that the barrel will move both above and below the zero line while the bullet is still inside. The starting movement, while usually down, can be either way and measurements show that 7 to 15 minutes of angle is common. (A degree is divided into 60 minutes and each minute into 60 seconds. In this respect, a degree is similar to an hour of time.) The .22 is more likely to make the initial movement down, than are the bigger calibers which may flex up.

With a rifle that shoots consistently in one place on the target, try a shot with the barrel end placed on something firm that will prevent the down movement. If this particular barrel flips down, then the bullet will be higher on the target because it will be prevented from doing so.

The whip, vibration and twist can be reduced by the addition of weight to the barrel. It is not uncommon for shooters trying for tight groups in target work to add weights that are adjustable as to position along the barrel. Careful test shooting at different positions will show where the best results are obtained.

One cheap method is to use two blocks of steel about 2" x 2" x 1". Place them together and drill a hole at the junction the same diameter as the barrel. If much taper is required, a metal lathe may be needed. Add a thin cloth or piece of rubber shim to prevent slippage and barrel scratches. Staying in the bulk alongside the barrel, drill bolt clearance holes in the top piece to correspond with taped holes in the lower piece. The position on the barrel is important. It will take testing to find the proper location. The moment of inertia and the balance point in relation to where the bending and twisting take place are critical.

Barrel weight is important in another way. The center of gravity of the gun is moved forward and this helps hold it down. Handguns in heavy calibers gain much in this respect.

One reason for different loads or different ammunition striking the target at a different place may not be the trajectory. If the trajectory is about the same, the bullet may be leaving the barrel when it is at a different place in its movement from whip or flex. **To hit the target at a consistent place, the bullet must always leave the moving barrel when it is at its same point in that movement.** The only way to be certain where a load will strike is by sighting and test firing. Change load or ammunition type and test again, even if the trajectory looks the same on paper.

The *action time* is the time between the firing pin hitting the primer and the bullet exiting at the muzzle. This time will not only change with different types of ammo, it will vary slightly from one shot to the next. The shots that have the lower velocity will take longer to exit the barrel and will leave at a higher angle. The lower velocity shots will, therefore, strike higher on the game or target.

Compensation is the word used to describe what we have been discussing. It can be called either favorable or unfavorable. It is when bullets with low velocity leave the barrel at a higher angle than the faster bullets. The vertical spread on the target is less than predicted. Favorable compensation is where higher velocity is low on the target in a small spread. Unfavorable compensation is where higher velocity is higher on the target in a larger spread.

While there are advantages to longer barrels, up to a point, the shorter barrels are not as influenced by the twisting and whip, etc. of the longer barrel. Heavy, thick bull type barrels are also less affected but heavy to carry. Most carried guns, including handguns, require a compromise by necessity. In a rifle, a thin, long barrel will be influenced the most. A long barrel will vibrate (or move, whip, bend or whatever) more than a short barrel. The long barrel may not necessarily be more accurate. It may be worse, unless it is very heavy and thick.

On the other hand, as was mentioned, a longer barrel will give a higher velocity up to a certain length. The powder will all burn and the gas pressure has time to give a longer push, but the velocity increase may be slight. (This covered in detail in Chapter 11.)

A .30-'06 cartridge that has a muzzle velocity of 2,600 f.p.s. in a 20" barrel goes to 2,700 f.p.s. in a 24" barrel. The same cartridge in a 30" barrel will produce 2,850 f.p.s. Of course, not all cartridges fit this exact pattern, but this one will gain 25 f.p.s. for each additional inch. The .22

long rifle is most efficient at between 18" and 22" in barrel length. All calibers and cartridges have a long length where the expanding gas and powder have done their job. (see Chapter 11) They will not be able to continue to accelerate the bullet against friction and drag and more length will have no benefit.

A sling can pull the barrel out of proper alignment by up to 3 minutes at a 10 lbs. pull. A proper fitting one piece stock helps reduce this.

It is interesting to note that many rifles have a small lateral jump (sideways displacement) during firing. The amount has been determined for some military rifles. For example, the .30-'06 bullet from a 1903 Springfield was found to be a little to the left at a 500 yard range. The gyroscopic drift (see Chapter 9) to the right from a right twist barrel is not enough to overcome it until a longer range.

PROOF: In the late 1950's and early 1960's, a large number of British No. 4 rifles in 7.65 mm were converted from the service barrel to barrels that were thicker and heavier. The original barrels weighed 2 lbs. 9 oz. The new barrel's weight was 4 pounds. There was a clear improvement in the group size at both moderate and long range.

The testing also showed that the barrel was moving upward while the bullet was still in the bore. Changing to a higher velocity ammunition caused the bullet to move down the bore faster and exit while the barrel was at a lower point. The faster velocity bullet not only had a flatter trajectory but a lower elevation.

Both of the two items pointed out about the No. 4 rifle prove items mentioned earlier.

RECOIL: In barrel movement, there is a jump or rise caused by recoil. This should not be confused with a much smaller change, whip, which is usually down and is a bending of the barrel itself. (Recoil is explained in detail in Chapter 6.)

ELASTIC PROPERTIES

Much of the barrel bulging and other changes are caused by the steels elastic properties. We tend to think of steel as being unyielding. It is hard to visualize steel, whether cold or hot, being flexible; but it is. It will flex, expand, stretch, spring, and change in many ways and has the ability to recover its original shape without permanent damage. Up to a point. All material has a limit, called an *elastic limit*, and Hooke's Law states that within the elastic limit, deformation produced is proportional to the stress. (or *force*, if you prefer that term.) Hooke's Law, by the way, is named after Robert Hooke, 1635-1703. He correctly stated that all materials, steels included, will be deformed and not return to normal if stressed beyond that limit. Worded in another way, the material, a gun barrel in this instance, has resilience; that is the ability to bounce or spring back as long as the maximum limit is not reached.

For barrels, much of the stress comes from the hot expanding gas. There are as many variables as there are barrels. No two are equal, even in the same model. For an example, consider an average rifle barrel that uses cartridges that are loaded to maintain a pressure peak at or below 50,000 p.s.i. For these barrels, pressures in the vicinity of 80,000 p.s.i. may cause a deformation that will be permanent. If the pressure reaches 140,000 p.s.i. a catastrophic rupture may occur. (These are fancy words that mean it will destroy the firearm and may remove a few much needed body parts.)

CARTRIDGE SHAPES

Not all authorities agree on the aspects of pressure and how it relates to different cartridge shapes. Muzzle velocity is easy to test and peak pressure, while more difficult for the nonprofessional, is still basically an easy task. Back in the 1940's, Olin Industries' ballistic laboratory in New Haven, Connecticut had the equipment and talent to take high speed x-ray photographs of firearms in use. 1/1,000,000 (one millionth) of a second exposure through a 360,000 volt x-ray machine made great photos of bullets and shot pellets as they moved down the bore. Still, we can not normally see what is happening in a chamber at ignition. We can only surmise and guess based on what facts we know about physics and chemistry and what actions seem logical. It is no wonder that people can not always agree, because a lot has to be taken on the mental acceptance of something unseen and unproven. Faith is fine in religion but falls short in science.

In tapered and bottleneck cartridge cases, not all of the total pressure inside is expended rearward. Some of the pressure is used against the tapered walls and case shoulder on the front end and attempts to hold the case to the front.

TEMPERATURE CHANGES

Heat can and does increase the pressure and give a higher velocity. The problem is caused when the pressure increase is more than safe. Hunters in the colder climates of the far north sometimes make up both winter loads and summer loads. There can be a problem with excess pressure in using winter loads in the summer. A very hot chamber from rapid firing can cause the same danger if a cartridge is allowed to stay in the chamber for a period of time before firing.

The heat from the hot expanding gas does very little to erode or burn away the bore. That it does cut away the metal has been a common belief for many years and it sounds logical and reasonable. The difficulty is that it is not true, at least not with the intent that is applied. It may remove a tiny amount of metal but the amount is so small that it is unimportant. If hot gas caused the metal erosion it would be where the projectile sealed the smallest and permitted the gas to escape past the bullet. That area is in the grooves but they erode the smallest amount. The lands wear the most starting at the chamber. Bullet surface friction is the primary source, although the heat from the gas will be a contributing factor and only a contributing factor. Nothing more.

That is not to say that heat does not have any effect. It can damage a barrel and change its accuracy. Machine gun barrels can have a very short life, with heat being a problem. Early barrels have even been water cooled. Machine gun barrels can also expand so much that the rifling has little influence and the bullets will have no gyroscopic stability. The gas escaping around the bullet also lowers the push and in turn the velocity and energy.

While most of us aren't concerned with long bursts from machine guns, the heat created from a normal weapon can influence accuracy and wear. The heat that is transferred to the barrel and action is lost and wasted. Only about 30% of the heat created is gainfully used to expand the gas and push the bullet. Another 30% is wasted on heating the barrel and the 40% balance goes to recoil, noise and escaping at the muzzle. With only about 30% being useful in creating kinetic energy, a firearm is not an efficient machine. (Note: Readers with good memories will notice that different figures were given a few pages back. That was energy and this is heat. Similar but different.)

The heat produced can exceed the melting point of the steel in the barrel. The short time the heat is influencing the metal is the only thing saving it from greater damage. The heat is dispersed by radiation into the air and by conduction into the stock and anything else touching it.

A heavy barrel absorbs more heat than a thin barrel and the larger surface area gives an increased heat transfer to the surrounding air. The heavy barrel has advantages in rigidity. They are usually inconvenient to carry hunting, but excellent for bench-rest and some other uses.

TIME

The time involved in interior ballistics is divided into 4 areas.

Lock time is the first, and it accrues before ignition. It is the time from when the trigger releases the sear until the firing pin hits the primer.

Ignition time takes over at the end of lock time. It is the time between when the firing pin strikes the primer and when the pressure of the hot gases is high enough to start the bullet moving from the case.

Barrel time is next and it starts where ignition time ended. It is the time from the projectile's first movement out of the case until it arrives at the muzzle. An average time for this action will be around .001 second. (One thousandth of a second.)

While it seems impossible to measure these short time intervals, it is almost easy with modern technology. Early science measured them accurately with methods that today seem crude.

It is possible to obtain an approximate lock time from a formula.

$$S = ft^2 / 2m \qquad t = \text{square root of } 2ms / f$$

Where: f = weight of the spring

S = length of travel for firing pin

t = time pin takes to drop

m = mass of the firing pin + 1/2 the mass of the spring

BARREL OBSTRUCTION

To prevent excessive pressure from damaging the gun and possible injury to the shooter, the bullet must be able to move forward out of the case and through the bore to the exit. It is easy to understand the dangerous situation created by a barrel obstruction. If the chamber neck is undersized or the bullet is much oversized or the case neck itself is big, especially with a bullet seated, use caution. The bullet will be held in place and the expanding gas will have no avenue of escape. The pressure will increase rapidly and the burning rate will increase as well. The two will "chase each other" until something blows out and the pressure is released.

BARREL LENGTH

A ballistics development is the Browning BOSS, which stands for Ballistic Optimizing Shooting System. This is a device that is threaded onto the muzzle end of the barrel and looks like a compensator, which in part, it is. It also can be used to adjust the barrel's length which, it is claimed, will tighten groups by reducing or shifting the vibration pattern.

As can be understood by a study of this chapter, a change in barrel length will also alter the barrels amplitude of vibration. Or simply put, it will vibrate or tremble differently. This will change the bullet's impact point, for better or for worse, or make it more consistent.

It has long been known that in some cases, the simple act of removing an inch of barrel and then recrowning it, can improve accuracy. Adding length would help at times, but this was not possible. All guns and barrels are slightly different, and what is correct for one may not be for another.

The public will see more of this innovative idea in the future. For ballistic studies, it proves some of the whip and vibration beliefs that a few readers may not have accepted. A lot of science and physics, whether for ballistics or any other subject, requires some faith. We have to prove and understand what cannot be seen by other things that can be seen.

NOTES:
♦ It is a wrong to believe that a fine grained and quicker burning powder always gives a higher pressure. Other factors must be considered.
♦ From ignition to exit at the muzzle, the time a bullet is in the barrel is called *barrel time*. A .22 Long Rifle in a 22 inch long barrel will be about .003 seconds. (Three thousandths of a second.)
♦ Guns operating at very low pressures have problems with holding a consistent velocity and with unreliable ignition. Low pressure also requires a light bullet to obtain ample velocity and also will develop low energy.
♦ A little basic math: For any enclosed bore or caliber, the volume is proportional to the square of the radius.
♦ More basic math: If we double the bore diameter, the bore area increases 4 times. (2^2)
♦ A Naval Academy text has the following statement: "..........a shot from a cold gun usually falls short................consistent shifts in deflection after the initial round have been observed in the case of some guns, and it is probable that such shifts occur for all guns, though they may often be small enough to escape notice." This is based on larger guns, but the rules are the same. It does not change accuracy as far as group size, but it will change group position.
♦ We have discussed all the heat expansion and twisting and bending of gun barrels. We have shown the advantage of a thick bull barrel for accuracy. It is interesting to note that the pressure exerted by the expanding gases can be contained in a thin barrel. Tests have shown that a barrel can be machined down amazingly thin and still hold normal ammo. Of course, this would be inaccurate as well as unsafe. After all, normal barrels have burst because of heavy loads, damage, or a factory defect.

CLOSING POINT

A gun barrel is subjected to lateral accelerations, stretch from gas pressure, expansion from gas pressure, recoil jump, gyroscopic forces from the bullet, counter rotation, whip up or down, friction from bullet and gas, high temperature expansion, and bend by combustion of a displaced mass, and pressure from the stock, shooters grip, sling strain, etc.

The end result will depend on how many of these factors cancel each other out or work together in the same direction. Another consideration is the effort to keep them all the same from shot to shot, so that it will remain the same. No easy task.

Chamber pressure is not just a subject to study in an academic manner. People with old weapons in poor condition or perhaps trying to find a cartridge that will work in grandpa's old Remington Rolling Block of unknown caliber, have to use caution. And hand-loaders know the risk they take in disregarding published figures. Don't they?

CHAPTER 6

RECOIL

In addition to being painful, recoil can be confusing if it is studied in detail. As with most fields of science, there is more to the subject than is noticeable on the surface. To properly understand recoil, some basic mathematics is required. As always in this book, the math will be available for readers that are interested and can be omitted by those that would just as soon disregard it. This explanation will nicely cover the subject even if the reader skips the mathematics. Let's try to break it down and explain it in easy terms.

THE BASIC FACTS

As we discussed in an earlier chapter, Sir Isaac Newton reported in the late 17th. century that *for every action there is an equal and opposite reaction.* Know as Newton's third law of motion, it is the basis for the rearward push of a gun. Sometimes called *kick,* yet kick is but a part of recoil, which is the best word to use. Kick is the feeling or sensation felt by the shooter and is different than the specific energy measured in foot pounds as recoil.

A quick way to visualize recoil is to skip a lot of the small details and consider only weight. That isn't to say that the other things are not important, for they are, but now we want just the barest basics. For an example, let us use a rifle that weighs 9 lbs. With 7,000 grains to a pound, the gun weighs 63,000 grains. If we use a 150 grain bullet, the gun will weigh 420 times more than the bullet. Recoil will be 1/420th. of the bullet's velocity. As stated, this is a simplified explanation.

The average military rifle weighs about 400 times more than the bullet. A shotgun will recoil harder because the weight of the shot is heavier in proportion to the gun's weight. The weight of the projectile affects recoil and so will the weight of the gun. A heavy gun will recoil less than a lighter gun if the projectile is of equal weight.

Recoil is caused by the expanding gas of the burning powder. It forces the bullet out of the case and down the barrel and exerts an equal force back against the rear of the chamber, breech, stock, grips, etc. The force is the same in both directions, but the gun is much heavier than the bullet so it is easy to understand why the bullet is moved much faster and farther than the gun.

Therefore, we see that everything else staying equal, a heavier bullet will increase recoil because it will require a heavier force to move it forward and hence a heavier force to the rear as recoil.

Recoil has a major influence on the impact point, because the barrel is moving up as the bullet comes out of the muzzle. This will be the same as long as the ammunition and load remain the same. Any change in ammunition will also change the point of impact. (While barrel whip is connected with recoil, it is explained in detail in another chapter.)

ADVANCED FACTS

The products of combustion generated at the time of ignition, mostly gas but including a tiny amount of liquid and solids, have the exact same weight as the powder itself. This travels out the gun barrel with the bullet and should be included in any weight that is used to figure recoil energy. A **black powder charge is heavier** than smokeless and it is noticeable in the heavier recoil of some weapons. Big revolvers made for large loads of black powder have a heavy kick and are especially good at demonstrating this fact.

A key word is *acceleration* because the recoil, as mentioned, begins the instant of ignition and continues as the bullet and expanding gases are accelerating through the bore.

Recoil begins the instant the bullet starts to move from the case and may move the barrel some in relation to the intended point of impact. This initial recoil is measurable but it is not a factor in aiming because if the weapon is sighted-in properly, the barrel movement will be the same with each shot. The change in bullet path will be the same. Of course, only with one type ammunition. A change in ammo may require a change in sight adjustment.

MUZZLE BLAST

Although it has little effect on the bullet and its flight, adding only about 1% to the velocity, there is an influence on recoil from the backward push of expanding gas **after** the bullet leaves the barrel. The gas pressure has already started the recoil process, and an extra push is given to the muzzle by the pressure of gas as it expands in all directions. The bigger the area of the muzzle, the bigger the push. The amount of force depends mainly on the expansion volume inside the barrel while the bullet is still there. This is one of the reasons a short barreled carbine with a high expansion factor may have a light recoil. Experts do not agree on how much of the kick is owed to this external push. They do agree that the expanding gas leaves the muzzle at a speed much higher than the speed of the bullet. This gives a reactive push to the muzzle. Of course, in the event of a barrel long enough for things to slow down inside, this would not be true, but barrels are not used in these extra long lengths. (What would be considered too long would depend on the cartridge and caliber and a few other things, as explained elsewhere.)

Measured by the U. S. Army on artillery at 4,700 f.p.s., the escaping gas will be less on small firearms. A general rule of thumb would be about one and a half times the projectiles velocity. An example would be approximately 4,000 f.p.s for a projectile with a 2,700 f.p.s. muzzle velocity.

This *after shock* is not felt as a separate kick, but perceived by the shooter as just a jolt. Actually it is several impacts rapidly combined together in what has been described as *waves*.

The maximum recoil or wave is reached after the bullet, powder and gas acceleration ends, which is always shortly after the bullet has left the muzzle.

Earlier in this book there was a brief discussion of a *secondary muzzle flash* which, while not a regular occurrence, sometimes does happen. This rapid gas expansion will increase the recoil a slight amount, although the increased sound and expanding air make it seem much larger than a measurement would indicate.

MUZZLE CLIMB

All recoil acts in a straight line to the rear of the barrel, but the apparent recoil may be different. The muzzle, in most cases, will swing up because the line of thrust, acting in line with the bore, is above the center of mass of the firearm. Also the bore axis, if extended, passes above the rifle or shotgun butt or handgun grip. In other words, above the resistance point at the hand or shoulder. This creates a pivot point at the gun butt for a long-gun and the rear of the grip on a handgun. The gun will then swing up from that location. And in many cases, particularly when shooting from a standing position, the bore axis is also above the shooter's mass center. The shooter and gun will both tend to pivot or swing away together. The recoil will be perceived differently with a straight stock (usually believed to be lighter in recoil-kick) than with a drop stock.

A simple test may help make this more understandable. Take a rifle or shotgun. **Make certain the gun is unloaded and keep your finger away from the trigger.** Place the butt of the stock against your shoulder and hold it in a normal position **except for your trigger finger.** Have someone push against the muzzle **after you have shown them the gun is unloaded.** Note how the gun rises and turns to the side as well as moves back. The side it turns toward will depend on the shoulder used. Right shoulder, right swing, etc.

TORQUE TWIST

Some weapons also have a noticeable torque twist. For readers that have never experienced it, the gun wants to turn or rotate around the barrel. This is experienced mostly with handguns like the .44 magnum that have a strong recoil and light weight rifles with high velocity cartridges. The torque twist is an opposite reaction to the rifling in the barrel which is forcing the bullet to turn one way as it turns the gun the opposite way. It would be proper to say that it was Newton's third law in operation once again.

The torque twist to a rifle is normally not noticed because it is absorbed by both the firearm and the shooter. A short explanation of why it cannot normally be felt is in order. This is a simple explanation that will not take into account such factors as the brief time interval that is involved and the delay caused by the greater mass of the rifle, which absorbs the torque.

That last sentence gave away part of the reason, *the greater mass of the rifle*. For example, a rifle that weighs 8 3/4 lbs. and a bullet that weights 100 grains. We already know from the math chapter that if dividing the 100 grains by 7,000 will convert it to pounds. In this case 0.0143 which is almost exactly 1 / 70 pound. The rifle will then weigh 612 times more than the bullet.

Since we were children and played on a see-saw, we have all had a basic understanding of leverage. We know what to expect from using a bar over a fulcrum to lift a weight. Also we know that as long as the bar has the strength, we can lift almost any weight by using a long enough bar with the fulcrum close or closer to the weight.

What does all this leverage business have to do with firearm rotation? A .30 caliber bullet will have a radius of about 0.15 inch and the leverage will be about 2 / 3 of the radius or 0.10 inch. We therefore have 1 / 70 pound multiplied by 0.10 inch. A small amount indeed. We can make an educated estimate (ok, a wild guess) that the leverage of the rifle will be one inch from the center line of the bore. The rifle, we figured, weighed 612 times more than the bullet and has a resistance to rotation 10 times as much. Therefore, it will take 6,120 times more effort to turn the rifle compared to the bullet.

AUTO-LOADING FIREARMS & RECOIL

Autoloading firearms have the same measured recoil, other factors being equal, but the kick felt by the shooter is less for some types of actions. The spring mechanism that is used to load the next cartridge distributes some of the shock that would be felt by the shooters shoulder or hand.

When checked with a recoil pendulum which measures total momentum, autoloading arms give the same readings whether they are used as autoloading or set to function as a fixed-breech weapon. This would be expected as a result of the law of conservation of energy.

The Springfield Armory M1A semi-automatic rifle has an *on* or *off* valve for the gas system. In other words, the semi-automatic mode can be deactivated. Felt recoil is distinctly harder when the valve is off and the gun in operated without the gas system. An interesting side note; a chronograph will show a 10 to 15 f.p.s. drop in muzzle velocity with the valve off.

Free recoil energy depends only on the gun's weight. The springs absorb energy when compressed and then release it as they return to their original length. The only energy lost is with heat in the spring.

To avoid confusion, remember that felt kick and recoil are not the same thing.

In the long recoil type of shotgun, the breechblock and barrel recoil separately from the remainder of the firearm. That is about 2 1/2 pounds or perhaps 35 times the shot in a trap load. This energy is picked up by these parts and will be transferred to the gun and the shooter's shoulder. The early Browning is an example of this system. Also the Remington Model 11 and the Winchester Model 1911.

In a blowback action, the breechblock is not locked in place, but held closed by spring pressure. When the cartridge is fired, the bullet goes forward and the breechblock moves backward. The bullet is lighter and not held by spring pressure, so its velocity is much greater. Some machine guns use a retarded blow-back action where the breechblock is slowed by lugs or a toggle.

In a short recoil and long recoil system, both lock the breech bolt to the barrel and the barrel recoils back with the breech.

In the long system, the movement will usually be several inches. The breech bolt and barrel stay locked together to the rear. The used case is ejected and as the breech bolt moves back forward toward the barrel, it picks up and loads a new shell. It is a good system and used in some Browning semi-automatic shotguns, Model 11, and Remington Models 8, 11 and 81. Also, it is found in Savage, Breda and Franchi. It should be noted that the recoiling barrel's weight adds to the felt kick and experienced shooters can distinguish two separate blows. This double kick is from the fired shell the breech bolt and barrel hitting the receiver. Friction rings and springs control the speed that the parts move back. Proper adjustment is important to keep the felt recoil at a minimum.

For an example of a *short* recoil system, one factory claimed their brand of autoloading shotgun would reduce felt kick by 20%. Winchester claimed this for their Model 50 in the 1950's. The barrel is fixed and not moveable. The breechblock and chamber move backward .090 inch from gas pressure. An inertia weight in a tube in the butt-stock carries the movement on backward and unlocks the breechblock and chamber after the shot has already left the muzzle. In most short recoil systems, the barrel only moves about 1/8" to 5/8". The Browning double automatic with its 5/8 inch short recoil and gas operated shotguns like the Remington model 58 have normal kick or felt recoil. The Remington takes gas from the fired shell to operate a piston which operates the action.

There are different types of mechanisms used in semi-automatic firearms. The long recoil type of system will make felt recoil worse. Other systems will make it feel the same or less as in the Winchester model 50 and a few others. Today most guns use a gas system instead of the recoil system.

KICK REDUCTION

While the actual recoil will remain the same, the felt kick can be reduced by the type of stock used on both a rifle and a shotgun. Perhaps it is not in the field of ballistics, but many things can help. The stock's shape, length of pull, shape of butt, use or lack of a recoil pad, even the shape and size of the forearm (forend) and the resultant grip are important. Also, holding the gun properly can make a big difference.

With handguns, the options are not as varied. Different grip size and shape, muzzle brakes and the way it is held can help. In some instances, the gun can be changed to one of a different design or frame size while staying with the same caliber.

From a ballistic standpoint, many firearms are too big for the job. It is true that the proper gun for either self defense or game hunting should be as large as needed and/or as large as the shooter can handle. No one would suggest a .22 caliber auto as a serious self defense weapon. A few .38 autos are small enough for concealment and pack a lot more stopping power, although they are still marginal. By the same token, a .44 magnum is too much for a lot of people. It is also more than most people need. Whether hunting or self defense, if the gun you are shooting has too much kick, perhaps a different gun or caliber would be easier to handle and still do the job. (See the terminal ballistic chapters, 24, 25 and 26. There is enough information in this book to make picking a suitable caliber an intelligent decision rather than a wild guess.)

SHOTGUN DOUBLING

Doubling, that is firing both barrels simultaneously from a double barrel shotgun, is not one of lifes' little pleasures. Yet if the gun is held firmly and properly, the shooter should not receive any permanent injury. The average double shotgun will weigh about 7 1/4 lbs. and with a 3 3/4 - 1 1/4 load, will have a 34.5 ft. lb. recoil. If both barrels fired together, the energy would be 138 ft. lbs. Notice that the difference is 4 times, not 2.

A doubling shotgun rarely fires both exactly simultaneously, even though it may appear so. There will be a small fraction of a second difference. The result is two recoils extremely close together. True doubling, that is both at precisely the same moment, can happen in a single trigger gun. In a double trigger it is not as likely, although possible.

If the second shot takes place as long as .003 second (three thousandths of a second) after the first, the shot charge from the first will have left the barrel. Therefore, it would be two successive kicks rather than one extra heavy. Although, to a shooter who has just felt this double kick, the technical points will not ease the soreness. Even if a trip to the doctor is not needed, doubling will call for a trip to a gunsmith experienced with double barrel shotguns.

At one time, Francotte, a Belgian gun-maker, made a 10 gauge double that had a third sear and could selectively fire both barrels simultaneously on purpose. (Talk about Sadomasochism.)

REVIEW NOTES

There are 3 main items that influence recoil that can be controlled by the shooter. **(1)** The weight of the gun. The heavier the gun, the less recoil, if all other things are equal. A reduction in weight of about 20% will increase recoil by a like amount. Other things to consider are: **(2)** the bullet weight, as a heavier bullet creates more recoil, and **(3)** the powder amount and type. The more powder amount and the more power of the type, the more recoil. The velocity is a factor, but it is a product of the others, in a manor of speaking. Also, remember the push from the departing gas, sometimes called muzzle blast.

Of course, any change in bullet or powder will also affect the performance of the firearm. If a lighter bullet is used, the recoil will drop but so will the performance. More powder can be used to increase the velocity of the light bullet and help the performance, but then the recoil goes back up again. They may not balance out exactly, but they will to a large extent. The powder used to increase velocity will always increase recoil. Not always a like amount, that is 20% to 20%, but enough to be important.

RECOIL INCREASE- BASIC MATH

Recoil increases as the square of the velocity. If the velocity is increased by 15% the recoil is increased by 32%. $(115 / 100)^2 = 1.3225$

Recoil also increases as the square of the bullet weight. Previously the comment was made that if everything else stays equal, a heavier bullet will increase recoil because it will require a heavier force to move it forward, thence a heavier force to the rear (recoil). With recoil energy squared, it increases steeply with an increase in bullet or shot weight. But, it decreases to only the first power with the guns weight so a change in the guns weight has much less effect. Or to put it in different terms, the bullet's weight and its velocity have a larger control over recoil than does the gun's weight.

RECOIL TABLE FOR RIFLES

Cartridge	Bullet wgt. (grs.)	Charge wgt. (grs.)	Muzzle Vel. (f.p.s.)	Gun Wgt (lbs.)	Recoil Impulse. (lb.sec.)	Recoil Vel. (f.p.s.)	Recoil Energy (ft.lbs.)
.22 Hornet	45	11.5	2690	7	0.74	3.4	1.3
.223 Rem.	55	27.0	3240	7	1.27	5.8	3.7
.22-250 Rem	55	36.0	3680	8	1.54	6.2	4.8
.243 Win.	80	48.0	3350	8	2.04	8.2	8.4
6mm Rem.	100	46.0	3100	8	2.19	8.8	9.7
.250 Savage	100	38.0	2820	8	1.93	7.8	7.5
.25-'06 Rem.	120	50.0	2990	8	2.48	10.0	12.4
.270 Win.	130	58.0	3060	8	2.79	11.2	15.7
.270 Win.	150	56.0	2850	8	2.89	11.6	16.8
.270 Win.	150	56.0	2850	7	2.89	13.3	19.2
.270 Win.	150	56.0	2850	6	2.89	15.5	22.4
7mm-08	140	47.0	2860	8	2.61	10.5	13.7
7mm R.Mag	150	79.0	3110	8	3.47	14.0	24.3
7mmR.Mag	175	66.0	2860	8	3.39	13.7	23.7
.33-30 Win.	170	32.0	2200	7	2.23	10.2	11.4
.308 Win.	125	50.0	3050	8	2.58	10.4	13.4
.308 Win.	150	47.0	2820	8	2.71	10.9	14.8
.308 Win.	180	50.0	2620	8	2.98	12.0	17.9
.30-'06	180	56.0	2700	8	3.15	12.7	20.0
.300 Win.M	180	74.0	2960	8	3.68	14.8	27.2
.300Wby.M	180	85.0	3245	9	4.10	14.7	30.1
.338Win.M	225	69.0	2780	8	4.00	16.1	32.2
.375H&H M	300	76.0	2530	9	4.72	16.9	39.8
.45-70	405	30.0	1330	8	2.92	11.8	17.2
.458Win.M	500	66.0	2040	9	5.70	20.4	58.1
.460Wby.M	500	130.0	2700	10	8.30	26.7	111.0

It is interesting to observe that there are three 150 grain .270 Win. cartridges with gun weights of 6, 7 and 8 pounds respectively. Notice the velocity and energy increase as the weight decreases.

RECOIL TABLE FOR SHOTGUNS

Gauge & Shell lgth. (ins.)	Dram Equiv. load	Shot charge (ozs.)	Muzzle vel. (f.p.s.)	Gun wt. (lbs.)	Recoil impulse (lb.sec.)	Recoil vel. (f.p.s.)	Recoil energy (ft.lbs.)
20-2 3/4	2 1/4	7/8	1155	7	2.36	10.8	12.8
20-3	3	1 1/8	1230	7	3.27	15.1	24.7
16-2 3/4	2 3/4	1 1/8	1185	7	3.06	14.1	21.5
12-2 3/4	2 3/4	1 1/8	1145	8	3.01	12.1	18.2
12-2 3/4	3	1 1/8	1200	8	3.15	12.7	20.0
12-2 3/4	3 3/4	1 1/4	1330	8	4.06	16.3	33.1
12-3	4	1 3/8	1245	8	5.25	21.1	55.4
10-3 1/2	5	1 3/8	1395	9	5.95	21.3	63.4
10-3 1/2	4 1/4	2	1210	9	5.75	20.6	59.1

These rifle and shotgun recoil tables, were compiled and written by Mr. William C. Davis, Jr. and originally published in the *American Rifleman*. They are reprinted here with their permission. The shotgun table, as it was originally published, contained some inadvertent errors. This corrected version was amended by its original author, Mr. William C. Davis, Jr. This author is indebted to Mr. Davis for his kindness in sending a newly corrected version.

SHOTGUN RECOIL ENERGY (in foot pounds)

Gauge	Load	Gun weight in pounds										
		5	5 1/2	6	6 1/2	7	7 1/4	7 1/2	7 3/4	8	8 1/2	9
.410 (2 1/2)	Max 1/2	6.5	6.0	5.5	5.0	4.5	4.5	4.5	4.0	4.0	4.0	3.5
.410 (3")	Max 3/4	13.0	12.0	11.0	10.0	9.0	9.0	8.5	8.5	8.0	7.5	7.0
28	2 1/4 -3/4	21.0	19.0	17.0	16.0	15.0	14.5	14.0	13.0	13.0	12.0	11.5
20	2 1/4 -7/8	19.0	17.0	16.0	14.5	13.5	13.0	12.5	12.0	11.5	11.0	10.5
20	2 3/4 - 1	24.0	22.0	20.0	18.5	17.5	16.5	16.0	15.5	15.0	14.0	13.5
16	2 1/2 - 1	25.5	23.5	21.5	20.0	18.5	18.0	17.5	16.5	16.0	15.0	14.5
16	3 - 1 1/8	38.0	34.5	31.5	29.0	27.0	26.0	25.5	24.5	23.5	22.5	21.0
12	3 - 1 1/8	34.0	31.0	28.5	26.5	24.5	23.5	22.5	22.0	21.5	20.0	19.0
12	3 3/4 -1 1/4	50.0	45.5	42.0	39.0	36.0	34.5	33.0	32.5	31.5	29.5	28.0

NOTE: This handy chart was compiled by The Remington Arms Company with ballistic pendulum figures rounded off to the nearest 1/2 pound.

RECOIL TABLE FOR HANDGUNS

Cartridge	Bullet Wgt. (grs.)	Charge Wgt. (grs.)	Muzzle Vel. (f.p.s.)	Gun wgt. (lbs.)	Recoil impulse (lb.sec.)	Recoil Vel. (f.p.s)	Recoil Vel. (ft.lbs.)
.22 L. R	40	1.1	800	0.5	0.16	10.4	0.8
.25 Auto	50	0.9	760	0.5	0.18	11.9	1.1
.32 Auto	71	2.3	905	1.5	0.33	7.0	1.1
.380 Auto	88	3.8	990	1.5	0.45	9.7	2.2
9mm L.	115	6.0	1155	2.0	0.70	11.2	3.9
9mm L.	124	5.2	1110	2.0	0.70	11.3	4.0
.38 Spl.	125	5.6	995	1.5	0.65	14.0	4.6
.38 Spl.	125	5.6	995	2.1	0.65	10.0	3.3
.357 Mag.	125	16.5	1450	2.1	1.10	16.8	9.2
.357 Mag.	158	16.0	1235	2.1	1.15	17.6	10.1
.41 Mag.	210	21.0	1300	3.0	1.58	17.0	13.5
.44 Mag.	240	22.5	1180	3.0	1.66	17.8	14.7
.45 Auto	185	7.4	940	2.4	0.90	12.1	5.5
.45 Auto	230	6.0	810	2.4	0.93	12.5	5.8
.45 Colt	250	8.5	860	2.8	1.10	12.7	7.0

This chart was compiled and written by Mr. William C. Davis, Jr. and originally published in *The American Rifleman*. It is reprinted here with their permission and kindness.

MATHEMATICS:

It is important to note that recoil is not the same mathematical figure as muzzle energy. A .38 special revolver will not deliver a recoil push or kick on the shooters hand of 278 ft. lbs. of energy (Winchester +P Super-X Lead Hollow Point). A rifle shooter will not have his shoulder hit with a 4,712 ft. lb. blow from firing a .458 Winchester Mag. with a 510 grain bullet. Although, in the last example, he or she will probably feel like it.

The foot pound is the unit of measure for the kinetic energy of recoil. It is, for the most part, a comparative and subjective figure that can and will affect different shooters in different ways.

◆ **Recoil is composed of both energy and velocity.**

As you know from reading earlier sections of this book, the energy and the momentum of a moving object are different. The momentum will be about the same in both directions; that is forward with the gas and the bullet and rearward with the gun. We have said earlier that momentum is the product of multiplying mass by velocity, which has a strong influence. If you remember the equation for kinetic energy, the velocity is squared.

$$\text{Kinetic energy} = 1/2 \text{ mass} * \text{velocity}^2.$$

This brings us to the main equation of recoil:

$$MG * VG = MC * C + MB * BV$$

The next 4 equations will help with understanding the basis for the main equation and explain the terms involved.

◆ *Free recoil* is the term used for the energy developed and used in recoil for the gun itself. It does not account for the shooters arm or shoulder which has to be considered. Free recoil velocity is only the gun as if it were hanging by a string and permitted to swing unrestrained.

$$VG = 32.17 * I / W$$

Where: VG = free velocity of firearm in f.p.s.
I = recoil of firearm in f.p.s.
W = gun weight in lbs.
32.17 = acceleration of gravity in f.p.s.

◆ The I in that formula is not easy to obtain, as is the weight. For I we can use this formula.

$$I = bw * bv + cw * c$$

Where: bw = bullet weight in grains
bv = bullet velocity in f.p.s.
cw = powder charge weight in grains
c = powder charge velocity constant

◆ The actual energy of free recoil in foot pounds is found from:

$$RE = W * VG^2 / 64.348$$

◆ This is considered the most important formula for foretelling how much kick to expect from different guns and ammunition.

$$EG = W * VG^2 / 64.348$$

eg = energy of free recoil in ft. lbs.

The 64.348 is the acceleration of gravity times 2, carried out to 3 decimal places.

◆ This covers 2 of the 3 basic elements of recoil mathematics. Remembering Newton's laws and with the momentum of a body the result of its velocity multiplied by its mass, (explained in detail elsewhere in this book), we have this formula for a 3 rd. **It is the basic equation of recoil.**

$$MG * VG = MC * C + MB * BV.$$

Where: MG = Mass of the gun
VG = velocity of gun (usually the required quantity)
MC = mass of the powder charge
C = velocity of the powder charge
MB = mass of the bullet
BV = velocity of the bullet

The $MC * C$ comprises the powder charge and the $MB * BV$ comprises the bullet. The $MG * VG$ comprises the gun and is also called recoil momentum or recoil impulse. The equation for this was given as I.

The constant C or *velocity of the powder charge* is simply the velocity of the expanding gases as they leave the muzzle. It is also called the *effective escape velocity*. If the velocity of the gun is known by use of a recoil pendulum, C can be figured easily as it will then become the only missing number in the equation.

For many years, and based on military cannons, a 4,700 f.p.s. figure was used for C. Modern tests show the best number to use for C, if it is unknown, is 4,000 f.p.s. for most standard guns using smokeless powder. Use 2,000 f.p.s. for black powder because of the heavier loads. Black powder will not necessarily be lighter with recoil. As with all situations of this type, the exact number may be higher or lower, but this is a good average figure in the absence of the correct number.

Testing at Aberdeen Proving Ground provided numbers for C from 3,710 f.p.s. to 4,115 f.p.s. The U.S. Army Ballistics Research Laboratories came up with numbers from 3,880 p.s.i. to 4,010 p.s.i. These figures are based on smokeless powders that are burned up while the bullet is still in the bore. If a powder was very slow burning for the bullet weight, it would not compute properly at a C of 4,000 p.s.i. This would be rare.

Mr. Wm. C. Davis, who has kindly helped with permission to use charts, has pointed out that some formulas are in error by using a constant for the velocity of the powder gases. The approach is flawed because it implies that shortening the barrel, which would lower the muzzle velocity, would also lower the escape velocity of the gases which is not true. Shortening the barrel will increase, rather than lower, the pressure of the gases at the muzzle. This, then, will increase instead of lower the velocity when they leave the bore.

The weight of the gun is easily measured. The powder charge and bullet weight may be known, but if not, they can be measured. Bullet velocity can be taken from a chart or measured with a chronograph. Even the velocity of the gun can be measured with a pendulum as explained elsewhere in this book. See, for readers afraid of math, it isn't as hard as it first appeared to be.

Previously it was mentioned that the gun part of the formula, the $MG * VG$, is also called the recoil impulse. A formula was given for finding its value. Also, using 4,000 f.p.s. for C with smokeless powder loads.

◆   The standard formula for recoil impulse.

I = WB * BV + 4000 * WC / 225,400

As with much of ballistics, it is based on Newton's laws. "The momentum of a body is equal to the impulse of the force that produced the momentum."

◆ Here is one last formula, included without explanation, for the readers who like math. For recoil energy:

$$RE = 1/2 \ GW \ (bw \ bv + cw \ C / 7000)^2$$

Where: RE = recoil energy in ft. lbs.
W = weight of the gun in pounds
C = constant
G = gravitational constant 32.17 ft. / sec. / sec.
bw = bullet weight in gr.
bv = bullet velocity in f.p.s.
cw = powder charge weight in gr.

SHOTGUN MATH.

◆ $$Vg = (Ws + Ww + 1.5 \ Wp) \ Vm / 7000 \ Wg$$

$$Eg = Wg * Vg^2 / 2g$$

Where: g = gun
s = shot in grains
w = wad in grains
p = powder in grains
m = muzzle
E = energy in foot pounds
V = velocity in f.p.s.
W = weight

NOTE: combine as required. Wg would be the guns weight and Vm would be velocity at the muzzle (or more commonly muzzle velocity), etc.

Recoil in shotguns, or any other gun for that matter, is not governed by bore size but by the weight and velocity of the shot.

For recoil figures, it is important to remember to add the weight of the wad to the weight of the shot pellets.

As the reader may have surmised, not every item that influences the felt recoil, or kick, is included in the equations. As mentioned earlier, their are differences between fixed breech firearms and autoloaders. Items such as recoil pads, shooters grip, stock shape and size, etc. What the shooter feels may be different than the equations would have us expect.

CAUTION

Some people want to figure their recoil on half of the powder charge. They base this on the opinion that before the projectile reaches the muzzle, the expanding gas is moving at half the speed of the projectile. **Do not use only half of the powder charge,** as this will give a wrong answer.

NOTES ON RECOIL & KICK:

- An ordinary rifle of .30 caliber with normal loads will recoil back about .060 inch. (sixty thousands) before the bullet leaves the muzzle.
- The shape of the cartridge case will have no effect on recoil. Different size head areas will cause different thrust amounts on the breech mechanism from equal pressure. A larger head area will cause more pressure but the recoil will not be different.
- The lighter guns cause less fatigue during a long hunt where walking is involved. They also kick more. What one gives, the other takes away. One consolation is most hunters are under enough excitement at the moment of kick that it goes almost unnoticed.
- A great rifleman, Col. Towsend Whelen, said most shooters can handle up to about 15 ft. lbs. of recoil without problems. Guns in the class of the .458 Winchester, which has a 60 ft. lb. recoil, are not for everyone. They also are not for all day target shooting. (Unless the shooter enjoys intense pain. In that case, a psychologist or psychiatrist may be needed more than a gunsmith or a knowledge of ballistics.)
- It is _not_ true that a slower burning powder will reduce recoil. The few milliseconds difference in burning will not change recoil.

Both of the following have been stated earlier, but they are important enough to repeat with slightly different wording.

- The gases produced by the powder's disintegration have the same weight as the original powder. The gas is expelled from the bore the same as the bullet and their weight must be included in any calculations.

- Recoil is the technical and mathematical amount of rearward movement that results from the projectile's forward movement. Kick is the amount that is felt. They are not the same. For most purposes, especially in casual conversation, the words can be interchanged and no one will know or care. Nevertheless, in the science of ballistics it still is necessary to understand the difference.

CHAPTER 7

BASIC STEEL PROPERTIES AND APPLICATIONS

In modern firearms, the choice of steel and hardness is crucial to safety, dependability and rust resistance. Modern steel properties and application is a complicated science. Many engineers have degrees in this specialized area known as metallurgy. No study of interior ballistics is complete without a basic knowledge of steel and heat treating.

Early gun makers had nothing but crude iron to use in their guns. Also, they did not have an acceptable method to drill the long hole through the solid iron. They used a long round iron bar called a mandrel and wrapped metal strips, heated to color in a forge, around the mandrel. Pounding with a hammer not only created by force the proper shape and appearance, it also flowed the adjacent strips together. This is called hammer welding and is still used in rare instances. This method of making rifle barrels was still in use as late as the 1850's by companies as big as Springfield Armory. By then, the hammering was done by machine power. Shotgun barrels were made of twisted steel or steel and iron twisted together in combination on into the 20th. century. (Damascus)

It is important to remember that these firearms were all made for black powder. There seldom was a breech pressure over 25,000 p.s.i. and bullets were soft lead.

Around the turn of the century, along with more powerful smokeless powder and jacketed-bullets, came better steel. Some was called nickel steel because nickel was added for strength. A bigger advance was ordnance steel which contained .45 to .55% carbon, 1.0 to 1.3% manganese, .05% phosphorous (maximum), .05% sulfur (maximum) and .25% silicon. Note the decimal point positions. These are small amounts but similar to some steels in use today.

MODERN STEEL CLASSIFICATIONS

Before we discuss modern steel; a brief background. By cooperative effort of the Society of Automotive Engineers (S.A.E.) and the American National Standards Institute (A.N.S.I.), a numerical system is used to classify both tool and standard steels. Tool steel is used to make the dies, jigs, fixtures and cutting tools used in the manufacture of guns and gun parts and accessories. It is readily hardened by heat treating. Standard steel is not as easily hardened, except on the outer surface, and is used for many other things. It has a classification system based on a 4 or 5 digit number.

The first figure indicates the class of steel, such as nickel or carbon. The second figure indicates whether an alloy is present and the approximate percentage of the predominant alloying element. The last two figures indicate the average carbon-content in hundreds of 1%. Any steel with the last two figures below 30 is considered low carbon. The higher carbon grades are less ductile and harder to form and weld than the lower-carbon steels, but they also have improved strength.

We occasionally see steel numbers that are prefixed with WD. This stands for War Department and was used by the military to allow tighter control over the specifications of steel used in military weapons.

Alloy steels contain added elements for the purposes of modifying their behavior during heat-treatment to result in improved mechanical and physical properties.

The addition of **nickel** increases hardness and tensile strength. Used in railroad rails, armor plate and ammunition, among other things. **Vanadium** adds the ability to resist repeated stress through an increased elastic limit. **Tungsten** adds air hardening qualities to steel and is usually used in conjunction with nickel or chromium or both. **Manganese** adds toughness in proper quantity. In a small amount (1.5 to 5.5%), the steel is very brittle and can be broken with a hammer. Additions up to 12 per cent make the steel ductile and hard. **Molybdenum** is usually used in conjunction with other alloys to improve high temperature service and wear resistance and improve hardening qualities. **Chromium** improves toughness, stiffness and hardness. The percentage used varies over a wide range from low to high chromium steel. Chromium also helps in resistance to acids, heat and nitriding. (Iron nitride). The addition of **sulfur, lead** or **selenium** improves the ability to machine or cut the steel in the manufacturing process. These additions that help the machining characteristics also weaken the grain structure. Most of the advantage gained in one area is lost in another.

The most common steel used in modern firearms is called **chrome-moly**. Properly designated as 4140 by S.A.E. grade, it is considered an ultra-high strength steel that, if properly oil quenched and tempered, will have a tensile strength of up to 180,000 p.s.i. Just plain cold drawn and annealed, its tensile strength is 98,000 p.s.i. The last two figures of the S.A.E. no. (40) indicate a carbon content between .38 and .43%. It also contains chromium (.80-1.10), manganese (.75-1.00), molybdenum (.15-.25), phosphorus (.040), sulfur (.040), silicon (.20-.035) and no nickel.

4140 steel is used by some manufacturers for barrels, actions, revolver and pistol frames, etc. It is an excellent steel, tough and strong and easy to machine and heat-treat.

HEAT-TREATMENT
HARDENING

Hardening by heat-treatment is an old skill that is now highly refined. There are two basic steps to the process. The first involves heating the steel at least 100° F. above its transformation temperature, the preferred term, but sometimes called critical point. If the steel is not heated to this temperature so austenite forms, no hardening action can take place.

The transition temperature is the point at which two solid phases exist in equilibrium. This temperature must be reached to the center of the metal piece without the outer surface being overheated. This takes care and time. The austenite is a solid solution of one or more elements and unless otherwise designated, the solute is generally assumed to be carbon. (Another would be nickel austenite.)

The second step involves rapid cooling (quenching) of the steel at a rate faster than the critical rate to produce a martensitic (very hard and brittle) structure. The hardness will depend heavily on the carbon content.

The heating can be as simple as a torch or forge or as sophisticated as a temperature and atmosphere controlling furnace. The quenching is determined by the type of steel and can be in oil, water, air or brine. The heating temperature will vary by carbon content from 1310° F. to 1550° F. and in salt baths up to 2400° F.

TEMPERING

To this we can add tempering (drawing) which is a process of reheating quench-hardened or normalized steel to a specific temperature that is lower than for hardening, holding it for a "soak", and then cooling it at a controlled rate. This also can involve oil and salt baths. The purpose of tempering is to reduce brittleness and remove internal stress.

STRESS RELIEVING

Stress relieving is similar to tempering and the two are frequently done in one operation. It is actually a separate process requiring different temperatures and conditions. Sometimes it is required when hardening and cold-work stress are induced from welding or from machine work.

After W.W. I, *autofrettage* has been occasionally used for prestressing barrels. In this process, a high hydraulic pressure is used to permanently enlarge the bore about 6% and the outside about 1% . The elastic strength can be almost doubled by the joint effect of the residual compressive stress near the bore and the higher elastic limit.

CASE HARDENING

Case hardening is where just the outer surface of low carbon (.30 per cent or less) steel is hardened. The surface is impregnated with carbon through cyaniding or nitriding or another of the seven common methods of

case hardening. This involves high temperature followed by quenching and controlled cooling.
SOFTENING
Hardened steel can be softened by a heating and cooling process called annealing.
NORMALIZING
Forged pieces have a coarse grain structure and normalizing refines the grain. It is done at a temperature usually about 100° F. higher than the regular hardening process. Cooling is done in still air at room temperature. Hardness is not desired because it is done before machining.
HEAT TREATING IN GENERAL
This has been an oversimplification. Heat-treatment of metals is a very exacting process and no reader is expected to attempt it from this brief section. The purpose was for familiarization. For proper results, most heat-treatment should be done by a company that specializes in heat-treatment for industry. If the result is important, as it is for gun parts, have them do a Rockwell or Brinell test after they have finished. There are 10 other methods of measuring hardness, but those two are the most common.

Whichever method is used, it will have a number which is based on resistance to indentation under the conditions imposed by the particular test. This number is meaningless unless it can be given significance by comparison to the hardness of familiar objects. Cutting tools, such as taps, require a hardness of about 63 on the Rockwell C scale. Below 61 and they will be too soft and wear excessively and over 65 they will be hard and chip. Brass can be from 40 to 95 Rockwell, but it is measured on a different scale; the B scale. As you can see, it is not an easy subject. Anyone having heat-treatment done should do some research to determine the proper hardness for the job at hand.

Hardness affects tensile strength. If the 4140 steel that was discussed earlier is 20 on the Rockwell C scale, it will have a tensile strength of 110,000 p.s.i. At 38 Rockwell C its tensile strength will be 180,00 p.s.i. The controlling factor is the temperature of the draw and quench. 4140 is considered medium at hardening ability and is frequently used "as is" at about Rockwell C-10.

Modern barrels are made of steels other than 4140. 4135 steel heat-treated to about 275 Bhn has been used with success. For a slight cost savings, cold drawn stress-relieved 1137 steel, with a minimum yield strength of 90,000 p.s.i. can be substituted. 1137 has less distortion than 4140 in hardening, but falls short in other areas.

The correct hardness for a barrel made with one steel grade may be too hard or soft for another grade. The early Springfield .30-'06 rifles with low serial numbers (below 1,257,762 or by Rock Island Arsenal below

285,507) generally had barrels that were too hard and brittle. They were usually above Rockwell C-50 at C-54 or C-55. They used a low grade steel that would be safer at C-30 to C-40. The barrels at C-40 had a tensile strength of 145,000 p.s.i. and were safe at 100,000 p.s.i. Above C-50 they were too brittle. Perhaps excellent wear resistance but a danger of blowup. Nickel steel Springfield barrels in the later models are safe and correct at C-45 to C-55

A hardness test on a firearm is normally useless. No, that is not an error. It is a waste of time and money. We discussed hardness testing and we discussed a few of the thousand grades of steel. A hardness number is worthless without knowing the exact steel used. Not a general idea but exact. Is it 4140 or 3140? Is it 1020 or 2330? For the hardness number to have meaning as far as firearm use and safety, the part would have to be destroyed by metallurgical testing to determine tensile strength, alloy, purity, etc. A Rockwell C reading of 45 can be good or bad depending on the steel. **It does not tell strength, only hardness.** The cheapest scrap steel can be heat treated or case hardened to give a Rockwell C reading anywhere on the scale between 20 and 80. It only has value when the steel designation is known.

Proof testing is the only way to know if an action is safe. Fasten the gun securely and fire it with a string. If the action is damaged, at least your body will be safe. An old tire is a common way of holding a long gun with the butt end inside the casing while the barrel or stock extending over the other side is tied in place. Don't forget to check where the bullet or shot will go and provide a safe backstop.

The hardness of a barrel's steel is a factor and like the composition, it is a trade-off. The harder steel has better heat resisting properties and more strength. On the other hand, it is more difficult and expensive to machine and as such, costs extra. As expected, a balance in the middle is obtained in the better guns while the cheap guns ordinarily use the softer steel with less alloys added.

A high percentage of gunsmiths do their own heat-treatment, and some are very good at it. Nevertheless, it cannot be recommended. Improper heat-treatment can result in quench cracking, uneven hardness, brittleness, distortion, and other problems.

Silver solder is a common method of adding sights, scope rings, etc. Modern silver solder will fuse at about 450 degrees. F. and some older silver solder requires around 650 degrees F. Silver soldering on a firearm requires both knowledge and patience to avoid damage.

CHROMIUM PLATING

The *hard* chromium used in bores is slightly different in its application than *decorative* chromium as applied to the outside for appearance and rust resistance. Decorative chrome is usually applied over a thinly added layer of copper or nickel.

A chrome plated bore can have good wear and corrosion resistance if done properly, but most have a thickness of only about .0005" to .00075". That is five to seven and a half ten thousands of an inch which is not much. It is sometimes as thin as .0002". Military plating is usually better at .001" to .002 in thickness. It is easy to understand why, even though it is extremely hard, it frequently does not hold up well.

Chrome plated barrels are not usually made to as close a tolerance as a standard barrel. Plating requires the use of an electrochemical bath which removes some of the steel. Sometimes called electro-polishing, it is needed for proper adhesion of the chromium plating (instead of the copper or nickel), then the thin chrome is added. Even under the best of conditions, extremely close tolerances cannot be held. Chromium plated bores are excellent for hunting, military and police use; anyplace where rust can be a problem. It is also very good at preventing metal fouling in barrels used for rapid-fire. For high accuracy in target work, it is not usually recommended.

Also, chrome plating will not increase velocity, as is sometimes believed. It is certainly no "slicker" than a good quality lapped barrel. Another misconception is that plating a bore will fill in and smooth out rough places. The plating follows the existing finish and all marks are still there only they are then in the plating. That is why the outside or visible parts of anything plated has to be polished so perfectly.

The plating wears away first near the chamber and then progressively on down the bore with the original thickness controlling how long it lasts.

Frequently, fouling does not stick as well to chrome as steel and it helps to control rust. *The American Rifleman*, a publication of the National Rifle Association, conducted tests with a .357 S & W Magnum. The barrel leaded about the same after chrome was added as before. Although, it was reported to be a little easier to remove the fouling from the chrome. It can be considered very good for rust prevention but only fair for other goals.

STAINLESS STEEL

Stainless steel is very popular for firearms despite the extra cost. Stainless steel was originally a trade name used for cutlery steel and patented in 1916. It was not legally maintained as a trademark and now the term is generally used for all steels that resist rust, corrosion, acids and high temperature scaling.

There are about 50 different compositions of stainless steel, each with its own unique properties of machineability, hardening, heat resistance, strength, stress resistance, etc. Some can be hardened and others cannot. As with many other things, the choice is always a compromise. For example, the best corrosion resistant steels have less strength and toughness.

Generally, to get the tensile strength required in firearms, the steel used will, under some conditions, rust and corrode. Stainless is also excellent at reducing barrel wear and fouling. The grade of steel used in manufacture and how it is machined and processed is the deciding factor.

The tensile strength of stainless steel is lower than properly processed A.I.S.I. 4140. For example, stainless A.I.S.I. type 410 is 65,000 p.s.i.

These steels always have chromium added, as little as 4% (A.I.S.I. 501) to as high as 27% (A.I.S.I. 446). Chromium percents below 10 are not technically considered stainless steels, but by common usage are called that. The common stainless steels contain from 11.5% to 27.0%. The iron content will be at least 50%. Alloying additions may include nickel, molybdenum, columbium, titanium, manganese, sulfur and selenium.

STELLITE

Stellite is used as a liner in barrels that are expected to receive extreme hard use. The lining is usually only at the throat or chamber shoulder and extended for 2" to 8". A close visual inspection will show a gap or a faint ring. The most common use of stellite in firearms is with machine gun barrels to reduce heat and wear resistance. The material is so hard that it is used to make cutting tools for machining metal and for valves in aircraft and high-r.p.m. automobile engines. Stellite has high impact and cantilever strength and heating and cooling can be repeated indefinitely without any loss in hardness. It is resistant to corrosion and oxidation and has good tensile strength. There are several grades with each composed of varying proportions of cobalt, chromium and tungsten. As you would expect, it is expensive to purchase and use.

TITANIUM

Titanium parts are costly to make, but they can hold tremendous pressure at half the weight of steel. With excellent corrosion resistance (almost 100%) and other long term advantages, titanium will probably be used more in the future; probably as parts or components rather than as a complete weapon or as an alloy such as titanium/graphite. Pure titanium has a melting point of $3,097°$ F. and a modulus of elasticity of 16,500,000 p.s.i. (No, that's not a mistake. It is in millions.) Even as an alloy, titanium has excellent qualities. By the way, modulus of elasticity is defined as the ratio of increase of unit deformation to the increase of unit stress within the elastic limit and is given in p.s.i. It can be measured in tension, compression and shear. Don't let the big words throw you. Modulus means a positive number expressing the measure of the function, which is elasticity (flexibility, resilience) in this case. The rest is the ability of the material to return to its original dimensions after the removal of stress.

PLASTIC

Plastics, composites, polymers; they are already on the scene in a big way; not for entire guns, but for component parts. They can be engineered and molded to be strong but light weight and corrosion proof. They are cheap to manufacture and have but one major flaw. They cannot handle high temperature, at least not at the time this is written. No doubt, the future will change the situation.

The Glock handgun was the first gun with a polymer frame to be widely excepted. (Although not without a huge fuss over whether it could be detected by airport security. - It could.) Other manufacturers are now following Glock's lead.

THE FUTURE?

Those who prefer a firearm made with fine wood stocks or grips, a deep lustrous blue finish on steel, and holsters and slings made of tanned leather, will have to make some adjustments for the future. The next century of firearms will be like nothing in the past. The guns of the future will be short on nostalgia but long on practical and worthwhile features. Right?

EROSION & CORROSION

This may be a good place to mention that many people use the terms *erosion* and *corrosion* interchangeably. In most cases, the choice of words would make no difference, but they do mean completely different things; both affecting gun parts and bores in different ways.

Corrosion damages by chemical reaction. Moisture and chemicals produced by primers, powders and saline agents are the main culprits. Rust is the most common result. Modern ammunition reduces the problem, compared to earlier powders and primers which caused corrosion because of their chemical composition.

Erosion is a gradual wearing caused by the friction of the bullet. Also, the heavy wear just forward of the chamber caused by the cutting action of the high pressure and high temperature gases.

Both damage the gun, but in different ways.

REFERENCE

For this section, refer to the bibles of the metal industry, *Machinery Handbook* from Industrial Press, Inc. and *Tool Engineers Handbook* from McGraw Hill Book Co. Inc. Also *Metals Handbook,* Vol. I and II by the American Society of Metals.

CHAPTER 8

GYROSCOPIC STABILITY

There is a strong relationship between the information in the chapters on gyroscopic stability, spin drift and rifling. Each chapter depends on facts given in the other two and they are all dependent on each other. To properly understand any of the three subjects, all three chapters should be read.

It is impossible to properly grasp external ballistics without a knowledge of gyroscopic action and yet there is a lot of confusion about the subject. It is tricky and difficult to comprehend. Perhaps it is comparable to religion because some of it has to be taken on faith. There is scientific proof, but unless someone is in the laboratory of a university physics dept., it is difficult to demonstrate. If Chapters 8 & 9 do not leave the reader with a crystal clear picture, perhaps there will be a basic understanding of the overall subject.

There are a few areas of gyroscopic action that are strange and confusing. Precession is one. But it is all fact, not fiction or wild theory. This section will go into many explanations and examples. The idea is not to bore you silly, although that may happen, but to try to build a solid background and understanding. **If at any point it becomes too confusing, keep reading. There are many explanations and examples to provide help.**

Gyroscopes appear to go against all the physical laws we have learned. A child's toy will stand up straight on just a single point. A gyroscopic instrument in aircraft, the attitude indicator (formally called artificial horizon), always remains in the same position to the earth, no matter how the aircraft is positioned. (Although earlier and cheaper instruments would *tumble* during extreme angle maneuvers.) Also in aircraft, there is the heading indicator which is freely mounted and utilizes the gyroscopic property of rigidity in space and the turn-and-bank indicator which utilizes the gyroscopic property of precession.

Ships use the same principal in gyrocompasses. They remain level and working correctly even when the vessel rolls and pitches from rough seas.

Any spinning object will have gyroscopic properties. This includes engine flywheels, fan blades, bullets and shotgun slugs from rifled bores. Technically, only a special wheel designed and mounted to utilize the properties is called a gyroscope.

The two primary qualities of gyroscopic action are rigidity in space and precession. There are several secondary actions and nutation is the most important in ballistics.

A gyroscopic instrument, as used in aircraft, has a spinning rotor which is universally mounted. That means it is mounted so that it is free to move on its axis to any position in space. It is said to have 3 planes of freedom and is free to rotate and turn into any position inside the case because of gimbals. Gimbals are devices consisting of a pair of rings pivoted on axes so they can swing one within the other. A ships compass is thus mounted so it will stay level as the ship tilts and rolls. Besides the common circular rings, gimbals may be rectangular frames or in flight instruments, a part of the instrument case itself.

Rigidity in space means that when it is spinning on its axis, it will remain in its original position regardless of how the case (attached to the aircraft in this example) is moved in relationship to the ground. This gyroscopic inertia is useful to the bullets flight path because of the stabilizing effect.

A rotor free to spin about its axis is the basic gyroscopic. Newton's first law of motion was discussed in an earlier section, but briefly it states that every body continues in its state of uniform motion (or rest) until acted upon by external forces. Therefore, in its simplest form, a spinning rotor will maintain its plane or rotation perpendicular to its axis, unless an external force is applied. For even simpler wording, this just says that it will stay in place if it is left alone.

Newton's second law states that the deflection of a moving body is proportional to the deflective force applied and is inversely proportional to its weight and speed.

A rotating body has an angular momentum (an energy of rotational motion, torque, moment). From a technical point, these terms are not 100% synonymous, but the reader needs to understand the basic power, to use still a different term, that exists in a rotating object. This angular momentum results from velocity, mass and distance from the rotational axis.

If we forget gyroscopes and bullets for a brief time, we could compare it to a wheel spinning free on an axle. If we want to stop or brake this free spinning wheel it is obvious that the more velocity, that is the faster it is spinning, the harder it will be to stop it. Also, the heavier the wheel the more braking effort needed. Distance from the axle will be important because if the weight is close to the axle, it will cause very little torque on the axle while the same weight swinging around at a distance will apply greater force and be harder to stop. Velocity, mass and distance from the axle (shaft, axis); all three determine the energy whether it is a spinning fly-wheel or a bullet spun by the rifling.

The most angular momentum is when the weight is farthest from the rim. The farther from the axis, the greater gyroscopic rigidity. A rotor that is cylindrical, as is a bullet, will have a good angular momentum for its weight if it is fat. With the same spin, the angular momentum would be less if it was long and thin.

To put it in other terms, the amount of gyroscopic effect depends on the mass of the spinning object, how far the mass is from the center of rotation, and its speed of rotation. All together it is the angular momentum or flywheel action which has both a direction and a magnitude (measurable size or quantity).

A few basic rules for all objects with gyroscopic stability, including bullets.

1. Weight. For a given size, a heavier mass is more resistant to disturbing forces than a light mass.

2. Angular Velocity. The higher the rotational speed, the greater the rigidity, gyroscopic inertia, or resistance to deflection.

3. Radius. The radius at which the weight is concentrated is important. Maximum effect is obtained from a mass when its principal weight is concentrated near the rim and rotating at high speed.

♦ For the mathematically inclined reader, torque is simply force times distance or ML^2 / T^2, where L is length, T is time and M is mass.

Energy in rotational motion, in cases of constant torque, is torque times angular displacement, measured in radians. Power in rotational motion is equal to torque times angular velocity, (radians / seconds). If we have rotational velocity in revolutions per minute, this must be changed to radians per second by multiplying by 2π. ($2\pi N$ = radians/seconds, where N is in revolutions per sec., not per minute.)

♦ To find the rim speed of a rotational round object, multiply the diameter by p and that answer by the number of revolutions per minute. That is, **rim speed = πD (r.p.m.)**

♦ The kinetic energy of a wheel, because of its angular velocity, is equal to $1/2\ IW^2$ where I is the moment of inertia and W is the angular velocity measured in radians per second.

PRECESSION

Precession can be very confusing, so let's go through it from several different angles and with several examples. Put simply, precession is the result of a force being applied to the outer edge of a spinning object and the following deflection. That part is easy, and more so as we word it in other ways. Now we shall throw in the part that drives grown men to do irrational and illogical things. **When the force is applied, the resulting action takes place in the direction of rotation and 90° away.** The force causes a pitch or yaw or both, depending on the exact location of the application of force. The exact effect of a pitch up or down or left or right depends on direction of rotation and location of the pressure point. But the result will always be found 90° from the point of pressure and in the direction of rotation.

Do not feel bad if you are a little bewildered. Following is more that will help. For example, consider the child's toy, a spinning top. When the top is spinning steady, it has gyroscopic stability and will stand up. If a gentle force is applied to the shaft to tilt it to the side, it reacts by first moving the shaft perpendicular to the force (90°) and then doing a spiral motion which is the precession. The top will then return to its normal upright position.

If the force that was originally applied to the shaft persisted until the top fell over, it would not fall in the direction the force was applied but almost perpendicular to the direction of pressure. (Again 90°).

In technical wording, the movement of the object with gyroscopic properties is not a direct one but a **resultant movement.** The axis is not displaced in the direction of the force applied but at a right angle and in a manner that causes the direction of the movement or pitch to assume the direction of the torque resulting from the applied force.

Study the drawings. The force as applied is **resisted** by the inertia and prevents it from being displaced as expected. However, **with the object spinning clockwise, the precession takes place 90° ahead in the direction of rotation.** The rate of movement is proportional to the force applied.

For those still confused, we will go through it more with a different sketch and more examples. It is important to the knowledge of a bullets flight path and should be understood.

The sketch on page 85 shows a gyroscope mounted where it has 3 planes of freedom. The axis are marked with X the rotor spin axis and Y the axis of the inner gimbals and Z the axis of the outer gimbals.

For example, when a force is applied upward on the inner gimbal, it is as if the same force were applied to the rim of the rotor at F. This is opposed by the resistance of the gyroscopic inertia, preventing the rotor from being displaced about the axis Y-Y. However, with the rotor spinning clockwise, the precession takes place 90 degrees ahead in the direction of rotation at P. The rotor turns about axis Z-Z in the direction of the arrow at P. The rate at which the wheel precesses is proportional to the deflective force applied (minus the friction in the gimbal ring, pivots and bearings.) If too great a deflective force is applied for the amount of rigidity in the wheel, the wheel precesses and topples over at the same time.

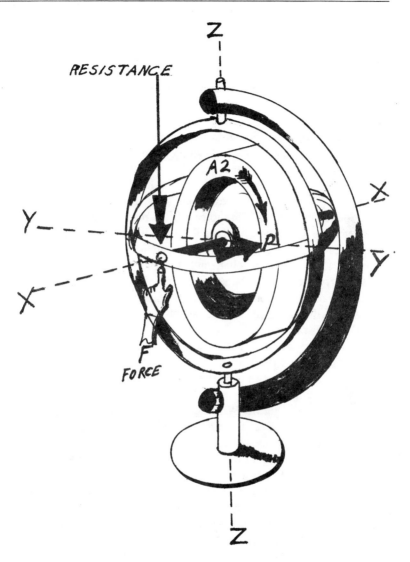

Please read and follow the explanation in the text at the bottom of page 84.

Force applied to a gyroscopic rotor mounted in gimbals and the direction of precession and the precessional movement.

If you are still confused, don't give up. Go back and read the last paragraph again as you trace the sketch with your finger.

The rate of precession is proportional to the deflective force applied and if too great a deflective force is applied for the amount of rigidity, the object precesses and topples over at the same time.

The problem is that a lot of intelligent and educated people, if their background is not in physics or engineering, have a problem with accepting it as a fact.

One more quick example.

Gyroscopic precession is well known to aircraft pilots. Here is a brief quote from the Federal Aviation Administration *Flight Training Handbook,* AC 61-21A, page 33. "The second factor that causes the tendency of an airplane to yaw to the left is the gyroscopic properties of the propeller." (Here we will leave out their explanation of gyroscopic precession as repetitive.) "That force will be particularly noticeable during takeoff in a tail-wheel type airplane if the tail is rapidly raised from a three-point to a level flight attitude. The abrupt change of attitude tilts the horizontal axis of the propeller, and the resulting precession produces a forward force on the right side (90° ahead in the direction of rotation), yawing the airplane's nose to the left. The amount of force created by this precession is directly related to the rate at which the propeller axis is tilted when the tail is raised."

Figure 1 shows a disk or wheel that is spinning rapidly about its axis A-B. The end of the axle A is constrained so that its only movement is in the slot between the two smooth fixed guides. If a force F is applied and the wheel is rotating, the pressure F will bring into effect a lateral pressure P against one of the guides. Which guide is pressed against depends on the direction of rotation. If the force F is regarded as caused by pressure of the flat board against the side of the spinning axle, the axle will try to move in the same direction it would if the board was rough. With rotation as in the figure, the axle end A is pressed toward the upper guide if the axle is pressed in F direction. If the guide was not present, it would move up as P. The magnitude of force P is proportional to the rotational velocity.

A bullet that is spinning has gyroscopic properties and precession is one of them. A bullet's deflection from branches is caused by a force on some part of the ogive. It will push on the bullet and precession will result. The direction will depend on where the force was applied. The amount of deflection will depend directly on the amount of force and inversely on the angular momentum of the bullet. The rules of gyrostatics will give an edge to a short fat bullet over a long thin bullet if the weight and the spin stability are equal. The short fat bullet will precess less and slower. (Brush deflection is subject with a lot of problems. It is covered in Chapter 10.)

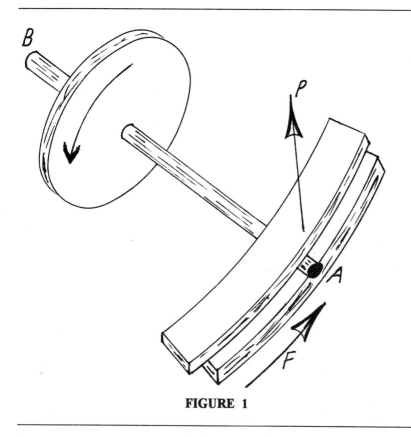

FIGURE 1

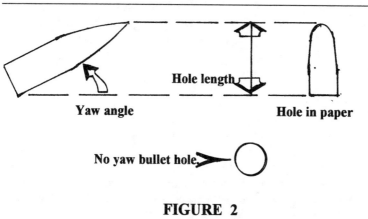

FIGURE 2

An object with gyroscopic stability is not affected thus by forces placed at the center of gravity but forces applied forward of it will tilt the axis. The same effect is noticed in spin deflection which will be discussed later.

NUTATION

Nutation is, by one definition, the act of nodding the head. For our discussion of gyroscopic stability, it is defined as the periodic inequalities in the motion of the axis and pole of a revolving object. A nodding of the spinning object, so to speak.

Something with gyroscopic action does not instantly go from zero r.p.m. to just the correct rotation speed and stability. There will be an initial axis wobble that is nutation, and under proper conditions, it will smooth out very quickly. It is not to be confused with the wobble of a bullet with down range decaying stability. That is different and it will be discussed later.

Nutation in projectiles is in the form of tiny loops which pass almost through zero yaw. (Yaw will be explained later.) In a gyroscope such as a top or a model for class-room study, the end of the axis may trace loops or waves or under some tight circumstances, even cusps (points).

The nutation would continue unending except for friction on a fixed gyroscope. On a bullet there are other forces such as aerodynamic drag that dampen it. (Study figure 3, page 92.)

The swinging point from nutation guides the bullet in a helical path like a spiral on the outer part of a screw thread. The sides are flattened at places as the yaw returns to zero. The radius of the helical path is normally a small fraction of an inch. That is worth repeating; they are a small fraction of an inch from nutation.

These nutants will dampen and disappear.

There are several methods of proving this is fact rather than theory. Of course, gyroscopes have been studied in laboratories since at least the early 1800's. Bullet's flight paths have been examined extensively. One of the more modern methods is by special photographs and spark shadow-graphs which show the attitude of the bullets as well as its place in space and time. This is limited to only the best laboratories.

The older method of shooting through cards or papers spaced apart provides helpful information. There is an advantage to the use of a special material such as photographic paper with the gelatin side toward the shooter. It will be cut with a clean, sharp hole that clearly shows the result even with low velocity projectiles. This is mounted in holders at a measured foot or two apart for perhaps 10 to 50 yards. The yaw can be increased on purpose to make it more visible by cutting away one side of the barrel at the muzzle. Another method is to use a barrel with a rifling twist that is too slow. In both cases, the condition is exaggerated.

The amount of yaw can be easily checked by measuring the longest distance across the punched hole. The paper will tend to close in the hole slightly, but it will still tell a good story. The amount of closure can be measured by checking the perpendicular and including it in the calculations. See figure 2, page 87 for more information.

Even a bullet that will quickly stabilize may leave the muzzle at a yaw as much as 5°.

Testing of .30 caliber match ammunition showed 4 nutations and .68 of a precession in the first 48 feet from the muzzle. That is typical. There are usually several nutations for each precession. The nutations are usually in the form of loops that pass through zero yaw. As we said earlier, this produces the path of a helix with flat spots.

It is sometimes said that rifle bullets gyrate around a large (?) circle after leaving the muzzle and then gradually stop spiraling and settle down to normal flight. For all practical purpose, this is false. There will be a spiral (helix is the correct word) that is very small, but it is small enough to frequently avoid detection. Partly because the precession that would cause the helix is modified by the nutations. It could be said that, at least initially, the bullets path is a tiny helix around the trajectory.

The nutations will die out quickly but the precession and yaw may continue for a longer time. They are, as explained, different actions. Somewhat similar but still dissimilar.

YAW

There will be a slight yaw and unbalance while the bullet is still in the bore. The confinement is relieved at the muzzle and the bullet departs in a slight yaw. This may be as small as a degree or two in a good barrel with the correct twist for the cartridge. Even a bullet that will become stable can have an initial yaw of up to 5%. Yet a beginning yaw of as little as a degree or two can have a substantial influence. It also may be extremely high with the use of poor and mis-matched equipment. In any case, the force on the bullet is not on the bullet's center of gravity, but in front of it. As discussed, this force will try to tumble the bullet.

The initial yaw outside the bore and the helix lead angle follows this interrelationship. (See figure 10, page 94.)

$$a = e \left[\frac{2B/A-1}{(1-1/S)\,1/2} \right]$$

a = amount of the first yaw past the muzzle (external)
s = stability factor
e = amount of yaw inside the muzzle (internal)
B/A = ratio of moment of inertia about the projectile's transverse axis (across)through the mass center and about its longitudinal axis (length).

The initial yaw as the bullet leaves the muzzle is governed mostly by stability. The direction of the yaw is random and can be anywhere around the 360° of the muzzle. This makes target dispersion larger in cases of considerable yaw with low stability.

Penetration tests conducted by the U. S. Army have shown less penetration at very short ranges than at moderate range. The lack of expected penetration at short range is blamed on initial yaw which prevents the bullet from striking straight on. When the bullet stabilizes, the depth of penetration increases. Then the penetration depth gradually drops again as velocity, energy and momentum decrease with range.

The initial wobble that is yaw and precession can last up to 200 yards before it settles down.

UNBALANCE

Bullets can be deformed in their travel down the bore so that their center of gravity is no longer in the center of form. Poor stability is the obvious result. This is not common because jacket and bullet material that is strong and hard is less likely to be deformed and usually, as a result, more stable and accurate.

A bullet that is unbalanced longitudinally may still be stable in yaw if the bullet is fired at the proper twist and the center of pressure is behind the center of gravity.

A bullet that is laterally unbalanced will be forced to conform while still in the barrel and its center of mass will revolve about its geometric center. After exiting the muzzle, the geometric center will begin to revolve about the center of mass. At 54,000 r.p.m. to 250,000 r.p.m., depending on velocity and twist, the centrifugal force can be tremendous. It will result in an outward or radial acceleration from the intended flight path and will try to get the bullet to rotate in a constantly growing helix. "Try to" are key words as other actions usually prevent a large helix and bullets are not normally that out of balance. While it is rare, it can happen.

Static unbalance is caused by the mass center being laterally displaced as in figure 4 and 8. In this type of unbalance, if there is no rotation and the object is free to turn on its axis, it will roll until the center of gravity is at the lowest point.

The centers of gravity may be divided on opposite sides of the rotation as drawn in figure 5 and 7. In these examples, the object will be in balance when stopped in static condition but during rotation it will be out of balance. This is called dynamic unbalance.

A third type of unbalance is shown in figure 9. In this example the centers of gravity are located so the forces formed by centrifugal force will create both static and dynamic unbalance acting together.

Bullets that are not properly balanced and follow a small helical path may continue that way to the target.

◆ For static unbalance we have: $D = \tan^{-1} 2\pi e / p$.

Where: D = angular deviation - lead angle
$\quad\quad e$ = eccentricity of mass center
$\quad\quad p$ = pitch of rifling

If you are not interested in math., don't worry about it, but the readers who can follow it will see that the deviation grows as the pitch of the rifling increases. While modern practice dictates that when in doubt, go for the faster twist, this can be overdone. Also noticeable is the very small amount of deviation required to obtain a one minute of angle error.

A different formula is used for dynamic unbalance and it reaches the same basic conclusion. That the deviation grows as the pitch of the rifling increases. Again, good pitch is necessary but can be overdone.

Mathematics can show that both static and dynamic unbalance can be improved by a slow twist. On the other hand, the stability factor is improved by a faster twist. A good balance is needed. (No pun intended.) See rifling chapter.

SUMMARY

To reiterate, **yaw** and **nutation** are different. **Nutation** is gyroscopic in origin and very small. **Yaw** is larger and while not caused by the same gyroscopic action, it still has a gyroscopic relationship. Both are affected by precession and interact together. If the bullet is not moving through the air exactly point on, for whatever reason, then the pressure is not through the center of gravity but before or ahead of it. This creates an overturning or upsetting action which, because of precession, has a result 90° from where it is expected. Further, the high speed of rotation is constantly changing the direction of upset, but always 90° from the force.

This section on gyroscopic stability and the next section on spin drift are difficult to comprehend. It is comparable to religion because some of it has to be taken on faith. If these two chapters do not leave you with a clear picture, perhaps you will at least have a solid basic understanding.

Review the sketches on the next 4 pages and also on the bottom of page 200.

The forces acting on a bullet in flight.

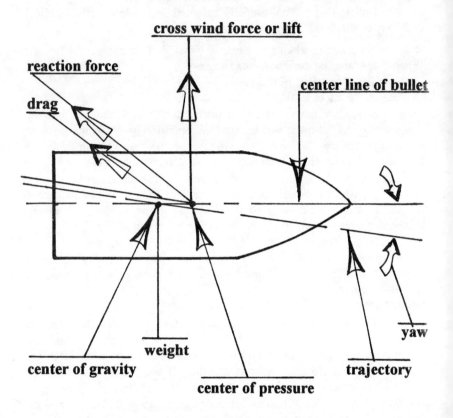

Direction of lift and reaction rotates through the complete 360 degree circle with precession, not the bullet's revolutions per second.

FIGURE 3

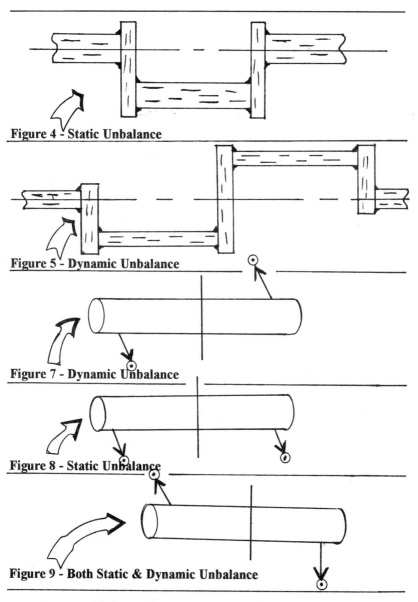

Figure 4 - Static Unbalance

Figure 5 - Dynamic Unbalance

Figure 7 - Dynamic Unbalance

Figure 8 - Static Unbalance

Figure 9 - Both Static & Dynamic Unbalance

For figures 7, 8, and 9, imagine the rotating object divided into 2 sections.

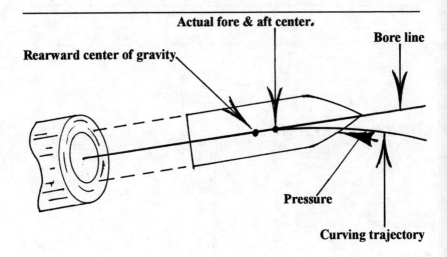

Figure 10

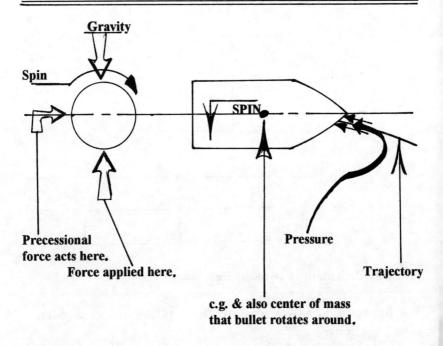

An example of pressure force that is not directed at the c. g.

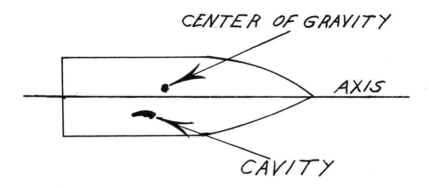

A cavity caused by any means such as an actual hole from an error in manufacturing or a thin jacket or other defect, shifts the bullets center of gravity off its axis. Bullet will deviate from its intended trajectory as soon as it leaves the muzzle because of static unbalance.

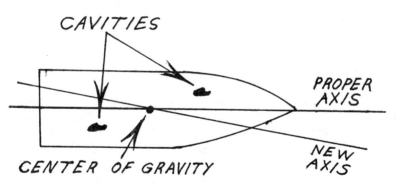

The center of gravity (mass center) is still on the proper axis but the mass is unevenly distributed. This dynamic unbalance will cause a wobble about a new axis after the projectile leaves the confines of the bore. Other factors and forces may dampen the wobble, but accuracy will be reduced.

CHAPTER 9

SPIN-GYROSCOPIC DRIFT

We will do the same as in the last chapter and go through everything several times in different ways to help in understanding. If the reader has skipped the last chapter, it should be read, or this one will make no sense.

Spin drift or gyroscopic drift is defined by Military Ordinance as, "The lateral deviation of the trajectory from the plane of departure caused by rotation of the projectile." Although we all use the term drift for both the action caused by the wind and gyroscopic action, deflection is the term for wind movement and drift for gyroscopic movement. Although the gyroscopic drift could be called a deflection as it deflects rather than drifts. Confused? If you are now thinking, "Who cares," you may be right. The problem is that over the years, ballisticians have started using terms that are opposite of their dictionary definitions. While it is best to know and use the proper word, as long as the person reading or listening understands what we are saying, then our thoughts and ideas have been successfully communicated.

Bullets drift to the right when fired from a right twist barrel and to the left from a left twist barrel. Frequently called gyroscopic drift, it increases at the longer ranges and forms a slight decreasing radius curve. We are not speaking now of wind deflection, so let us assume a rare no wind condition. As with many fine points of external ballistics, this is more noticeable at the longer ranges. Sighting in will correct it, but only for the range planned.

Gravity pulls the bullet in a gradually increasing and accelerating downward path. Most bullets will quickly take on a constant trajectory attitude with the nose pointing a little to the side of the direction of rotation and a little nose high. This position is sometimes called equilibrium yaw.

Spin drift is caused by gyroscopic action. As the nose moves around its small swinging arc, it moves farther to the one side on the down swing than to the other side as it swings up on the opposite side. *Cycloidal* is the proper term to describe the axis movement which is more and more down and toward the direction of rotation.

With the spin from the rifling producing gyroscopic stability, it incorrectly seems to some people that the bullet should remain nose high as it follows the trajectory curve. The bullet's nose does not remain at the angle to the horizon it had when it left the muzzle. The angle will decline as the trajectory drops. It will remain slightly above the flight path but not necessarily above the horizon. The nose position changes in flight so that it remains almost point first. In other words, if it is fired at an angle of 20° above the horizon, it will not hit at that angle down range. A properly stabilized bullet will still strike point first.

Air pressure from the atmosphere presses against the bullet on the front end, but very slightly on the under side rather than dead straight. This will be more pronounced as the projectile moves farther from the muzzle. This pressure is ahead of the center of gravity and will try to force the nose higher above the curve of the trajectory. The center of gravity being somewhat to the rear of center will intensify this condition because of the streamlined, tapered, or pointed front end. The phenomena that is involved with gyroscopic action causes the bullets nose to begin an arc to the right and down. The action is continuous and gradually diminishing. As air pressure pushes in one place, the point swings out in a different place around the circle of rotation as gyroscopic action dictates. This is with a right hand bullet rotation. In case of a left hand rotation, it would be to the left.

The bullet's nose will never move as high above the trajectory curve as it was on the previous turn because of axis shifts and changes in pressure. The nose then drops as does the trajectory curve.

Without the gyroscopic effect, the air pressure would pitch the nose further up around the center of gravity and tumbling would be a possibility on a long range shot. The shape of the hole in the target at long ranges, particularly under test conditions, backs up the result if not the reason.

One test conducted to determine if bullets stay point-forward at long range is especially interesting. The experiments were done by the U. S. Army on the .45 cal. U. S. military rifle with .500 grain bullets. Although conducted in 1880, the laws of physics applying to rifled barrels and projectiles are unchanged today.

At the extremely long range of 2,500 yards, the trajectory was low enough and the bullet point forward enough for it to make a cut in beach sand. The bullet was marked under the point in an area that showed a point-forward impact.

At 3,200 and 3,500 yards, the angle of impact was much higher and only a round hole was made instead of a cut. The impact marked only the point and was again, point-forward. The testing was carried out at ranges from 1,500 to 3,500 yards, and the bullets always hit point-forward. At the longest range of 3,500 yards (60 feet short of 2 miles), the angle of descent was 65°.

The slight yaw that a bullet has as it leaves the muzzle becomes a diminishing precession.

For a more detailed explanation, consider a bullet fired from a barrel with a right hand twist. A left twist will cause the same action only to the reverse or left direction. The bullet's front-end is lifted as the trajectory curves down and air pressure pushes beneath the nose. The bullet has gyroscopic stability and resists this moment upward by pointing its nose slightly to the right because of precession. The nose right is maintained by the upward moment and the stable flight attitude is nose high and nose right. A very tiny amount, of course, but present nevertheless. This *yaw of repose* tends to steer the bullet slightly to the right. The air pressure is giving a push to the nose on the left side.

To summarize in a few words, the curved trajectory causes pressure on the bottom of the bullet's nose and that, because of gyroscopic precession, causes a yaw and drift to the side in the direction of rotation.

Many people have blamed this side drift on the bullet "rolling on air". This theory states that as the bullet drops, it compresses the air below it and then the bullet rolls to the side on the compressed air cushion. It has been cleverly compared to throwing a spinning log or baseball bat on water. Modern ballistic laboratory tests have shown that the rolling on air theory is incorrect. It is still believed mainly because of two reasons. **First**, it is a simple and easy to understand idea that sounds reasonable. **Second**, it was embraced by Major (later General) Julian S. Hatcher. (1888-1963) He mentioned it in his book *Textbook of Pistols and Revolvers*, (bottom of page 142) published in 1935. J. S. Hatcher was an expert on the subject and highly respected. In this case he was in error but due to no fault of his own.

For people who do not want to move into contemporary times, the rolling on air theory may assist a very slight amount, but the fact that the drift increases with distance shows the help to be minimal. Remember, the rotational speed gradually slows, so the drift would not increase if caused by something that decreases.

The .30-'06 will be used once again for an example. Because of early and long time military use, a great amount of information has been gathered about this cartridge which does not exist for other ammo; thanks partly to the U. S. Army.

The gyroscopic drift of this bullet is about 12" at 1,000 yards. Long range artillery and machine-gun tables include this information.

For ranges below 1,000 yards, it is suitable to use the approximate relationship that the drift is proportional to the square of the time of flight, or to the third power of the range. For example, at 500 yards the drift will be approximately 1/8 that at 1,000 yards or 1½". At 300 yards it will be about 1/30 that at 1,000 yards or about 3/8". While it is excessive at ranges well over 1,000 yards, it is small enough to be of no concern at normal ranges.

Gyroscopes precess slower as their r. p. m. increases. This is a basic fact for any object having gyroscopic properties. Also, slower precession needs a larger yaw to maintain equilibrium as the trajectory curves. An increase in spin will increase the yaw. The extent of the equilibrium yaw is in a direct proportion to the spin. The faster spin, therefore, has little advantage over the slower spin.

THE EARTH'S ROTATION EFFECT

Knowledge of the earth's rotational effect has very little practical benefit, for most of us are not going to be shooting at ranges long enough for the earth's rotation to be noticeable. Nevertheless, it is an interesting subject and should be included in any serious study of ballistics.

Gaspard Gustave de Coriolis worked at the Ecole Polytechnique in Paris, France as an assistant professor of mechanics and analysis. In 1835, he wrote of the natural law of physics that makes moving things, such as flowing water and bullets, move to the right in the Northern Hemisphere and move to the left in the Southern Hemisphere. (For readers who missed that day in school, the imaginary circle around the earth, equally distant at all points from both the North Pole and the South Pole, known as the *equator*, forms the dividing line between the two hemispheres.) The Coriolis effect, as it is known, supposedly causes water to drain out of a bathtub with a counterclockwise swirl in the Northern Hemisphere and clockwise in the Southern. Weather masses take curved paths across the earth as they move from high pressure areas to low pressure areas because of this effect. Hurricanes, waterspouts and cyclones are likewise affected.

Very simply, this effect is caused by the earths rotation. It is moving east at about 750 miles per hour.

The British Small Arms Committee discovered back in 1886 that a bullet will deviate about 6" to the right in 1,000 yards from the earth's rotation. (Northern hemisphere.) Maximum drift occurs when the bullet is fired toward the south east and minimum drift when fired to the north west. This is in addition to the normal gyroscopic spin drift in the direction of the rifling twist.

Generally, the bullets curve will be about 1/10th of an inch in 125 yards in the central part of the U. S.

The drift effect on a projectile caused by the earth's rotation varies inversely according to proximity to the equator. Sighted-in rifles will remain the same unless moved an extremely long distance north or south. Even then, it would only be noticeable at long ranges like 1,000 yards. Of course, only the very best bench-rest shooter could detect such an effect.

In *The Book of the Rifle*, Freemantle writes of testing by Sir Henry Halford with the Metford .461 caliber match rifle at 1,000 yards. He stated that deviation from both spin drift and earth rotation together, using a right hand rifling twist, was 3' 6" to the right. Using a left hand twist of equal inclination, it was reduced to 2' 6" to the left. While these figures may seem high, remember that no two calibers and firearms are the same. Also, the actual amount is not as important as the knowledge and proof that this occurrence takes place.

After the projectile leaves the gun, its plane of motion is fixed in space, except for drift and deflection, **and the earth turns under it**. The linear velocity of the surface of the earth at the gun, due to both rotation and translation, is conveyed to the projectile. As long as this linear motion is constant in amount and direction, there is no effect on the trajectory. It is only the angular diurnal motion that influences the trajectory and it does this in both range and deflection.

Without going into astronomics or mathematics, there are many obvious results. Consider that this rotation can cause an angular lag or deviation of the projectile to the west as the vertical axis is carried by the earth's rotation to the east. The rotation about the axis perpendicular to the plane of departure causes a change in range.

During the flight of the projectile, it is at a constantly varying distance from the surface and hence from the center of the earth. These latter distances are all greater than the radii of the earth at the beginning and end of the trajectory. Therefore, the projectile will lose in angular velocity, due to its increased distance from the center of the earth. A projectile fired vertically at the Equator to a great height would strike the earth at a point to the west of the point of departure. The amount of the westward deviation would depend on the height of the trajectory and the time.

The surface of the earth at the time of shooting is the horizontal surface of projection, and the range is measured on this surface taken as fixed in space. By the rotation of the earth while the projectile is in the air, the surface of the earth to the west of the gun is elevated in respect to the horizontal surface of projection and the surface of the earth to the east of the gun is depressed. Thus fired westward, as the projectile approaches the surface of the earth, it will be moving with a greater velocity but at a lower altitude than would have been the case if the earth had not been in rotation. The earth comes up and meets the projectile and shortens the range more than the first-mentioned term lengthens it, except in case of very high angles of elevation. The rising of the western horizon makes the sun appear to set and the lowering of the eastern horizon makes the sun appear to rise.

The effect of the two terms in range may be visualized by drawing a trajectory of low elevation and another almost vertical. Now, keeping the origin fixed, rotate the horizontal and vertical axes counter to the direction of the trajectory, which will correspond to the rotation of the earth when firing west. It will be noticed that the range will decrease for the trajectory of low elevation and increase for the nearly vertical trajectory.

For artillery, the amount of the change in range due to the rotation of the earth is obtained by numerical integration. (Differential equations; meaning advanced mathematics, in other words.)

Vertical axis: The rotation around the vertical axis produces a deflection. It varies as the sine of the latitude and is independent of the azimuth of fire. A projectile fired from the pole of the earth would undergo an angular deflection proportional to the time of flight, and hence a lateral displacement approximately proportional to the product of the time of flight by the range. In the northern hemisphere this deflection is to the right while in the southern hemisphere it is to the left.

Horizontal axis: The rotation around the horizontal axis produces a term in deflection very similar to cant. The projectile is carried out of the vertical plane and the amount of the deflection is approximately proportional to the height the projectile goes as well as the time of flight. It is at a maximum when firing on a meridian and zero when firing east or west. When firing northward in either hemisphere the deflection is to the left and when firing southward it is to the right.

Perpendicular axis: As explained a few paragraphs earlier, a rotation about the perpendicular axis from the point of departure, causes a change in range.

VERTICAL DRIFT

An interesting side-note to wind drift is the very slight vertical drift, either up or down, that is caused by a cross wind. At first that sounds impossible, but it is a proven fact and not a theory. One reason it is little known and seldom mentioned is the vertical movement is so small. Also, it requires extra laboratory skill to prove because several details that can change the trajectory have a normal variation larger than the vertical drift. In other words, ammunition and other variables may cause a larger change in performance. This gives added meaning to the statement, *all other things equal.*

The cross wind causes the bullets front end to precess slightly and slowly. A right hand twist and spin and a right cross wind will raise the nose and the impact point. A right twist and a left cross wind will lower the nose and impact point. A left spin will cause the opposite reaction. Wind from the right will lower the nose and from the left will raise the nose.

The effect is noticed in extra long projectiles where the center of pressure from the drag is ahead of the center of mass. This includes most rifle bullets.

The vertical drift is in the region of 1/10th the side drift. The Ballistics Research Laboratory at Aberdeen Proving Ground considers the

effect so small that it is not listed in firing tables for any gun smaller than an 8" bore.
LATERAL JUMP
It is interesting to note that many rifles have a small lateral jump during firing. That is an angular displacement sideways. The amount has been determined for some military rifles. For example, a .30-'06 bullet from a 1903 Springfield was found to be a little to the left at a 500 yard range. The gyroscopic drift to the right from a right twist barrel is not enough to overcome it until a longer range. This lateral jump may not exist for some guns and may be different for each firearm that has it.
LONG RANGE WOBBLE
Gyroscopic stability is what holds a child's spinning top erect until it slows down and begins to wobble and eventually falls over. A spinning bullet can do the same thing for the same reason. If the range is long enough and especially if the bullet is not as stable as it should be, it will wobble and change position. Bullets must have complete stability, not just marginal stability. After all, a long and streamlined bullet will tumble almost at once if given no spin from a smooth bore gun. As a projectile moves down range, the velocity decreases faster than the spin which helps it to maintain stability at range. From that it is obvious that a faster twist will not usually help long range stability. Still, gyroscopic stability is vital and over stabilized bullets are more practical than previously believed. The most popular theory today, which is backed up by testing, is when in doubt, go for the tighter twist.

This is opposite some people's opinion that for long range work, a bullet should leave the muzzle at minimum stability. That is false. It is also false that firing a bullet at minimum stability is necessary to keep the point on the trajectory as the bullet arches over on a long range shot. This falsehood is explained in several areas of this book. Much more on correct twist in Chapter 12.

Normally, a bullet will be about 1½ times more stable at 500 yards than at the muzzle. Part of this is due to the initial yaw which normally corrects itself.
MATHEMATICS
For readers who like mathematics, consider the following: Stability can be expressed by the military conceived formula:

$$S + A^2 N^2 / 4B_u$$

Where: S = stability factor which must be greater than 1.
  N = spin.
  A = Moment of inertia of projectile about its longitudinal axis.

B = Projectiles moment of inertia about the axis going through the center of gravity and perpendicular to the longitudinal axis.

u = moment factor which includes air density, projectile speed and diameter and a moment coefficient which is based on the projectile's shape and velocity relationship.

For mathematically inclined readers, there is a benefit in studying this formula, but you are not expected to work it. Not enough information is given. The reader cannot look up the moment coefficient as he or she can the ballistic coefficient in another section. Nevertheless, with a knowledge of mathematics and how numbers and terms interrelate and function together, some basic information can be deduced.

There is a big advantage to expanding the diameter of a projectile as is shown by the moment of inertia (A) which is squared in the formula. The other moment of inertia (B) has its effect in the opposite direction as it is opposite in position. But, notice B is not squared in the formula so its effect is only to the first power. Therefore a projectile that is short and fat should, everything else equal, be very stable.

When the center of gravity and center of pressure are moved apart, the stability is reduced. This is logical and basic. It also increases the moment factor in the formula. The longer projectiles usually have the two points the widest apart and also have the poorest stability. **(There are advantages to long projectiles, but that is not the subject at this time.)**

The spin created by the rifling is a major detail and it is squared in the formula. Velocity affects spin and stability.

HOLLOW POINT BULLETS

In a hollow point bullet, removing the center portion removes material close to the axis which has a small radius of gyration. The design shifts some of the mass out from the axis. The average radius of gyration of the rest of the bullet is increased. This improves the stability and helps slow down rotational deterioration. The greater the radius of gyration the stronger the gyroscopic effect and the stronger the stability. The hollow point also transfers the center of gravity farther to the rear as does a nose radius and streamlining.

Moving the center of gravity back separates it farther from the center of pressure. This hurts and makes it harder to stabilize. Some people believe hollow points are more accurate because they are better stabilized while some think the reverse is true. In most cases, the two things probably cancel each other out and the end result is a tie.

The changes in the formula listed are easy to visualize and the increase in stability may be small.

If the hollow is in the base, as on a shotgun slug, the increase in stability will be improved as the center of gravity is moved forward.
VELOCITY
If other factors remain the same, changing the bullets velocity changes the stability but not in a direct way that is easy to work mathematically. Generally, the stability increases a large amount at 200 f.p.s. or 300 f.p.s. above the speed of sound. The speed of sound and the transition range above and below it are discussed in detail in Chapter 13.

The moment factor in the stability formula takes into account projectile velocity. While a velocity increase will increase stability, it is a small amount. The ratio between the velocity and the twist is a constant. The gain in stability will be from a small decrease in overturning moment.
BULLET STABILITY DEFINED
A technical definition of a bullet's stability: it varies directly as the square of the axial moment of inertia and inversely as the first power of the transverse moment. At first reading, that is overwhelming, but it isn't really difficult. To vary directly as the square means exactly (precisely) the quantity times itself. The axial moment of inertia is the rotary force about the axis, (the mass). Inversely means opposite or inverted as one quantity is greater or less according as another is less or greater. The transverse moment is the force operating across (crosswise) of the axis which tends to throw a rotating object out of position. Finally, as mentioned at other places in this book, a force in rotary motion is called a moment.

Let's repeat the technical definition. A bullet's stability varies directly as the square of the axial moment of inertia and inversely as the first power of the transverse moment. See? Now that we know what the strange words mean, it makes a lot more sense. It follows that a bullets stability is directly proportional to its density. In other words, if the size and design are the same, a heavier bullet will be more stable. Light metals are hard to stabilize and also poor in other ballistic areas.

The transverse moment of inertia can be increased by added weight at the axis by using a longer bullet or long point, boat-tail etc. This decreases the axial moment and will not help stability. In fact it may decrease it. Hollow points, blunt noses, etc. increase the transverse moment and increase stability. (Once again, there are advantages to long projectiles, but that is not the subject at this time. As is frequently the case, a compromise is required in choosing or designing a bullet.)
RIFLING TWIST
That rifling twist is important to stability is obvious. Gyroscopic stability varies as the square of the rifling twist. This means that a very slight change in twist will have a disproportionately big result. Twist is covered in detail in Chapter 12.

CHAPTER 10

RICOCHETS AND BRUSH DEFLECTION

These two subjects are closely related and they will be discussed together. As with all of the subject areas of ballistics, this field of knowledge is intertwined with other information. If it has not been read lately, it would help to review Chapters 8 and 9 on gyroscopic stability and spin drift.

RICOCHET

This is a word with a french background. It is as hard to get experts to agree about ricochet details as it is to spell correctly. Ricochet is defined as the glancing off of a projectile from a hard surface. One writer said it was, "a phenomena of low velocity rather than of high velocity." This is not correct, as a high velocity projectile will ricochet as much as a low velocity one will. The thinking, of course, is that at high velocity, all projectiles will break apart. This is also incorrect. Of course, more will break up at high velocity than at slow velocity, but not all of them. Not by a long shot. (No pun intended.)

For about as long as people have thought of the subject, it has been believed that high velocity prevents ricochets. The idea is that a bullet at high velocity - the required speed usually not specified - will break apart on hitting any object. Even leafs and tiny twigs are supposed to destroy a high velocity bullet.

If we stop and think for a moment, we realize that a bullets velocity deteriorates. In most examples, only near the muzzle is the velocity high enough to believe in this statement with any certainty and then only with something more substantial than a leaf or a twig. While it is true that ricochets can be reduced by high velocity, it is not a "given".

Soft-point and hollow-point bullets are safer from ricochet. They still ricochet, but it is frequently in pieces because they are more likely to break apart on impact.

It is also important to be aware the a bullet's velocity will drop very little from a ricochet. Tests conducted by Winchester show that a .22 Super X bullet ricocheting from water, lost on the average, only about 45 f.p.s. Bullets hitting at 1240 f.p.s. grazed off at 1195 f.p.s. The angle has some influence, but it is obvious that much care must be taken. Other calibers will be affected in a similar way.

Regardless of the type of surface, a bullet of any caliber will ricochet at the angle of incidence. In other words, a shallow angle strike will ricochet at a shallow angle and very little energy will be lost. With a steeper angle of incidence, the angle of ricochet will be steeper and the energy lost will be greater.

A safe backstop should be used even when shooting at water.

Under some conditions, even a soft point bullet at high velocity can ricochet off a sand pile if it hits a tiny rock at the correct angle.

Penetration testing on steel can be dangerous. The bullet can ricochet straight back toward the shooter or friends standing nearby. This would not be a ricochet in the true sense of the word. Perhaps spring, bounce or rebound would be a better word choice. Most people think in terms of glancing at an angle down range, not bouncing back. It the bullet cannot penetrate or knock out a slug, then it has the potential to come back. Wear eye protection and use caution.

Of course any bullet can ricochet and if you are in a situation where you are afraid it might, then don't shoot. Remember that high velocity bullets loose velocity over long distances. Also, the shattering ability of high velocity bullets is perhaps overrated. Even so, the best will lose velocity to where it can ricochet the same as any other slower bullet.

BRUSH DEFLECTION

Deflection from brush and twigs is a big concern when hunting in the woods. Numerous experiments have been conducted to determine which cartridge and bullet type will handle brush with the least problems. Unfortunately, the tests don't agree with each other. Some promote a heavy slow moving bullet (the theory for years) while various other test results speak of great things with light high velocity spitzer point bullets.

Some tests have shown velocity is not a factor in brush deflection and other tests have shown that the higher velocities are better. Many hunters prefer low velocity cartridges, partly due to the popularity of the old .30-30.

One thing is certain. The testing methods leave much to be desired when the results are so wide spread and divergent.

KNOWN FACTS

The evidence tends to lean toward a high velocity being better than a low velocity. (Bullet weight and other factors being equal.) Also toward heavy bullets without excessive length.

It was just reported in the chapter explaining gyroscopic properties that a bullet's deflection from branches is caused by a force on some part of the ogive. The pressure will cause precession and the direction will depend on where the force is applied. The amount of deflection will depend directly on the amount of force and inversely on the angular momentum of the bullet. The rules of gyroscopic action will give an edge to a short fat bullet over a long thin bullet if the weight and spin stability are equal. The short fat bullet will precess less and slower.

Bullets hitting twigs and brush are not only deflected from their trajectory, they yaw and lose gyroscopic stability. They can be knocked into a helix spiral and because of gyroscopic precession, the travel will be in a different direction than the direction of original force. Testing has shown maximum yaws as much as 50° are common from heavy brush. A bullet with high stability can become stable again if the range is long enough, although it will be in a new direction.

Good brush deflection requires a bullet from a barrel with a good twist. If the stability is marginal, it will take very little to cause it to tumble. Although tumbling and deflection are two different things, they are related. A yaw is not tumbling, but if the bullet cannot stabilize itself again, it may tumble if the range is long enough. As previously mentioned, the direction will be changed.

Testing conducted for military purposes at Aberdeen Proving Ground showed that bullets break up easily on obstacles. When the bullets held together, the least deflection was by bullets that possessed the most density, the most velocity, and the most gyroscopic stability.

This is the opposite of the older theory promoting slow bullets.

It should be kept in mind that all bullets can and are deflected by brush, tree limbs and branches; high velocity, low velocity, short and fat or long and slim.

A lot of benefit can be obtained by using the proper bullet. Much of the trouble with hunting in brush is not deflection but as was just mentioned, disintegration. They simply fly apart when they hit something. Out of the countless cartridges and bullets manufactured, look for strength as well as weight, shape and expansion.

In the chapter on gyroscopic stability there was a discussion of the use of screens or papers placed down range to check for yaw and spiraling etc. The same technique is useful in testing for brush deflection. Screens can be placed in such a position as to intercept the bullet at different points after it has deliberately hit a peg or dowel.

Some testing has shown that a bullet's shape has little effect on deflection. That is probably true in many cases, but a careful analysis of the physics involved would indicate otherwise. Most likely the differences are so small as to be difficult or impossible to determine in all but the best laboratory testing.

SUMMARY

All bullets can and are deflected by brush, tree-limbs or Aunt Martha's favorite rocker. (If you or she are in the wrong place.) High velocity, low velocity, short and fat or long and slim; the result can be the same.

The amount or angle of deflection is a direct result of impact angle.

Metal cases and strength helps to prevent fragmenting.

Big, heavy bullets are better in brush than small, light bullets.

Velocity is still a matter of debate, but most research leans toward more velocity instead of less.

Last and most important, if possible, avoid shooting at animals through forested growth.

◆ A well know sportsman and gun expert, Col. Charles Askins, believes there is no easy and accurate solution to using bullets through brush. He said in the March 1982 issue of the NRA's *The American Rifleman*, "Our sporting bullets are pretty much incapable of slashing through the limbs, the brush and the second growth with any degree of assurance of thereafter killing the game."

Amen.

17th century Italian physiologist Giovanni Borelli said, "The perpetual law of nature is to act with a minimum of labor......avoiding, in so far as possible, inconveniences and prolixities." This is sometimes called, "the principle of least action," and it applies to both static and dynamic equilibrium as well as gyroscopic action, trajectories, etc. The list can go on to almost infinity.

CHAPTER 11

BARRELS

A barrel consists of a lot more than a metal bar with a hole down the center. Much of the science of firearm ballistics is influenced by the barrels length, material, rifling, choke, etc. They are frequently changed or modified with the hope of receiving better performance.

All ballistic subjects are so intertwined that some barrel topics are discussed in other chapters. The index should prove helpful.

MANUFACTURING

Rifle barrels are subjected to considerable stress during manufacturing. The steel blank has to be deep-hole drilled through its entire length with the resulting chip clearing problem. The hole is left about .010 inch below the finished bore size. Several reamings are required to bring the bore to the final dimension. The outside of the barrel has to be machined to size and concentric to the bore, if possible. Other operations include the chamber, sight mounting holes and threads on the breech end, and we must remember the addition of the rifling grooves. (Rifling is covered in detail in Chapter 12.)

Chambers and barrels are not always made as accurately as they could or should be. Diameters can be too big or too small, tapers inadequate and in the wrong place, lengths long or short. Misalignment of the chamber to the axis of the bore, both in straightness and concentricity, is not unusual and very difficult to detect.

STRAIGHTENING

Straightening is an important detail, and one that requires much human skill and experience. Two different methods are in use. **Ring straightening** involves sighting down the bore toward a light source and rotating the barrel while watching the light rings and how they appear in relationship with each other. The barrel is then placed upon two lead blocks and struck with a lead or bronze hammer at the high point. The barrel will be inspected after each blow until the rings maintain a concentric position while the barrel is rotated.

Line straightening involves the use of a back-lit frosted glass with a thin opaque horizontal crosspiece at its center. A straightening jack, which is similar to an arbor press, is 18 feet away. When sighting through the bore at a point just below the cross piece, two shadow lines will converge at the muzzle end and form an angle with the vertex at the bottom of the bore. The lines will extend back and up the sides and appear as an inverted V. The lines will extend about half way down the bore, as viewed through the breech end. If the barrel is straight, when it is rotated the lines will keep their symmetrical position. The barrel is straightened as required with blows from the jack.

As you can easily understand, it takes great skill, experience, and a lot of patience to be a good inspector and barrel straightener.

All of the machining and cutting places a lot of internal stress in the grain structure of the steel. In cheap barrels, little correction is done and whip and other barrel movement is greater than it needs to be. Best quality target barrels are carefully heat treated and stress relieved. It is interesting to note, as many old time machinists know, that the hammering involved in the straightening process aligns the grain structure and relieves stress. The more straightening a barrel needs, so long as it is done properly, the less internal stress the steel will have.

A perfectly straight barrel is extremely rare. This should not be confused with concentric, which is a separate topic.

Testing a barrel involves fitting the barrel with two concentric and accurately machined rings. Each is tightly placed around the barrel at about 20 percent of the total length from each end. The barrel is then placed in a special accurate and smooth V block that is long enough to reach both rings. Two V blocks can be used, but getting them positioned and mounted so they are dead in line is not easy and requires both instruments and skill. The long block is (or blocks are) securely fastened to something extremely rigid or heavy and stable. It must not move even the slightest amount.

Both of the rings must be held firmly in the V groove as the barrel is fired at a target at least 100 yards down range. A no-wind day is best because winds are not steady and constant. The barrel is rotated and fired at different points. The four 90° points are satisfactory. Don't be frugal with ammunition. Fire at least five shots from each position and use only ammo from the same box. As always, an occasional "flier" can be expected. But if the barrel is straight, it will shoot a normal group of 2 inches or 4 inches or whatever the barrel is capable. If it shoots 4 separate groups at 90° points, the answer is obvious.

The "fliers" that were mentioned are hits that do not follow the normal tendency of the firearm. Some people believe that nothing can be done to prevent them. That is not always true. Several suggestions are in this book. The section on gyroscopic stability and rifling have some. Other sections also have tips in this area.

LENGTH

Barrel length is an important factor in bullet efficiency as it relates to chamber pressure and the expansion of gas. The length of the barrel is important in velocity figures and generally, the longer the barrel the higher the velocity, up to a point. The word *generally* indicates this is not always true, as will be explained.

When the powder in the cartridge is ignited, it expands as a hot gas which in the open air, would extend in all directions. This expansion, differing with various powders, can be as much as 800 to 1300 times in the free air. The pressure accelerates the bullet down the barrel because it has no place else to go, discounting the "leaks" in some firearm types.

The longer barrel does not help because the powder is still burning in the extended barrel but because the gas is still expanding and pushing the bullet to a higher velocity. There may be some unburned powder found in the barrel after firing, but in that case, it probably would not have burned in any length of barrel.

As the bullet moves down the barrel, the area behind it that is being filled by the expanding gas increases and the pressure will eventually begin to drop. There will be a point in an excessively long barrel where the drop in pressure would not be able to continue acceleration against the friction required to push the bullet through the rifled bore. (Or a smooth bore, for that matter.) A shock wave can also develop at high gas velocity. At this point the velocity will begin to drop. Few firearms are made today with barrels excessively long, although weak ammunition can create this problem in an otherwise excellent gun.

Example: The velocity of a .22 rimfire long-rifle bullet will begin to slow at about a 16 to 18 inch barrel length, depending on ammo type and firearm. At 16 inches, there is a gas expansion of about 37 times its original volume. The .22 has a tiny chamber and this gas expansion is extended to 44 if the barrel length is increased to 19 inches. The .22 short cartridge has an even smaller chamber and would give an even higher expansion in the same barrel length. Years ago, testing by Eric Johnson of Hoffman Arms Co. showed an 18 inch barrel length gave best velocity for the .22 long rifle cartridge. An interesting side point: The very high velocity of about 4,000 f.p.s. is hard to exceed in a .22 caliber regardless of powder charge because of the barrels resistance (friction) to the movement of both the bullet and the gas. Larger center fires have a slightly higher limit.

The barrel should have nothing in it except the bullet and the push of the expanding gas. The combustion of the powder should all be in the cartridge case, not the barrel. That is why the statement that long barrels are best because they let all the powder burn is wrong. Barrel lengths are best at a length (if practical) that lets the gas expansion be most efficient. As stated, combustion is best in the cartridge.

The requirement for longer barrels was stronger in black powder days than it is with modern powders. Today, the need for a long barrel for gas expansion is a fact but in earlier times this need included length for proper burning. Occasionally this persuades people in the present time to believe they need a barrel much longer than is required.

EXPANSION RATIO

We mentioned expanding gas and the barrel length. This "expansion ratio" needs more consideration. The technical meaning is the ratio that we obtain if we divide the volume of the powder chamber with bullet seated into the volume of the powder chamber and bore added together. Or, put in simpler terms, how much room the expanding gas has to expand into before the bullet leaves the bore. This ratio is a determining factor in muzzle velocity. Also involved are peak pressure and the distance the bullet has moved.

◆ To work this out mathematically, we must first find the bore volume. For this, we can use a formula that assumes the lands cover 1 1/2 per cent of the bore area. (The cross section as seen from the muzzle.)

$$B = L * D_2 * .773$$

Where D = groove diameter of barrel in inches
L = length traveled by the base of the bullet from seat to muzzle in inches
B = bore volume in cubic inches

◆ The bore volume, along with the powder chamber volume, can determine the expansion ratio.

$$R = B + U / U$$

Where: R = ratio of expansion
B = bore volume in cubic inches
U = volume of powder chamber in cubic inches

The .30 caliber M1 carbine uses a cartridge with a .081 cubic inch capacity. When fired, the gases will have expanded to double their volume when the bullet has moved approximately 1.2 inches down the 18 inch barrel. At the muzzle, the gas will have expanded approximately 15 times.

By comparison, the .30-'06 cartridge has a capacity of 2.51 cubic inches and will therefore require 3.4 inches of bullet movement to double its volume. At the muzzle end of a 24 inch barrel, the gas expansion will be only 7 times.

In the two examples just listed, the longer barrel is not as efficient as the shorter barrel. The cubic capacity of the cartridge case is the controlling factor. One expands 15 times to the others 7 times. Bullet weight is less in the carbine and it receives only a 26% loss of velocity (compared to the .30-'06 with a 24 inch barrel) with 70% less powder charge by weight and a 25% shorter barrel.

Another point that has to be considered is the ratio of bullet weight to powder-charge weight. Faster and slower powders or a heavier bullet will bring about the least change in velocity. Hand-loaders with a chronograph will find this easy to test for themselves.

VELOCITY AND BARREL LENGTH

Both ballistic laboratories and gun-magazine writers have conducted many experiments comparing barrel length to velocity. Handguns, shotguns and rifles have all been subjected to the hacksaw as inches were removed and the gun test fired. Then another inch or two is removed, the muzzle is properly deburred and crowned, and another series of test shots and chronograph readings are taken.

For the most part, the testing confirms what the industry has known for years. The shorter barrels produce less velocity. Interestingly, in most calibers the drop in velocity and accuracy is small enough to be of little concern for hunting at normal range. Hunters that seldom shoot over 100 or 150 yards will be about as well-off with a 20 or 22 inch barrel as a 24 or 26 inch. The loss in velocity will, in most cases, be small. There will be increased muzzle blast, but as with the velocity loss, it will be small.

For example: testing by Springfield Armory with .30-'06 cal. produced 2,710 f.p.s. with a 24" barrel: 2,775 f.p.s. with a 28" barrel: 2,830 f.p.s. with a 30" barrel: and 2,848 with a 32" barrel.

If rifle barrels in calibers of .30-'06 and larger are shorter than 22" to 24", the muzzle blast can be disturbing. From a ballistic point of velocity and energy, the shorter barrel may be satisfactory. It will be fast handling and light to carry. From a practical point of view, a rifle that has to be carried over any distance may as well have a short barrel. The shorter barrel will probably have only about a 25 f.p.s. drop in muzzle velocity for each inch of barrel lost. Of course, with all general rules, it will be just an estimate or approximation. As always, a compromise is required for the performance area the shooter believes is most important.

Note: A 2" difference in barrel length may have little effect with a rifle or a shotgun, but with a handgun the difference between a 2" and a 4" barrel can be dramatic. (See Chapter 22 on handguns.)

Originally, in black powder days, long barrels were used for three reasons. (1) They were helpful in burning the powder. (2) The long distance between open sights was a help in aiming. (3) It was safer for bayonet use as it kept the opponent farther away. (If the soldier was both good and lucky.)

Another section of this book discussed barrel whip, flex, warp and the movement and stress created by the hot expanding gas and projectile movement. (Chapter 5.) Technically, this physical motion will be less in a shorter barrel than a longer barrel, all other things being equal. Once again, the amount of reduction will be small.

As long as the barrel length does not exceed the optimum length, (which it seldom will in any cartridge except a .22) the bullet will continue to accelerate as long as it is in the barrel. After exiting the muzzle, the deceleration begins and the velocity starts to drop. The highest speed is at the muzzle, with the noted exception of an excessively long barrel.

OCTAGON SHAPE

Octagon barrels, besides their traditional attractive appearance, are generally stiffer than round barrels. While stiffness and accuracy are related, they do not always go together. In the mid 1800's, when barrels were made in large quantities by both hand as well as machine, the round barrels were accurate because they were more likely to be machine made. In modern times, the accuracy depends on the quality of work, not the barrel's shape. But, that's the same reason round barrels were better in the 19th. century.

MEASURING LENGTH

A revolver's barrel length is the distance from the rear of the forcing cone to the muzzle. The cylinder length is not included. It can be measured with a rod from muzzle end to a shim placed in the gap between barrel and cylinder or directly with the cylinder out of the way.

For rifles and shotguns, the chamber length is included. Notice the difference with revolvers where this is not included. The barrel length is from the muzzle to the back as far as a rod can be inserted with the action closed. Federal law limits barrel lengths on shotguns to a minimum of 18"and on rifles to a minimum of 16". Over all length of 26 1/2 " minimum. This may change as laws are being altered to abolish gun ownership.

A shotgun barrel that is too short cannot be made legal by adding a choke device. A Cutts Compensator or Poly Choke, to name just two, will add length at the muzzle end, but it is not considered barrel length by the Federal laws. Many of these products work well, from a ballistic point of view, but they cannot change an illegal barrel into a legal barrel.

If there is a problem over the legality of a barrel length, the actual measurement is used. Not a figure in an advertisement or from factory literature. Caution: Police officers can't be expected to know everything and some do not know the correct way to check barrel length. Most importantly, that the chamber is included on a shotgun. If you disagree, be polite and cooperate. The proper answer will come out in due time.

CALIBERS FOR LENGTH

The military occasionally list a barrel length in calibers. It is used very seldom and mostly for artillery and heavy cannon. It has a practical use in showing if a barrel length is suitable for a bore diameter.

For example, a .30 caliber bore 22 1/2 inches long would be 73.3 calibers in length. A 5 inch cannon with a 100 inch barrel would be 20 calibers in length.

CALIBER	Bullet weight in grain	Charge weight in grain	Powder type IMR	MV,f.p.s. in 24" barrel	18" to 20"	20" to 22"	22" to 24"	24" to 26"	26" to 28"
.17 Rem.	25	22.5	3031	4015	87	76	67	61	54
.22 Hornet	45	11.5	4227	2504	31	28	24	22	19
.222 Rem.	50	20.5	4198	3111	54	46	42	37	33
.223 Rem.	55	25	3031	3165	59	52	46	41	37
.22-250	55	35.5	4895	3645	80	71	62	56	51
.243 Win.	100	43.5	4350	3033	68	60	53	47	42
.250 Sav.	100	34.5	4064	2875	57	49	43	39	35
.25-'06	120	48.5	4350	2950	71	62	55	49	44
.270 Win.	150	54	4350	2955	67	58	52	47	41
7x57 mm	145	47	4350	2650	55	47	42	37	34
7mmRemMag	150	66.5	4831	3055	75	66	58	52	47
.30-30 Win.	170	32	3031	2214	35	31	28	24	22
.300 Sav.	150	38.5	3031	2575	46	40	36	32	28
.308 Win.	150	45	3031	2850	52	46	41	36	32
.30-'06	110	56	3031	3391	70	61	54	48	43
.30-'06	150	52	4064	2908	60	52	47	41	37
.30-'06	180	57	4350	2772	57	50	45	38	35
.30-'06	220	52	4350	2489	52	45	39	35	32
.300 WinMag	180	71.5	4350	3100	73	64	57	51	45
.338 WinMag	300	64.5	4350	2392	50	45	39	35	31
.35 Rem	200	37.5	3031	2122	32	27	24	22	20
.375 H&H	300	78	4350	2601	54	47	41	37	33
.44 Mag.	240	23.5	4227	1693	17	15	13	12	11
.444 Marlin	240	47	4198	2323	34	30	26	23	21
.45-70 Gov't	405	51.5	3031	1787	23	21	18	16	14
.458 WinMag	510	72.5	4895	2089	32	28	24	22	19

The last 5 columns give the difference in velocity in f.p.s. for each 2" change in barrel length with the barrel length and muzzle velocity established by equation. Barrels of intermediate lengths that are not listed can be estimated by interpolation between the lengths given. The effect of barrel length on velocity depends on bullet weight and powder charge.

This chart is the work of William C. Davis, Jr. and was originally published in *The American Rifleman*. It is reprinted here with their kind permission.

EUROPEAN BARREL LENGTH

European barrel lengths are given in centimeters. To convert centimeters to inches, multiply by 0.3937.

Here are a few common lengths for readers who hate math.

50 centimeters	equal	19.7	inches
55	"	"	21.7"
60	"	"	23.6"
65	"	"	25.6"
68	"	"	26.8"
70	"	"	27.6"
71	"	"	27.9"
75	"	"	29.5"
76	"	"	29.9"

All lengths are described in the U.S. as the next longer number in inches. For example, both the 70 cm. and 71 cm. would be called 28 inch.

DIAMETER AND SHAPE

For safety, the barrel diameter at the breech end should be at least 2 1/3 times the diameter of the cartridge case. For example, the .308 Winchester and the .30-'06 have a case diameter of .470 times 2.333 equals 1.096 or rounded off to 1.1 inch.

Barrels can be straight as with most heavy bull barrels or tapered with the taper beginning at any point in the first half after the chamber. There can be a steep taper in the first part and a smaller taper in the last section or a tapered first part and a straight final section. Or about any combination on the list.

WEIGHT

The barrel is the heaviest part of most big caliber rifles and careful planning should be done in building a custom job. The action can be easily weighed, probably around 2 3/4 to 3 lbs. Normal stocks will weigh between 1 3/4 and 2 1/2 lbs. and special target stocks can go up to 3 1/2 lbs. Scopes can vary from 5 1/2 to 16 ounces and in a rare case, to 24 ounces. A good average scope weight would be around 10 or 11 ounces. Total everything up and if your custom rifle needs to be kept below, say, 8 1/2 pounds, the barrel will become very important for weight as well as accuracy.

BARREL LIFE-DURATION

Barrel wear is first noticeable in the area immediately in front of the chamber. Farther down the bore, the grooves will be roughened. Even when the wear is quickly noticeable, the accuracy may still be as good as ever. As wear continues, there will be a velocity loss from gas escaping between the bullet and the bore. Extreme wear will cause yaw and poor stability.

The wear in the bore can be visually obvious. Target dispersion is another noticeable detail. If a barrel which has had heavy use will not group as it did in the past, and everything else is unchanged, check for barrel wear. The wear can be measured by instruments that check the bore itself. Any competent gunsmith can make the inspection for a gun owner.

Some riflemen believe that boat-tail bullets cause excessive barrel wear. Tests by the Finnish State Cartridge Factory indicate this is not true. Tests conducted later by the U.S. Army, backed-up the earlier study. There is no evidence to show excessive wear by the use of boat-tail bullets. The idea was probably started with the theory that the hot gas was more likely to go past the bullet or at least into the boat-tail area.

CONCENTRIC BORE

Some barrels have a special problem where the barrel appears straight on the outside but the bore is crooked. As these barrels heat up, the impact point changes and a good group is impossible. It is usually caused by errors during manufacturing. If the outside of the barrel was turned after the bore, with a lathe center in the muzzle end, that end would appear concentric. (Inside and outside diameters on the same center with equal wall thickness around the circumference.) The outside would be straight even if the bore was not. A too heavy machine cut in turning, too much heat during machining from a lack of coolant or dull tool or heavy cut, no follower rest or adjusted incorrectly, improper straightening; these are among the possible reasons. It is sometimes seen when a barrel is cut apart. The incision exposes a place where the bore is noticeably not centered properly.

HEAVY BARRELS

Heavy barrels, called bull barrels, whip and bend less than lighter barrels and the variation between shots will be less. This can be caused by internal heat. The impact point can differ between a hot and a cold barrel. The heavier barrel is affected less because the greater surface area transfers more heat to the atmosphere and more is absorbed by the bulk.

The heavy barrel may help some people shoot better because it will be influenced less by wind and recoil. All of these advantages of a heavy barrel, even when added together, can be small. For a long field hunt where the gun is carried by hand, it may not be worth the additional effort. If that sounds familiar, it is a similar situation with barrel length.

CHOKE

Choke in rifle barrels? Yes. It is rare and unusual, but not new. A book, *The Improved American Rifle*, by John Chapman, discussed it in 1848. He called it "freeing" and later, by the 1880's, it was called "choking". This was about the same time shotguns started to have a choke and the same word was used for both. The difference was in the amount and the purpose. A full choke shotgun can be restricted as much as .040 inch. For rifles, the idea first started in muzzle loading times when the bullet had to be put in from the muzzle end, and only .001" or .002" was used.

The barrels were finished by a lapping process which left it a uniform diameter from end to end as well as very smooth. The final lapping was done with a stop positioned so that the last 1 1/2" was of a smaller diameter with another 1 1/2" taper to the main bore diameter.

Today, some muzzle-loaders still believe that a choked rifle barrel is more accurate. They require skill and time to make, and are therefore expensive and rare.

BLUING

Occasionally it is said that bluing a barrel will affect its accuracy. There is no logical basis for this opinion. If you have a barrel blued and the accuracy is not as good as it was, the reason will be somewhere else. Something was damaged or changed during the process. None of the proper bluing methods use a temperature high enough to warp, distort or change a barrels molecular structure. Notice the term *proper method*. It is possible that someone may use a hot bath method at too high a temperature, but not very likely. Also, bluing does not add a thickness or plating as does hard chrome. If by some method it did add something to the bore, it would not be tough enough to survive over a few shots.

MORE ON LENGTH VS. VELOCITY

This chart is a fine way to demonstrate the effect and advantages of barrel length. It is a comparison of the same cartridges used in both a short barreled rifle (a carbine with an 18" barrel) and a revolver with a 4" barrel.

Muzzle Velocity		Muzzle Energy		50 yard Velocity		50 yard Energy	
18"	4"	18"	4"	18"	4"	18"	4"
1750	1235	1075	535	1520	1104	810	428
2140	1450	1270	583	1750	1240	850	427

The first line is with a 158 grain Jacketed Hollow Point. The second line is a 125 grain Jacketed Hollow Point. **All data by Federal Cartridge Corp.**

Standard guidelines on approximate velocity changes for each 1" change in barrel length. Based on a 20" to 26" barrel.

Muzzle velocity	up to 2,000 f.p.s.	change expected, about	5 f.p.s.
"	"	2,001 - 2,500 f.p.s.	"	"	"	10 f.p.s.
"	"	2,501 - 3,000 f.p.s.	"	"	"	20 f.p.s.
"	"	3,001 - 3,500 f.p.s.	"	"	"	30 f.p.s.
"	"	3,501 - 4,000 f.p.s.	"	"	"	40 f.p.s.

The word used above, *approximate*, is important. This is just a general guide based on a variety of firearms and calibers all averaged together.

CHAPTER 12

RIFLING-TWIST

HISTORY

Rifling is the helix grooves inside the bore of a handgun or rifle. (It is usually called a *spiral*, although technically speaking, that is incorrect. A spiral has a constantly increasing or decreasing radius in a single plane.) Although not common, rifling is also found in shotguns made for slugs.

The grooves began so long ago that the exact date, location and inventor are unknown. An Italian Arms inventory of 1476 refers to "one iron handgun made with spiral rifling." Some people question the translation so it is not confirmed. A rifle that was reported to be made for the ruler Maximilian I between 1493 and 1508 appears to have helix rifling but it is so corroded and worn it is hard to determine if the grooves are straight or curved. Some historians believe it was started by Gaspard Kollner of Vienna in 1498. Others place it later at between 1500 and 1520 by August Kotter of Nuremberg. It definitely was in use by 1544 because the National Museum of Zurich, Switzerland has rifled firearms from the Zurich Arsenal with the inventory list dated 1544. It was common enough that by 1563, shooting competitors in Bern, Switzerland were complaining. They said it was unfair to use rifled barrels against their smooth bores.

PROPER TWIST

The amount of twist is more important to accuracy than is generally known. A perfect example is the experience of G. David Tubb, the gentleman who earned his sixth N.R.A. High Power Rifle Championship in 1993. His .243 Winchester rifle would not shoot with consistent accuracy using a barrel with an 8 1/2" twist. Reportedly, a change to an 8" twist stopped the *flyers* and improved the accuracy. Just a 1/2" change in twist can make a difference, for better or worse.

A good quality barrel and a good quality bullet can do wonders with the proper twist. The problem is, few people agree on what is the proper twist. Some people want an over stabilized bullet from a fast twist. They claim best accuracy at all ranges. Other shooters believe a fast twist builds pressure and heat and they want a slow twist for minimum stability with claims to back their theory. The modern thinking is that while there is an ideal twist for each caliber, bullet and barrel length, if in doubt, lean toward the fast side.

For hunting and all-around guns where a variety of bullet weights and lengths will be used, the twist should be chosen for the biggest and heaviest bullet. For bench-rest and target work, the twist should precisely match the bullet.

Before getting into a long and technical discussion of twist, it is important to remember something. A knowledge of twist depends on a basic understanding of gyroscopic stability, spin drift and the related subjects. If you have just picked up this book, you may want to read the two chapters that cover these subjects. (Chapters 8 & 9.)

GROOVES AND LANDS

The area without the grooves, with the smaller inside diameter, is called the lands. (See page 125.) The groove may twist or rotate to either the left or the right similar to a bolt. Most U.S. makers use a right twist. An exception is the Colt revolver. Normally, the barrel will be threaded to fit the action with the same direction of twist. The torque of the bullet spinning through the grooves tightens or holds the barrel in place. The opposite arrangement will hold if the work is done with great precision. The Enfield Model 1917 is an example. It had 5 grooves with a 1: 10 left twist.

If the grooves progress in a clockwise direction, the twist is called right-hand. A counter clockwise progression is labeled left-hand.

The twist is expressed as slow if the barrel length needed to complete one turn is long. The twist is called fast if the barrel length needed to complete one turn is short.

If the rifling grooves make one complete turn in 10" of barrel length, it is called a ten inch twist and written as 1:10. The barrel may be any length, 2" or 30", it would still be a 10" twist. For example, a Winchester Model 94 in .30-30 caliber has a twist of 1 turn in 12" with a normal barrel length of 20". The bullet will make 1 2/3 turns between chamber and muzzle. If the barrel was shortened or a longer barrel was added with the same twist, the number of turns to the muzzle would change, but it would still be a 12" twist.

The twist can also be given as the angle between the grooves and the bore. With its helical nature, like a bolt thread, it is correctly called a helix and the angle a helix angle. To convert from a twist to an angle or from an angle to a twist requires math a little harder than some in this book. But it really is not as rough as it sounds, especially if the reader has a scientific calculator. If not, you will need a book of natural trigonometric functions. From any angle, no pun intended, the calculator is the way to go.

◆   From inches per turn to helix angle:

$$A = \text{arctan} (3.14 * B / T)$$

Where: A = angle in degrees
B = bore in inches
T = twist in inches per turn

Arctan = arctangent. A trigonometric function in reverse. It is the angle whose tangent is known. Check the directions on your scientific calculator. Many will require the use of only the INV and TAN keys on the number in the display.

♦   From helix angle to inches per turn.

$$T = 3.14 * B / TAN\ A$$

Where: B = bore in inches
          T = twist in inches per turn

TAN = Tangent is a trigonometric function. After the angle A is in the display, most scientific calculators require only the press of the TAN key. Check the directions for your calculator.

Twist can also be given in number of calibers per turn. While not used very often, this method is handy for a feeling of proportion. Example: One turn in 12 inches in a .30 caliber would be the bullet diameter of .308 divided into the 12 for 38.96 calibers.

BORE DIAMETER

The subject of bore diameter in a rifled barrel can be confusing. It is complicated by some printed material that is either wrong or worded so poorly that it is hard to understand. It is further complicated by many commercial cartridges named by one method while others are named a different way. **There is no consistency.**

Generally and correctly, the caliber is the bore of the rifle or handgun barrel measured from land to land. The land is the part of the barrel left between the groves. In other words, the bore diameter is before the grooves are cut in the barrel. After the grooves are cut, this larger diameter is the bullet diameter. Normally. Exceptions exist and that is the problem. (See figure 21.)

The basis is historical and stems from the days when bores were smooth. Rifling was added and the bullets were made bigger to fit the opening.

In the U.S. and England, the diameters are given in decimals of an inch. On the European Continent, it is measured in millimeters (mm).

The bore and grooves must be completely sealed by the bullet so the pressure from the gas cannot escape until the bullet leaves the muzzle. Lead bullets are usually slightly oversized and forced into the grooves.

Military arms designate caliber by bore diameter but the bullet is bigger to fill the grooves. Some commercial names are just that, names. The caliber or size may be different than expected. Some are named to bore dia. and some to groove dia. Caliber is sometimes given for bullet diameter. In other words, while there is a standard to follow, much is not as expected.

For example, the bullet is normally slightly larger than the caliber. Most .30 cal. bullets are .308 dia. yet the .38 Special uses bullets with a dia.

of .357 to .359. Therefore the .38 is not really a .38. The correct caliber would be the bore diameter; a number slightly smaller than .357. The .357 magnum is based on bullet diameter and is the same except about a tenth of an inch longer. Incidentally, the term *magnum* was started in Britain.

NUMBER OF GROOVES

For most of the early history of rifling barrels, no standard was established and makers used, normally, six, seven, or eight grooves. Harpers Ferry settled on 7 grooves in 1803 and other companies followed suit. It was believed that the hollow based bullets in use at that time would be more accurate if a land forced the bullet into an opposite groove. In other words an odd number such as seven. In 1852, a British report said their experiments showed 3 grooves to be far superior to 4 grooves. The U.S. military used odd number rifling (except for a few special 6 groove target barrels) until the 4 groove Krag in 1892.

2 GROOVE

Springfield 1903 A3 rifles were made in 2 grooves, 4 grooves and 6 groove rifling. 4 grooves was standard. There has been some controversy about whether higher pressures will be developed in the 2 groove barrels. The well known hunter and gun writer, Elmer Keith, said that he fired hundreds of both 2 and 4 grooved barrels at Ogden Arsenal and could detect no difference in pressure. As he said, it would take an accurate pressure gun to test for a disparity. Also, some people have complained of bullet distortion from 2 groove barrels.

Technically, there would be a pressure increase, but it would be a very slight amount. The Springfield 2 groove barrels were usually broached and not as smooth or as accurate as a 4 groove barrel. (A broach is a cutting tool consisting of a bar with a series of cutting edges which increase in size so each succeeding tooth removes more metal. In rifling, sometimes one tooth is used per groove and adjusted larger with each pass.)

Remington made thousands of barrels with 2 grooves to speed production during W.W. II. They were the same as the 4 groove in dimension and shape, less 2 grooves. The lands cover 5/8 of the bore and work well with cast bullets that have a groove body size and the rest bore size.

4 GROOVE

4 broad grooves and lands that are only 1/3 as wide and use 1/4 of the circumference are used on M1903 and M1 Garand and M14 barrels. With the lands only covering 1/4 of the bore circumference, they are too narrow to serve as a pilot for anything riding on them such as a cast bullet nose. For good performance, the bullets must have groove sized bodies of favorable length.

5 GROOVE

This was originally British in origin and used on the 1917 Enfield, British .30-'06 sporters and .303 Lee-Enfield, etc. The most noticeable characteristic is the equal width of the lands and grooves. Each occupy one half of the bore. Cast bullets should be the same as used in 2 groove rifling.

Used in handguns with Iver Johnson revolvers and some Smith & Wesson handguns.

6 GROOVE

Common among commercial sporting and target rifles. As with the 4 groove Government Springfield, the lands cover only 1/4 of the bore circumference. For best performance, cast bullets should be groove diameter as in 4 groove barrels and long bodies of at least 2 calibers.

Handgun use is common on a large variety of guns including Colt, Harrington & Richardson and some Smith & Wessons.

7 GROOVES

Used in handguns on some .32 and .380 semi-automatics.

8 GROOVES

Some target barrels use 8 groove rifling and it is common enough to be considered conventional. Lands normally cover about 1/4 of the bore. Results are excellent with proper bullets and loads. Cast bullets, for best performance, should be of a similar type as for 6 groove barrels.

MULTI-GROOVE 12-16-22-24

Marlin Firearms Company has been the leader in multi-groove rifling since it was adapted in 1953. As with some others, the lands occupy only approximately 1/4 of the bore circumference. Marlin uses a bore diameter a little larger than normal. A .30 caliber barrel will normally have a bore of .300" to .301". The earlier 16 groove Marlin Micro-Groove barrels were about .305" to .306" The newer 12-groove barrels are about .302" to .303". The grooves themselves are .308" to .3085", as they have to be to fit the bullet. Marlin uses 22 grooves for the .22 Magnum. The increase in the number of grooves is to provide the same grip or firm hold as a barrel with deeper grooves. Cast lead bullets can *strip* if the lead is soft or they are pushed to a high velocity. Lead works fine if used properly.

The factory claims their Micro-Groove rifling prevents gas leakage which reduces muzzle velocity and the muzzle flip which destroys accuracy. They say deep grooves may distort and unbalance the bullet. There is evidence to show they are correct in their opinion, although some shooters claim the barrels provide less guidance to the bullet.

GROOVE DEPTH

Modern groove depth is normally in the area of .004 to .006 inch. Deep groove rifling is not produced today and is almost always found on genuine old guns. The shallow grooves can be found in both old guns and reproductions. They are preferable because the bullet will not fill deep grooves and seal in the pressure. If it did seat in the deep grooves, it would use a lot of pressure (energy) to accelerate the bullet at a loss in velocity and an increase in pressure and temperature.

ANTIQUE RIFLING

Several types of rifling are found in old Kentucky type muzzleloaders and other firearms of the period. **(1)** Deep wide grooves with narrow lands usually are the best. Groove depth of .006 to .010 inch and land width from 1/4 groove width to full groove width. **(2)** Followed in order of efficiency by shallow wide grooves and narrow lands. Groove depth of .005 or less. **(3)** Then next in ability, deep narrow grooves with wide lands. Groove depth running .010 to .015 deep. **(4)** Finally, narrow shallow grooves. Some had shallow grooves that were cut straight with no helix. While these guns were occasionally more accurate than the smooth bore guns, they were normally not better because the grooves did not impart a spin.

An average twist for early rifled barrels for patched round balls was one turn in 66 inches. When conical bullets became popular, (see Chapter 16 on bullets.) a faster twist rate of about 48 inches was required. Conical bullets, if the caliber is the same, are naturally longer and heavier.

UNUSUAL TYPES OF RIFLING

Rifling comes in many forms. Today, as well as in the past, manufactures have different ideas on what is best.

Danish Rasmussen and British Metford are types of rifling which have rounded grooves. It was claimed to have less fouling than normal grooves. Some Krag rifles can be found with 6 groove Rasmussen rifling.

The Ballard rifle, which was popular as a long range gun in black powder days, used a rifling that was similar to Marlin's shallow grooves of today. Most rifling is about .004" deep and the Ballard's was from .002" to .0025". Wide and shallow, there were usually 6 grooves with the land and grooves the same width.

Mannlicher grooves are wider at the top. Mannlicher claims this method helps to stop fouling in the grooves and requires less pressure to spin the bullet. This leaves more pressure for acceleration. These barrels have a worn-down look when in good condition..

A Metford type of rifling was used in the Japanese Model 99 rifle. In 7.7 mm, it had grooves .006" deep and a right twist of 1 turn in 9.5".

Oval rifling was used for a time in both the U.S. and Europe. one type was known as the *Lancaster oval bore*. The oval spiraled down the barrel and gave a spin to the projectile. Its lack of success should be obvious. The bullet had to be literally squeezed down the bore.

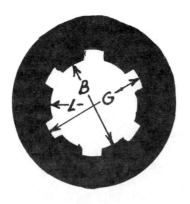

NORMAL RIFLING GROOVES
B is the BORE DIAMETER
G is the GROOVE DIAMETER
L is the LAND LOCATION

DEEP GROOVE STYLE

PARABOLIC

**OVAL - X is .010 to
.012 bigger than Y.**

MANNLICHER
(cut away view)

➔**ALL SKETCHES EXAGGERATED FOR CLARITY**⬅

Parabolic rifling was tried for a short time in the 1920's. It was unusual in shape and design and was difficult to manufacture. Not as proficient as normal rifling, it is in some Newton rifles of about 1919-1920.

A SAMPLING OF FACTORY SPECIFICATIONS

◆ Smith & Wesson M & P revolver -- Caliber .38 Special -- 5 grooves, right twist -- Bore Diameter .346 to .3472 -- Groove Diameter .3555 to .3572 -- Land width .1034 -- Groove width .114

◆ 1903 A3 Bolt action rifle -- Caliber .30-'06 -- 2 grooves, right twist -- Bore diameter .2995 to .3015 -- Groove diameter .3075 to .3095 -- Land width n.a. -- Groove width .1767

◆ Winchester Model 52 bolt action rifle -- Caliber .22 Long Rifle -- 6 grooves, right twist -- Bore diameter .217 -- Groove diameter .2223 to .2229 -- Land width n.a. -- Groove width .085

◆ Colt Government Model 1911 Semi-auto pistol -- Caliber .45 ACP -- 6 grooves, left twist -- Bore diameter .442 to .444 -- Groove diameter .449 to .451 -- Land width .069 to .073 -- Groove width .156 to .160

PRESSURE AND TWIST

Pressure is covered in detail earlier in the book. (Chapter 5.) Here, we should mention that twist has an influence on pressure. Of course, testing to determine the effects of twist variations on pressure is difficult because all of the other items influencing pressure must be controlled.

Testing was conducted by one manufacture under very strict guidelines. Without going into details of the test and moving to the result, they found that the velocity difference between a fast twist and a slow twist, all other things equal, was only 1/4 %. The pressure difference was about 1%. This is so small that ammo tests using the same barrel will usually have more variation because of differences in the ammo itself. The conclusion; moderate changes in twist create such a small change in pressure they can be ignored.

GAIN TWIST

Gain twist is where the rifling twist gains or increases in twist as it moves down the bore. Put another way, it changes and magnifies. The chamber end may start the twist at, for example, a slow twist like 1-30. It will increase to the proper required twist of say 1-14. The projectile starts into the rifling with only a slight rotation and is up to full revolutions per second at the muzzle. The basic theory is sound and it works good, but is seldom used. While soft lead can hit normal rifling so fast that it strips instead of revolves, jacketed bullets do not have this problem. It is also more difficult and costly to produce. The idea is not new. Some of the best and most expensive barrels in the 1800's had a gain twist.

Harry M. Pope (1861-1950), a famous New Jersey maker of top quality barrels, used gain twist in some barrels. It can be very efficient in some cases. For example, with bullets that have short bearing surfaces or of lead. The gain twist places less torque strain on the bullet and yet it can leave the muzzle at the full desired spin. Long bullets will not perform well in a gain twist because they cannot adapt to the change without slippage, which is just the problem that it is supposed to correct. This can be solved in bullets with a long bearing surface by the addition of one or two narrow circular grooves in that area. If, at this point, you are wondering why this would even be wanted, recover an undamaged bullet fired from a modern revolver. Notice how it may have slipped as it started into the rifling.

DECREASING DEPTH GROOVES

For a short period of 11 years, the U.S. military used rifling that had shallower grooves at the muzzle end than at the breech. It was thought to improve accuracy and was based on studies conducted by the French.

Springfield Armory, in 1855, declared that, "The grooves of all small arms be three in number, equal to the lands in width, and rounded in shape; that the twist be a uniform spiral, one turn in six feet for the long musket barrels; and that the depths of all grooves be cut; uniformly decreasing, commencing in the musket at the breech with .015 inch, and ending at the muzzle with .005 inch and in the pistol, at the breech with .008 inch, and ending at the muzzle with .005 inch; or in other words, that the slope of the grooves of all small arms shall be the same, commencing at the muzzle with .005 inch in depth." This was approved by Jefferson Davis, Secretary of War, on July 5, 1855. That rifling method was dropped and replaced by the modern system of uniform grooves with the Model 1866 breech loading musket.

RARE EXTREMES OF TWIST

For extremes of fast and slow twist, according to the N.R.A. publication *The American Rifleman* of March 1976, we have the 6 mm Lee Navy (.236 U.S.N.). Remington Arms Company barrels for this cartridge had a twist of one turn in 6 1/2" and Winchester barrels had a slightly slower twist of one turn in 7 1/2".

One of the slowest twist rates was in converted U.S. cal. 58 percussion muskets with a rate of one turn in 68". In modern times, Winchester .50-95 W.C.F. barrels had a twist of one turn in 60".

FREE-BORING

Free-boring is also called jump, free travel, free run and throating.

Should the bullet be kept away from the rifling when in the chamber or is it acceptable for it to touch or even be pressed into the grooves before firing? Tests by many groups, including the U. S. Army, have found that if the bullet touches the rifling, the accuracy will be improved and the pressure and velocity higher. Of course, as with most technical things, it is not as simple as that. The changes will, in most cases, be from 2% to 10% on pressure and only up to 1% on velocity. The accuracy, as stated, may improve.

On the other hand, free-boring is the practice of boring the barrel ahead of the chamber and bullet up to 2 1/2 inches. It is usually done to cut out worn or damaged places. Some cartridges and powders permit the ignition to push out some of the powder with the bullet. This causes *combustion erosion*. The burning powder damages the starting (lead) section of the rifling. It causes eventual damage from the washing effect of the high pressure gas and intense heat. (This is not always as it appears. For more detail see internal ballistics earlier in this book.) If the lands are worn some at the area behind the chamber, it may even help so long as the bullets fit the bore correctly and seat down in the grooves. The lands, though, must guide and align the bullet and they cannot do this with any looseness.

A few advocates believe free-boring helps the ballistics. The theory is based on the bullet leaving the case straight and by overcoming its inertia, reaching a higher velocity before encountering the rifling. While that sounds good in theory, in practice it does not work very well.

For one thing, it is much harder to have a snug fit in the free-bore area and any escaped gas is a loss in pressure and velocity. Other problems come when the bullet *slams* into the rifling instead of being *pressed* into it.

The free-bore travel itself is not the only uncertainty. A problem arises from bullet deformation, which in turn causes unbalanced stability and yaw. A bullet that is pressed, albeit suddenly, into the rifling will pick up and follow the rotation with less shearing. A conservative twist can be a help in this regard. It is easier for the bullet to pick up and this also will help to hold down deformation and shearing. Of course there is not a major difference between twist angles, as the previous mathematics shows. Therefore, this can usually be disregarded in favor of the most effective twist. As with many areas of ballistics, gains in one area are pushed aside by gains in others.

Most people will recommend free-boring to cure a bad place in the bore if it is not to large. But don't expect it to improve performance beyond what it was before the damage occurred. Most likely, accuracy and barrel life will suffer in the process.

For normal use, about 1/8 inch or 1/4 inch of free-bore is the maximum; the less the better. Generally, all calibers from .17 up to the big bores will perform well at a free travel of .020 down to zero. Not every one agrees on the best space, but most experts lean toward the shorter the better. The only way to be certain about a particular weapon is to test fire it on the range. It also should be mentioned that the throat will wear. Over a period of time an increase in free travel will occur.

The distance can be controlled by hand-loaders, but problems can surface. Adjusting the free run by tuning the bullets seating depth in the cartridge can be a self defeating move. For example, deep seating will increase pressure due to increased loading density while the increased free run will lower pressure. The final result will depend on so many variables such as type of gun, condition, cartridge size, etc., that it is impossible to predict the outcome. Usually, though, if the gun already has a long free run, the change will be an increase in pressure instead of a stalemate. Of course, cartridges kept in magazines may be limited in useful length or suffer point damage during recoil.

It is strange that something as simple as this can be so controversial. The owner of a popular bullet manufacturing company has said that the bullet should always be kept back from the rifling and "given a running start at the rifling, and the longer this is, the easier it is to start the bullet down the barrel and the less the pressure." With the exception of the lower pressure, this opinion is in the minority and wrong.

To carry the subject one step farther, there is frequently an area of transition just forward of the chamber where it is tapered to the bore. Usually called a throat, it is occasionally called a forcing cone, although the latter is usually reserved for the forward part of a shotgun chamber. The throat can be either beveled, tapered, counter-bored or rounded or a combination of two or three. This transition space is also called a lead or leade, with an e on the end. While no major dictionary gives the e ending as correct, common usage makes it proper.

The free travel and therefore, the bullet seating depth, will change when a hand-loader changes bullets. Different bullets have various ogive shapes and lengths and fit the throat in assorted ways. Even in the same gun. Of course, we all know that not all guns are throated even close to the same. For hand-loading, this can be checked **by first being certain things are safe and the rifle cannot be fired.** Remove the firing pin if it is the only way to be positive. Set the seating die so the bullets are seated long and adjust down until the bullet just touches the lands in the bore. The marks should show on the bullet. If a different bullet type or brand is used, the check should be done again.

Some people incorrectly use lead, leade or throat for free-bore, but that is incorrect. Free-bore is the space that may or may not exist between the bullet (usually the ogive, but not always) and the leade or throat. Note that this is not intended to be overly critical, but this is a text-book and it is necessary to be precise.

For accuracy, the best practice seems to be a gradual taper of about 3° to 5° with the bullet just or almost touching the rifling. All diameters should be closely held to minimum size tolerance with all of them, including the bore and groove diameter, closely concentric. Even with the best lathe, proper pilot size and perfect technique, concentricity is not as easy to hold *dead on* as it may seem. Top gun-shops do it regularly. Others.....? One problem with concentricity is that errors cannot be seen by the eye and special skills and equipment are needed for inspection.

For ordinary shooting, free-bore, throat and concentricity will make little difference. For utmost accuracy, everything should be perfect. No exceptions.

MEASURING TWIST

It is easy to measure the twist in the barrel of a handgun or rifle. All that is needed is a vise or clamp that is padded, a cleaning rod, a patch and a fine tip felt marker.

Clamp the barrel or gun in a horizontal position. Don't forget to pad it against scratches and damage. Make extra sure the cleaning rod sections and tip cannot become lose or turn even the slightest amount. The patch should be a tight fit so it will track in the grooves.

Start the patch and rod deep enough to have a good grip on the grooves. A half inch to an inch should be enough, especially on a handgun where there isn't much barrel length to work with. A rifle barrel should have length to spare.

Place a vertical mark on the muzzle end and a matching mark extending down the top of the rod. Also make a cross mark on the rod at the muzzle. On a long barrel, push the rod into the barrel, permitting it to turn with the helix. Stop when the long mark down the rod's top is again lined up with the vertical mark on the muzzle. At this point make another cross mark on the rod. Remove the rod and measure the distance between the cross marks. This will be the distance for one complete turn.

Hand gun twist can be measured but will require one extra step. For example, a popular .45 semi. automatic has a 5" barrel length with a twist of 1 turn in 16". If we didn't know the twist, we couldn't go 16" to find the answer. Carefully, we would have to go perhaps a quarter turn and mark the rod. After measuring the distance, multiply the number by 4. If done with caution, the result will be accurate.

MEASURING BORE & GROOVE DIAMETER

The inside diameter on barrels with an odd number of grooves is harder to check because the lands and grooves are not opposite each other. If just an approximate size for either the bore or groove depth is satisfactory, it can be done by holding a micrometer or caliper at the muzzle and *eye balling*. For proper accuracy, time and effort are required.

The bore diameter is determined first. If a machinists ball gauge in the proper size is available, it can be adjusted so that it is a snug fit, then removed and checked with a micrometer. A ball gauge is circular and when expanded it will not go into the grooves to give a false reading. Another method is to use a lead bullet or lead slug and press it against the muzzle and turn it by hand. If care is used, a short area of lead will take on the bore diameter and can then be measured.

The rifling can be checked with a piece of soft lead that is about the bore diameter. Mushroom one end to a little over groove diameter, lubricate it, and force it through the bore with a wooden dowel and mallet. **Caution:** use wood and soft lead so no damage will be done to the bore.

Measure the diameter of the lead piece over a groove to land section and subtract the bore diameter to obtain the land depth. Bore diameter plus 2 groove depths will be the groove diameter.

ROTATION SPEED AND MATHEMATICS

This is a good place for some mathematics. Relax. This is easy.

◆ **Rotational speed** is an important part of ballistics.

$R = (12 / T) * V$

Where: R = Revolutions per second of bullet. Note second, not minute.

T = twist of rifling in inches (example 10" or 16")

V = Muzzle velocity in feet per second. (f.p.s.)

For **example**, a .22 long rifle with a muzzle velocity of 1,200 f.p.s. and a barrel twist of 16", the bullet rotation would be 900 f.p.s. People are more familiar with revolutions per minute (r.p.m.) so multiply the answer by 60 to obtain 54,000 r.p.m.

A bullet at 3,000 f.p.s. from a 10" twist would rotate at 3,600 f.p.s. or 216,000 r.p.m. The high gyroscopic stability and centrifugal force are obvious. This also shows why a bullet can disintegrate if used at a velocity and rotational speed higher than what it was designed for. Rotational speed will decay, but it will not be enough to be of any concern.

◆ **Circumference:** The spinning bullets surface speed is the circumference of the bullet times the number of turns in the time involved. In most instances, the largest diameter would be used, but any part of the bullet could be calculated.

Circumference = $\pi *$ **diameter**

Example .30 * 3.1415927 = .942" ÷ 12 = .0785 feet.

◆ **Surface speed.** To find surface speed of bullet.

$S = C * R$

Where: S = surface speed in feet per second, circular.

C = circumference in feet

R = revolutions per second.

Another route to the same answer.

$S = (R \; p \; * 2 * T) / 12$

Where: S = surface speed in feet per sec., circular
  T = turns of bullet per sec. (revolutions)
  R = radius of bullet

◆ **The spin** is integral with the forward velocity.

$$RS = \sqrt{V^2 + S^2}$$

Where: V = muzzle velocity in f.p.s.
  S = surface speed in ft. per sec.
  RS = resultant speed

The spin is imperative for trajectory and accuracy but adds, on the average, about 4/10 of 1 percent to the kinetic energy and therefore has almost no effect on tissue destruction and killing power. This is covered in terminal ballistics.

◆ **Energy of rotation (spin)**

$E = 1/2 \ Iw^2$

Where: E = Rotation (spin) energy
  I = moment of Inertia about the bullet's axis
  w = angular velocity

◆ **Bullet kinetic energy**: As explained in the internal ballistics section: **kinetic energy** = $MV^2 / 2$ where M is mass and V is velocity.

For bullets: $Et = WV^2 / 450400$

Where: W = weight of bullet in gr.
  V = velocity in feet per sec.
  Et = translation energy in foot lbs.

This is frequently stated as muzzle energy, but that is not a fully true statement. It does not take into account the small amount of energy required to rotate the bullet.

◆ **Energy to rotate the bullet.**

$ER = IS^2 / 2$

ER = rotational energy in foot lbs.
  I = axial moment of inertia in slug feet ^{2}.
  S = rotational speed in radians per sec.

For I, the moment of inertia is generally in gr.-in.^{2} (grain-inches squared). To convert to slug-feet ^{2} we should take 3.084×10^8 times the gr.-in.^{2}.

For S, know that a radian is a unit of measurement for angles used in calculus. Although confusing to people not familiar with its use, it is both natural and useful. Radian = 75.4 times the muzzle velocity divided by the rifling twist in inches per turn.

That is a lot of mathematics and work to show and prove the energy required to rotate the bullet is very small. Extremely small. If an example is worked out, it will fall in the general area, depending on cartridge, twist, etc., of between 1/4 % and 1% of the available energy. The loss in velocity, compared to a barrel with straight rifling (no twist) would be even less.

◆ **For rotational kinetic energy**, we can put this all together with:

$ER = (8.766 * 10^{-5} AV^2) / T^2$

A = axial moment of inertia in grain inches squared
V = velocity in feet per sec.
T = twist of rifling in inches per turn
ER = rotational kinetic energy

A variation for kinetic energy for bullets is to multiply bullet weigh, times velocity, times velocity, times 0.000002221 for an answer in f.p.s. This is stated as $E = W \times V^2 \times .000002221$. (Note in Chapter 2, exponential notation.)

This formula for kinetic energy is basically the same as dividing by 450,240, but multiplication is easier, particularly if a calculator is not handy. (Note that 450,400 is preferred as explained in Chapter 2. The difference is negligible.)

For math fans among the readers, another formula is interesting.

$E = 1/2 \, IA^2$

E = spin energy in ft. pounds.
A = angular velocity
I = moment of inertia about the long axis.

GREENHILL FORMULA FOR TWIST

Professor (Sir) Alfred George Greenhill, M.A. (1847-1927) was a lecturer in mathematics at Emanuel College, Cambridge, England from 1873 to 1876. Then for 30 years, from 1876 to 1906, a professor of mathematics to the advanced class of artillery officers at Woolrich Military Academy. He received a knighthood in 1908. He made original contributions to the mathematics of dynamics, hydrodynamics, and elasticity, and worked on special problems in aeronautics and ballistics. It is the latter that concerns us here with his *On the Rotation Required for the Stability of an Elongated Projectile.*

The Greenhill formula is based on his rule; "The twist required in calibers equals 150 divided by the length of the bullet in calibers." Greenhill was aware that this application was limited because it applied to fairly blunt-nosed, jacketed lead bullets with a specific gravity of 10.9. Lead alone has a specific gravity of 11.35. The number is changed for other materials in the ratio of the square root of their densities. (This is explained shortly.) There is a big margin of stability involved, and the number 150 was well chosen. Most bullets will stabilize at less than 200. The formula will also work for jacketed, sharp point bullets.

The rule does not account for nose shape, base shape, jacket or jacket thickness, density of material, etc. These can be used in a computer program for twist.

The wonderful advantage of the Greenhill formula is that it is very simple. While it is not always dead accurate on some modern bullets, it will always give a useful approximation and in many examples it will be exact. It is very good for its age and simplicity.

◆ **To use**, first divide the length of the bullet by its caliber in inches. For example a .30 caliber is .308 diameter, divide that into its length, say 1.125 to obtain 3.65 calibers long. Divide the 3.65 into 150 to obtain 41.09 or one turn in 41.09 calibers. To change to inches, multiply the 41.09 by the diameter of the bullet in inches which in this case is .308. The answer is 12.655 or 1 turn in 12 5/8 inches. (.625 = 5/8")

At first it would seem that velocity is ignored in the formula, but that is not the case. It is included in a hidden form in the spin segment. For explanation, let us use a barrel with a twist of 1 turn in 12 inches. If a bullet fired at 1,500 feet per second. has a spin of 1,500 revolutions p.s., then one fired at 3,000 feet p.s. would have a spin of 3,000 revolutions p.s. The faster velocity gives a faster spin, therefore velocity is involved in the formula even though it is not declared as such.

Note that while this formula is old and an excellent starting point, it is not the definitive answer. Bullet shape and occasionally velocity need to be considered. Bullet weight, most experts believe, is not of much importance in determining twist. A few opinions state that extreme long range may require a different twist. The majority also believe range is unimportant.

Some people believe that the Greenhill formula is still good at all velocities. They do not recommend a different twist unless past experience indicates it is acceptable for the cartridge and bullet in question.

Another view is that faster than 2,200 f.p.s., a slightly slower twist is required. They also suggest at velocities over 3,200 f.p.s. to decrease the turns about one turn per inch for each 350 f.p.s. increase over 3,200 f.p.s

Velocities over 3,200 f.p.s. are not common. A lot of shooters, hand-loaders in particular, believe that a velocity of 3,000 f.p.s. is a special number that will promise great accuracy. Sorry, but in a general sense, it is not true. While there are very accurate cartridges of about 3,000 f.p.s., there are also excellent cartridges at both higher and lower velocities.

Of course, it is understood that the Greenhill formula is designed for projectiles moving in air. In the rare event that someone wants to calculate twist rates for a bullet moving in some other material, it can easily be worked out. To stabilize a bullet in another material, work the spin out as always for an air environment. Then multiply the answer by the square root of the number obtained by dividing the density of the new material by the density of air. A denser material would require a faster twist. For example, water would need about 30 times the twist necessary for air.

CONCLUSION ON GREENHILL

In analyzing the last opinions, we can safely say the formula is good for bullets up to at least 2,200 f.p.s. For higher velocities, add some research on similar cartridges. Keep in mind that the modern theory is to go with a higher twist when in doubt, and that is what the old Greenhill formula does. All considered, it still works very well.

SPECIFIC GRAVITY

It is fairly easy to find the specific gravity of a particular bullet for the Greenhill formula or for any other reason. Hand-loaders may have balance scales, so hang the bullet by a thread from one side of the scale so that the weight can be read. Put a container of water in a position where the bullet hangs in the water and again find its weight. (It may be necessary to raise the scale up by placing it on a few books or other object.) Subtract the water weight from the dry weight. The specific gravity will be the result of dividing the weight of the dry bullet by the difference found in the subtraction.

The Greenhill formula was based on a specific gravity of 10.9. The correct spin can be found for other specific gravities by dividing 10.9 by the new or adjusted specific gravity and finding the square root of that number. Then multiply it by the spin supplied by the original formula.

NOTES ON THE .22 CAL. & COMPROMISE

The twist per inch will usually be a compromise. The .22 short works best at about a 20" twist with its short 30 grain bullet. (Some years ago, 24" was used.) The .22 short can be surprisingly accurate if fired from a gun with the proper twist. With the wrong twist, it will do poorly, as will all other cartridges.

The .22 long rifle cartridge with a 40 grain bullet works best at one turn in 16". The .22 long, if tried at a 10" twist, gives a larger group size and considerable flyers. At a 20" twist, it has little stability and wobbles and hits the target in a position other than point first even if its static and dynamic balance are near perfect.. Instead of a round hole it may be oblong, hence the term *keyhole*.

The longer bullet requires more stability because it has more of a tendency to tumble in flight. Therefore, it requires more twist to obtain

more spin. If several kinds of cartridges will be fired from the same gun, it is impossible for the twist to be corrected for both. A compromise is required.
CANNON
Large caliber projectiles, such as cannon shells, require less spin to stabilize than smaller bullets. The basics of gyroscopic stability dictate that the larger the radius of gyration, the more flywheel effect for greater stability. For example, a howitzer of 240 mm with a velocity of 2,300 f.p.s. has a rifling of 1 turn in 25 calibers. This gives about 6,960 r.p.m. (116 r.p.s.). A 16 inch diameter shell with a 2,650 f.p.s. velocity is fired from a barrel with a twist of 1 turn in 21 calibers. The spin is almost half the previous example at 3,720 r.p.m. (62 r.p.s.). Theoretically, the stability required and possessed would be about equal.
CAST BULLETS
Cast bullets have been known to strip through the rifling grooves, not damaging the bore but tearing off the outer layer of lead. This will become an unstabilized projectile that will not group well and may keyhole. With normal loads, lead has enough strength to withstand the stress involved in the forced passage through rifling. It is usually caused by too heavy a powder charge and that, plus proper bullet alloys and lubricants, can control the problem. It is almost unheard of in a jacketed bullet. (Note the use of the word *almost.*)
NOTES AND CONCLUSIONS
For proper twist, we must consider: **(1)** The longer the bullet in proportion to its diameter (caliber) the more twist to stabilize it. **(2)** A very long nose is good for drag reduction and aerodynamic efficiency but is harder to stabilize. (See note following) **(3)** Density of the bullet is also a factor. **(4)** A longer and heavier bullet can be stabilized at the same twist by increasing velocity. **(5)** The diameter of a bullet in proportion to its length is a factor to consider. **(6)** A low velocity bullet requires a faster twist to stabilize it.

◆ The main reason a long pointed bullet is harder to stabilize is the slight errors in manufacturing. It needs to be properly balanced, concentric and symmetrical. The point will cut drag and because it can sustain velocity better, it will give a flatter trajectory. If the bullet is not perfectly made, problems show up with the center of gravity and other forces as discussed in a previous chapter. (See figure 3 on page 92.)

◆ A plain bullet, reasonably short and flat at both ends, will work best with a slower twist. A bullet with a very long point will need a faster twist.

◆ If two bullets start at the same velocity and spin, the heavy long bullet (strong sectional density), will hold its speed of rotation better than a lighter and shorter bullet. The momentum of the heavier bullet also helps it maintain velocity, among other things.

♦ While not common, any barrel with a bad groove which drags a piece of displaced metal into a fin will have poor accuracy. If all grooves in a barrel leave fins at the base of the bullet, wayward steerage can be caused if the projection is more pronounced in one area. (Which it certainly will be.)

♦ High velocity will stabilize a bullet with a slower twist than would be satisfactory for a lower velocity.

♦ The best rifling twist is governed by the bullets length, not its weight. (The two are, of course, connected.)

♦ Barrel manufacturers have tolerances that are held in regard to twist. An ordinary barrel will be held to about one inch of required twist. If a 16 twist is specified, it will be within 15 to 17. More expensive barrels from the custom makers are held to closer tolerances, such as 1/2 or even 1/4 turn. Dead accurate barrels can be made, but as in everything, precision work is expensive.

♦ A conservative twist will help the bullet pick-up and follow the rotation with less shearing. This will help to hold down deformation which can create and unbalanced bullet with stability and yaw.

♦ Because of tumbling, some experts believe that when the M-16 rifle barrel twist was changed from 14" to 12", the killing power was reduced.

♦ Increasing the twist for a faster spin will not give a flatter trajectory.

♦ It is not true that more twist should be used for long range stability. Changing barrels may help, but if it does, it is caused by factors other than twist. (Assuming it was correct to begin with.)

♦ For many years, standard .30-06 rifles used a 10 inch twist. This works fine for bullets with a weight over 125 grains but is insufficient for lighter ones. The modern trend is to use 12 or 14 inches for better long range performance with any weight bullet.

♦ A good method to determine proper twist is to follow the practice of commercial and military firearms.

♦ It is common for rechambered rifles not to shoot as expected. Frequently this is because if the rifling twist was correct for the earlier bullet and load it will not be correct for a new bullet. Most noticeable if the bullet weight is changed either up or down.

♦ When switching to a heavier (longer) bullet, the rifling will frequently not be fast enough for the increased weight, if it was correct to begin with. Manufacturers occasionally put a notice on an ammunition box to use the cartridges in firearms with a specific twist. This has nothing to do with safety but may have a lot to do with accuracy.

◆ Gyroscopic stability varies as the square of the rifling twist. This means that a very slight change in twist will have a disproportionately big result. Or to put it another way: Changing twist on your rifle? A small change will make a big difference. Perhaps more than desired.

EXAMPLES OF TWIST AS USED BY MODERN MANUFACTURES

Firearm	Cartridge	Rate of twist	Direction & grooves
RIFLES			
Winchester 94	.30-30 Win.	1 in 12"	right hand
Winchester 94	.375 Win.	1 in 12"	right hand
Winchester 94	.22	1 in 16"	right hand
Winchester 70	.222 Rem.	1 in 14"	right hand
Winchester 70	.243 Win.	1 in 10"	right hand
Winchester 70	.270 Win.	1 in 10"	right hand
Winchester 70	.30-.06	1 in 10"	right hand
Winchester 70	.308 Win.	1 in 12"	right hand
Winchester 70	.300 Win. Mag.	1 in 10"	right hand
Winchester 70	.375 H&H Mag.	1 in 12"	right hand
Winchester 70	7 mm Rem. Mag.	1 in 9 1/2"	right hand
Winchester 70	7 mm Mauser(7x57)	1 in 8 3/4"	right hand
Winchester 70	.458 Win. Mag.	1 in 14"	right hand
HANDGUNS			
Ruger speed six	.357 Mag&.38 Sp	1 in 18 3/8"	right hand 5 groove
Ruger old army	.44 blackpowder	1 in 16"	right hand 6 groove
Ruger semi-auto	.22 L. R.	1 in 14"	right hand 6 groove
DanWesson revl	.357 Mag&.38 Sp	1 in 18 3/4"	right hand 6 groove
DanWesson revl	.41 or .44 Mag.	1 in 18 3/4"	right hand 8 groove
DanWesson revl	.22 L.R.	1 in 12"	right hand 6 groove
DanWesson revl	.22 Mag.	1 in 16"	right hand 6 groove
1911-A1(Sprfld)	9 mm	1 in 16"	left hand 4 groove
1911-A1(Sprfld)	45 ACP	n.a.	right hand 6 groove
Desert Eagle	.357 Mag.	1 in 14"	n.a. 6 groove
Desert Eagle	.44 Mag.	1 in 18"	n.a. 6 groove

CHAPTER 13

VELOCITY AND DRAG

Velocity and speed appear to be words that are synonymous. Velocity seems to be a fancy word for speed but actually they have slightly different meanings. Speed is a rate of movement. Velocity is a rate of movement in a particular direction in space. The word magnitude is also used in connection with speed and velocity and it is the measurable quantity as in *magnitude of the velocity*. It is used for the purpose of comparison with other quantities of the same category. Velocity changes with respect to time and may be constant or changing. **Definition:** velocity is distance with direction per unit time and may be uniform or varying.

$$v = ds/dt \qquad s = \int v\, dt$$

Where v is velocity, t is time and s is distance.

Acceleration is the rate at which velocity changes with respect to time. It may be uniform or varying and positive (acceleration) or negative (deceleration).

Estimated velocity is just that. Estimated without an instrument. Experience is helpful in estimating cartridges with similar characteristics. If two cartridges have the same bullet weight and qualities and the velocity of one is known, then a good estimate can be made by shooting tests. Find the trajectory of the unknown and match it with a known trajectory.

Of course, even the best estimate by a very experienced person is still just an approximation and should be used with that knowledge in mind. The best velocity figures, if used to compare cartridges, must consider bullet shape and weight, distance from muzzle and in many cases, gun type and barrel length.

Instrumental velocity is measured velocity whether measured by the old style pendulum or a modern chronograph. The actual distance from the muzzle is always an important consideration. The velocity of a bullet slows as it moves forward. It cannot and will not remain constant, so the distance is always an essential figure.

Remaining velocity is at any point in the trajectory. It may be measured or estimated.

Striking velocity, sometimes incorrectly called terminal velocity, is the speed at impact.

Summit velocity is the velocity at the highest point in the trajectory.

Terminal velocity is frequently used as the term for the velocity at impact (striking velocity). It is easy to see how this habit started because terminal can mean "end" or "final". Terminal velocity has an exact and specific meaning which is completely different. Terminal velocity is the greatest velocity which a body can acquire by falling freely through the air. The limit is reached when the increasing atmospheric pressure becomes equal to the increase of the force of gravity. With bullets, normally only a bullet fired straight up and then falling free back to earth would be subjected to terminal velocity. (Covered elsewhere in this book.)

Uniform motion is a constant velocity with no acceleration or deceleration. This is rare in ballistics.

MEASURING VELOCITY

There are several ways to accurately measure the velocity of a bullet. Here we will discuss three. The modern chronograph is a device that can be purchased at moderate cost and is the most practical of the three. It was only in more recent years that the cost and ease of owning such an instrument was within the means of the average person. As a matter of both curiosity and learning, first we will briefly discuss two much older methods.

BALLISTIC PENDULUM

This is the oldest known method for measuring the striking force of a projectile. The ballistic pendulum has a history that is hundreds of years old. Invented by Benjamin Robbins (1707-1751) of Great Britain in 1742, it is simply a block of wood of a known weight hanging by a piece of chain, also of known weight. When a bullet (also of known weight) is fired into the block, it swings away like a pendulum. A scale measures the distance the block has moved. In actual use, it is not that simple, but that is the basic principle. It measures kinetic energy instead of time so it can not be called a chronograph.

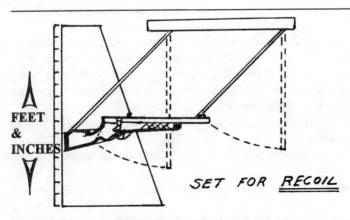

BASIC BALLISTIC PENDULUM

With the distance and the weights, calculations will determine the momentum or kinetic energy in foot pounds. From this, the velocity can be calculated. By firing at the pendulum at different distances, the velocity at the different ranges can be figured. It is still used today to measure recoil. The pendulum must be mounted from a strong frame and in an area protected from the wind.

The construction must be done with care using numerous spread out wires to hold the pendulum. Both ends of the wires should be mounted with bearings to reduce friction. The normal pendulum weight can be maintained during repeated use by having a container on the pendulum with bullets inside. For each bullet fired into the block, a bullet of equal weight is removed. A method must be used to catch and include any bullet fragments. A sliding marker is pushed rearward with the swing and will give the inches of recoil. It is easy, from either experience or math, to mark the scale so it will read directly in velocity (f.p.s.). Some have been made of all wood and some of iron with a replaceable wood face. Calculations usually included the pendulum weight plus 1/3 the weight of the wire supports.

Referring to Newton's 3rd Law of Motion, we see that if a moving object strikes a stationary object, the momentum (mass * velocity) of the moving object before the strike is equal to the momentum of the stationary object after the strike. For an equation, we put the weight and velocity of one object on one side and the weight and velocity of the other conversely.

$$wv = WV$$

◆ For example, if an 8 lb. block of wood is attached to a 2 lb. swing, the total moveable weight would be 10 lbs. If we fired a bullet at 2,000 f.p.s that weighed .01 pound into the block, it would move away at 2 f.p.s. (The V is equal to 2.)

$$.01 * 2,000 = 10 V$$

The swing and block, at 10 pounds, weigh 1,000 times more than the .01 pound bullet so they are given a velocity of 1/1000 the bullet's velocity. This does not account for energy lost through friction, heat, etc.

This formula is from the *British Textbook of Small Arms* of 1909. "The velocity of the recoiling parts when the angle of swing is small and when V = foot seconds of recoil, a = inches of recoil and n = swings to and fro in one minute is as follows:

$$V = (2 \pi n / 12 * 60) * a = (na / 114.7) \text{ F.S.}$$

The weight of the recoiling parts is the weight of the bob (W lbs.) plus the weight of the bullet (w lbs.) embedded in it, so that with V as the striking velocity $wV = (W + w) na / 114.7$ which is the accurate working formula for the ballistic pendulum. The value $2 \pi n / 12 * 60 = n / 114.7$ is a mere numerical constant which can be evaluated once and for all by observing the number of complete swings (to and fro) made in one minute."

This is complicated, so let's try to make it easier.

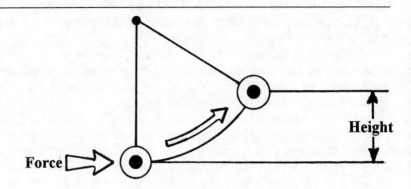

If a pendulum swings backwards from any force, as it moves rearward it will also swing (arc) upwards. Besides the rearward distance which is horizontal, there will also be an upward or vertical distance. It is known from Galileo's studies that a swinging pendulum's velocity is related to the height it will swing (or rise above the horizontal).

$$V = \sqrt{2gh}$$

where: V = velocity
 h = height
 g = gravitational constant

REVOLVING DISK

The revolving disk measures the time the bullet takes to travel a known distance so it can correctly be called a chronograph.

Two disks with calibration marks similar to the 360° marks around a compass are secured on each end of a shaft at a known distance apart. The marks or lines on the disk must be identical and beginning witness marks on the separate disks must be directly in line with each other. The disks should not be too close together, not less than four feet, and not so close to the muzzle as to be affected by the blast of escaping gas. We must know from a tachometer or other method the exact speed (revolutions per minute) of the shaft and disks.

The gun is fired with the bore of the gun parallel to the shaft so the bullet enters the disks near the outside edge. The bullet will enter the second disk, not directly behind the first disk but to the right if the shaft is rotating

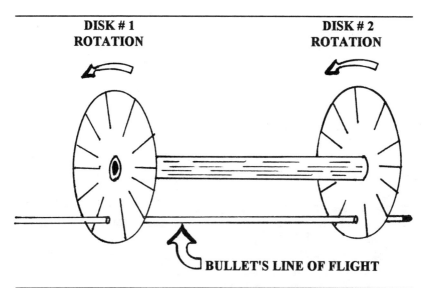

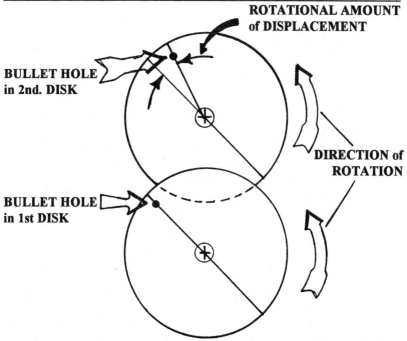

VELOCITY MEASUREMENT BY REVOLVING DISKS

to the left. This is because of the movement of the disks while traveling between them. Make the disks of a material that will not slow the bullet's velocity and do all of the steps carefully and accurately. The measurement of the lag or distance the second disk moved and some basic mathematics will give the velocity. For this method to be successful, both disks must be moving at the same speed. Their exact r.p.m must be known, and the lag angle between the hole punched in the first disk and the hole in the second must be measured with accuracy. (See page 143.)

CHRONOGRAPHS

A chronograph measures only time, or in this case, elapsed time. An electrically operated timer is started automatically as a bullet is fired through an electrical screen or photoelectric device. The bullet travels a set distance and stops the timer at the second screen. The elapsed time will be very short and measured in milliseconds. The instrument will calculate the time and convert it into a reading in feet per second. A remarkable accomplishment, when we consider the bullets millionths of a second passage has to start and stop the device with accuracy.

For many years, velocity was measured by the Boulengé chronograph. This was the invention of a Belgian Army Captain, Paul Emil le Boulengé. It originated in 1874 and was still in use in the late 1940's.

It consisted of 2 screens held by frames and placed a predetermined distance apart. Through both electric and mechanical means, a timer is started as the first screen is penetrated and stopped as the second screen is penetrated. The timer is not a clock type of system that we would all be familiar with, but a falling tube that is marked and the time is determined from the mark's location.

The first screen was placed at 3 feet from the muzzle and the second placed for a 150 feet spacing. This gave a midpoint of 78 feet from the muzzle. (75' + 3') If the screens were set at 100 feet apart, the midpoint was 53 feet and handguns were measured at 50 feet distance for a 25.5 feet distance with the first screen at 6 inches.

The 78 feet distance just mentioned, was the military standard for many years, and listed in the charts and papers as "Instrumental at 78 feet." This was the distance from the muzzle to the midpoint but a few people mistakenly thought that it meant the maximum velocity was at 78 feet. The maximum velocity is always at the muzzle because the bullet begins to decelerate as soon as it leaves the muzzle. Technically, in some cases, there may be an inch or two before the air resistance stops the acceleration, but certainly not much.

Electronic timers were invented that measure time by an oscillator that, for example, may oscillate at a frequency of 100,000 times each second. With a modern method of counting the oscillations from screen to screen, the time could be computed with a high degree of accuracy.

Improvements in screens and methods of starting and stopping the time were also made. A gate circuit, a sandwich screen, a solenoid screen and a photoelectric screen have all been used.

Now-days, computers take chronograph figures and print out a standard ballistic table. Modern equipment is cheap, easy to operate, and the two screens are set close together on a short bar. Because of battery operation, the units can be used any-place where a gun can be legally and safely fired. (Have spare batteries. They always go dead when you need them the most.)

Modern electronic chronographs are wonders of technology. Easy to use and very accurate. One and a half percent accuracy is common. Many features are usually included which make their use a pleasure. Most models will store in their memory the results of a string of shots as well as displaying the last shot velocity. Calculation features include reporting the highest velocity, standard deviation, extreme spread, average velocity, etc. All of these items will be explained in detail in this book so chronograph owners will understand better the information they have available. With a computer, a chronograph and this book for information, an amateur can accomplish things that not long ago would have created problems for a complete ballistic laboratory.

MUZZLE VELOCITY

Muzzle velocity is just what a reader would expect; the velocity of the projectile as it leaves the gun barrel. It is usually estimated from a chronograph test made from a short distance away. This is to prevent the blast created by the hot escaping gas from damaging the valuable equipment.

Air density affects velocity and Army Ordnance uses 0.64 f.p.s. per foot for the distance from muzzle to mid-point between screens. (Note the term *mid-point.*)

♦ If d = distance between muzzle and start screen
and s = distance between screens,
$$D = d + 1/2\ s \quad \text{then} \quad MV = D * C + IV$$
Where: MV = muzzle velocity
D = as above
IV = instrumental velocity
C = air density factor Rho (or use 0.64 f.p.s.)

Technically, muzzle velocity varies inversely as the mass of the bullet and directly as the square of its caliber. In other words, for highest velocity, use a light bullet with a large diameter. This is primarily for acceleration while the bullet is still in the bore. Of course, a compromise would be needed because the ballistic coefficient which maintains that velocity outside of the bore requires more weight and less diameter. (It varies directly as the mass of the bullet and inversely as the square of the caliber.) To help avoid confusion, remember that inversely means opposite or inverted order. As one quantity is greater or less according as another is less or greater. Directly means exactly or precisely.

SPEED OF SOUND

"The speed of sound is the rate at which small pressure disturbances will be propagated through the air and this propagation speed is solely a function of air temperature." This is a quote from an *Aerodynamics Training Manual for Naval Aviators*, NAVWEPS 00-80T-80. Many people believe that other items enter into the situation, but they do not. The speed of sound slows with increased altitude, but this is due to the lower temperature instead of lower air density. Most people, including educated individuals, believe that the thinner air is the reason for the drop. This not correct. Temperature is the one and only reason. This error is further enforced by many gun writers who don't bother to research the subject and say in their articles that it is a result of thin air. Now you-all know better. For example, the speed of sound at sea level is 661.7 knots or nautical miles per hour (761 statute or common miles per hour) at a standard temperature of 15° C. (59° F.). At 60,000 feet the speed of sound drops to 573.8 knots. This is based on a standard temperature at that altitude of -56.5° C. (-69.7°F.)

In firearm ballistics we usually give velocities in feet per second instead of knots or miles per hour. Many people still use an out-of-date figure of 1,080 f.p.s. for the speed of sound. This would be correct at about 24° F. For a number closer to general use, 1,120 f.p.s. is best and is used in this book. Rounding off the figures slightly, this is based on about 60° F. (The standard previously given of 661.7 Knots at 59° F. is 1,116 f.p.s.) Chapter 2 has more information on this subject.

To find the velocity of sound with the air temperature in centigrade, multiply by 1,088 the square root of one plus the temperature divided by 273. The answer will be what we want in ballistics, feet per second. Remember to use centigrade.

$$V = 1088 \sqrt{1 + \frac{C}{273}}$$

This formula works well. 59 degree F. converts to 15 degree C. divided by 273 = .0549 + 1 = 1.0549 The square root of which is 1.027 times 1,088 = 1,117.4 feet per second.

The speed of sound increases as the temperature increases and conversely will decrease as the temperature drops. The change is about 2 ft.

per sec. for each degree of centigrade. Above the speed of sound is the supersonic range and below it is called subsonic. The speed of sound is important in both bullet design and performance as well as the louder noise, sometimes described as a *crack*, that the supersonic bullet makes as it travels through the air. The same noise as an aircraft sonic boom, it is a product of a pressure shock wave sufficient to create an audible disturbance.

For the study of aerodynamics and air flow at velocities below the speed of sound, air is termed incompressible. At the higher speeds, there are large and important changes in air density and compressibility.

As the bullet moves through the air mass, pressure and velocity changes create turbulences in the surrounding airflow that move at the speed of sound. These pressure changes move in all directions. Even ahead of the bullet's front at below the speed of sound, but not above it. At over the speed of sound, there is no apparent change preceding the front edge of the bullet. This is all similar to the situation of high speed aircraft.

The example of surface waves on water may help clear up a misunderstanding. Although the speed is slower, a surface wave on water is simply a propagation of a pressure disturbance. A boat moving at a speed less than the wave speed (which varies) will not form a bow wave. A boat moving over the wave speed will form a bow wave which will become stronger as the speed increases.

◆ The speed of sound is termed Mach 1.0. Mach is the ratio of velocity to the speed of sound. Therefore: **Mach number** = V / a where V = true velocity and a = the speed of sound. This is a simplified formula which does not consider the temperature.

Compressibility is the main difference between subsonic (below Mach 1) and supersonic (above Mach 1) air flow. At supersonic velocities, all changes in air flow direction, pressure, density, etc. change suddenly. There are 3 types of waves formed at supersonic velocities. **(1)** A compression shock wave that is wasted energy and called *oblique*. **(2)** A normal compression shock wave. **(3)** The no shock *expansion* wave. Without going into the technical details on the wave forms, it should be added that they create a *wave drag* on the bullet. *Drag,* meaning to hinder or slow down.

The shock wave generated by a supersonic bullet can be seen on the range, under certain conditions of light and heat. The wave is compressed air which is more dense than the surrounding atmosphere. Light rays are bent as they travel from one density area to the other as with a prism. Mirage, the bending of light rays by different temperatures rather than pressures, can help in seeing the shock waves. Occasionally, under ideal conditions, even the bullet can be seen as it streaks toward the target, followed by an expanding round vortex.

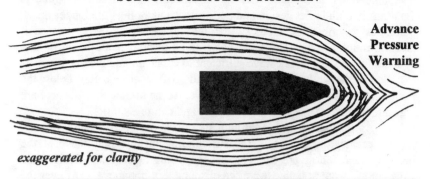

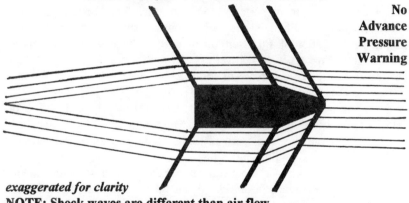

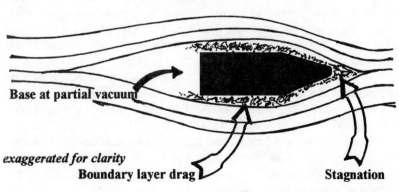

While the speed of sound is about 1120 f.p.s., there is a *transonic* range that begins at 0.75 Mach and extends to 1.20 Mach. This is the speed where the actual projectile is not at the speed of sound but if below it, some of the airflow will be pushed to Mach 1. If the projectile is slightly over the speed of sound, some of the airflow will be retarded and lag behind and the full supersonic effect will not take place. This range is roughly from 840 f.p.s. to 1,344 f.p.s. The figures are approximate because the speed of sound is only a precise amount at a known temperature and the transonic speed is lacking in precise limits.

A lot of phenomena occur during this transonic range. Most of the difficulties are associated with shock wave induced flow and pressure changes. A boundary layer forms between the supersonic airflow and the subsonic airflow. This area is always accompanied by a shock wave. There will be a static pressure increase behind the wave that the boundary layer may not have sufficient kinetic energy to withstand. As the speed increases in the transonic range, the shock wave moves farther back and at Mach 1 and above, a bow wave forms.

For long range rifle work, velocities in the transonic range should be avoided. Handgun calibers that operate in this velocity region, such as the .45 ACP, are not used at long ranges to avoid the effect.

Some muzzle-loading experts can accurately estimate the velocity of their bullets and powder required by the sound. Generally, the sound of the powder explosion and the gasses escaping the barrel will remain about the same, but as the bullet approaches the speed of sound, the bullet will begin to make a *crack* sound. This will increase in volume until it covers the original explosive noise. The increase in volume will be rapid until the top of the transonic range. Beyond this velocity, the increase in volume continues, but at a much reduced rate. (**Caution. For safety, do not use too much powder.**)

This small sonic boom created by the bow shock wave can be heard from any bullet with a velocity at or above the speed of sound. Below that velocity, the sound is no more than a light *sigh*. Down range and safely to one side is a good place to hear this event. Under some conditions, it can also be heard bouncing back as an echo from poles and buildings. It is interesting to note that silencers can reduce and almost stop the sound of ignition, but will have no effect on the supersonic sound wave of a bullet in that velocity range. The *crack* will still be as loud as ever:

SOUND AND LIGHT OBSERVATIONS

It was written in a magazine article that the .220 Swift was so fast that with a 50 grain bullet leaving the muzzle at 4100 f.p.s. (Norma factory ballistics.), a varmint would die before the sound arrived. While this is true, it would also be correct for any bullet that had an average velocity above the speed of sound. This assumes, of course, an instant kill.

With the speed of light instead of sound, the reverse is true. It is interesting to note the time involved for a bullet to travel a long distance and the huge difference between any bullet's velocity and the speed of light. (186,330 miles per sec. in a vacuum or $2.9979250 * 10^8$ meters per sec. using scientific notation.) For extreme long ranges, if a potential victim moved upon seeing the flash of an opponents gun, he could be out of the way before the bullet arrived. **Don't try it.** The key is *extreme* **long range. Also, the shooter could be off of the target (miss) in the same direction the intended victim moved.**

◆ A person who can see a firing gun's muzzle flash can estimate the distance to the gun by counting the seconds between seeing the flash and hearing the "bang". Multiply this by 1,130 for an answer in feet.

INCREASED VELOCITY

Increased velocity has disadvantages. In hunting, for example, if the bullet is not designed for the velocity it can disintegrate on impact with tissue and not expand or penetrate properly. Barrel and throat wear will be increased and usually at a much higher rate than the percent of increase in velocity.

DECAYING VELOCITY

If the range is long enough, velocities that are well over the speed of sound can decay down to the transition range and become subsonic. For distance work, usually targets at 600 yards and beyond, bullets should be chosen that have enough weight to hold on to their velocity as well as having enough to start with. If a bullet's velocity decays to the speed of sound, it is influenced by all the pressure and shock waves that are in that speed realm. For a brief example, the 7.62 M852 cartridge leaves the muzzle at about 2,600 f.p.s. The velocity will drop to the speed of sound, a loss of over half its velocity, at about 750 yards.

It is sometimes presumed that velocity drops at a rate whereby a bullet that losses 9% of its velocity in the first 100 yards will continue to loss 9% of its remaining velocity in each succeeding 100 yards. This theory does not work well for most cartridges. For example, a .30-06 bullet of 150 grain will fluctuate in each 100 yards from .889 to .928. Some will check out worse and some better, but the results are not good enough to consider the rule valid for anything requiring accuracy. Of course, **very few formulas consider the variables.**

WHY VARIATION IN VELOCITY

The ammunition is the leading cause of differences in velocity under what appear to be similar conditions. Ammo that is made as carefully as possible will still vary in velocity. There are several causes for this difference. **(1)** Variation in powder charge. **(2)** Variation in powder location in case. **(3)** Variation in bullet weight, roundness, diameter, hardness, etc. **(4)** Variation in case capacity, hardness, etc. **(5)** Variation in ignition caused by; **(a)** variation in flash-hole size and shape, **(b)** variation in hammer blow force, **(c)** variation in primer cup hardness and thickness, **(d)** variation in primer compound.

DRAG

To begin a discussion of drag, we need to acquire three basic facts. **(1)** Drag is the resistance of the air to the passage of an object through it. **(2)** Drag acts in a direction to the rear of the intended path. **(3)** Drag is the net aerodynamic force parallel to the relative trajectory.

Let's discuss these in more detail to clear up any foggy areas.

Air pressure at sea level is 14.7 pounds per square inch. Called static pressure, it is the result of the mass of air supported above it. This pressure is in all directions, not just down toward the earth. The bullet does not slip through the air without some resistance from the air as it touches the sides. *Parasitic drag* is the friction of the air molecules as they slide along the outer edge of the bullet. This is also called *skin friction*.

The air is retarded (slowed down) as it sticks to the outer surface of the bullet. The retardation is greatest at the surface of the bullet where the air particles are slowed to a <u>relative</u> velocity of near zero. Above this area, the particles experience successively less retardation until at a distance away, the velocity reaches the full air stream velocity. This area is called a boundary layer. The gas molecules can slide easily over each other but they cannot slide easily when they are pressed against the moving object.

$$D_{p2} / D_{p1} = (V_2 / V_1)^2$$

The velocity of the bullet has a powerful effect on parasitic drag. In this formula, D_{p1} is parasitic drag corresponding to an original speed V_1. D_{p2} is parasitic drag corresponding to a new speed V_2. The formula shows the powerful effect of velocity on parasitic drag.

In other words, the drag increases almost as the square of the velocity. If we double the velocity, the drag increases 4 times. Triple the velocity and the drag increases 9 times. This formula holds for the subsonic velocities. **At higher than the speed of sound, this effect is more rapid and pronounced.**

The compressed air at the nose or front of the bullet is tremendous. The flow created as the bullet pushes and forces into the air creates what is called a *bow wave*. If you put your hand out the window of an automobile moving at 55 m.p.h. (only 80 ft. per sec.) you can easily understand the force on a bullet at 1,000 or 2,000 or 3,000 ft. per sec.

Below the speed of sound, the air piles up in front of the bullets nose and air density increases. A pressure wave travels ahead of the bullet. We could say the air ahead of the bullet is being warned of its advance.

At velocities over the speed of sound, the pressure waves are not able to advance ahead. The foreword air is not aware of the approaching bullet until it is hit by the air particles pushed immediately in front of it. This causes a major increase in air density and air pressure.

As the bullet parts the air with its passage, it leave a void behind it. This area at the base is sometimes called "a hole in the air". It is created because the flowing air does not close in behind the moving projectile until it has moved on a short distance. This area with no air may not be a complete vacuum, but it has almost the same effect. If there were air at the back, it would exert the normal 14.7 lbs./sq. in. of pressure forward because the pressure is in all directions. The lack of pressure at the back does not hold the bullet back, as is occasionally believed, nor does it increase the pressure on the nose, but the consequence is the same.

The total drag is a combination of all types of drag added together.

Bullet design and the ballistic coefficient are all factors in this subject area and are explained in another section of this book. (See Chapters 14 and 16.)

AIR DENSITY

Air density is the term for the air's thickness or compactness. Technically, it is the ratio of mass to volume. The air gets thinner as we rise above sea level and this reduces the resistance (drag) on the projectile.

The density at a given altitude over a given place will show large variations do to the changing distribution of high and low pressure. The density of the air at any given altitude is obviously dependent upon the barometric pressure at that point. Variations in barometric reading due to cyclonic areas of high and low pressure passing over a given spot give a noticeable increase or decrease. The best that can be done for a simple constant law is to seek an average.

The standard atmosphere, as defined by the NACA, is an approximation of the average. For the U. S., it is based on sea level and a pressure of 29.921 in. Hg., mass density 0.002378 slug per cu. ft. and a temperature of 59 degree F. The temperature change is fairly uniform to the beginning of the isothermal region (stratosphere). That is -67 degree F. at 35,332 feet.

The air density drops with altitude, and as such, offers less drag (resistance) to the bullet's flight. The percent of sea level density is about .94 at 2,000 ft., .88 at 4,000 ft. .83 at 6,000 ft., .78 at 8,000 ft. and .73 at 10,000 ft. It would be a rare hunter that would have a need for the percent above this altitude.

A bullet's flight trajectory and ballistics are affected, but not as much as it would appear at first thought. One of the main reasons for the small change in performance is that the flight time, while reduced, is changed very little. If all the air were removed to a complete vacuum, the result would not be too dramatic. (In most cases.) A .180 grain .30 caliber bullet with a muzzle velocity of 2,700 f.p.s. at 10,000 ft. and a 300 yard range would require a lower elevation, compared to sea level, of only 1/2 minute. If this same bullet could be fired in a complete vacuum, the elevation change would only be a 7/10 minute difference. Remember, a minute is 1/60 of a degree. (More on density in Chapter 14.)

COEFFICIENT OF DRAG

The coefficient of drag increases strongly in the transonic velocity range. It peaks at about 1,500 f.p.s., which is actually past the transonic velocity. Depending on the projectiles shape, at higher velocities it will drop to the level it had at the subsonic velocity. This will be close to Mach 2. (About 2,240 f.p.s. at 60° F.)

SEMI-AUTOMATIC VELOCITY LOSS

There is a velocity loss in semi-automatic, self-loading firearms that is so small it is inconsequential. It is so slight that only a very careful test with good chronograph equipment will prove it.

A good explanation is reminiscent of a discussion on the effects of shooting forward or backward from a moving object. For example, let us say we are firing a gun that is solidly attached to a heavy object so it can not move in relation to the earth. The chronograph screens also cannot move. If a bullet is determined to move at 2,000 f.p.s., that will be in relation to the earth, as everything is mounted solid on it. If the same gun were fired at the same 2,000 f.p.s., but the gun was free to recoil at 15 f.p.s., the chronograph would show 1,985 f.p.s. The bullet is still moving 2,000 f.p.s. relative to the gun barrel but 1,985 f.p.s. relative to the earth and the chronograph that is fixed on it.

The same principal is in effect for a self-loading firearm where part of the mechanism is free to recoil.

Guns that use a small portion of the gas to fill a gas cylinder can have a slight velocity loss if the system is the type that permits leakage. In these guns, the velocity loss will vary from zero to perhaps a half a percent.

Recoil operated guns have parts that move back as the bullet moves forward. These parts may weigh up to half the total gun weight. The velocity loss is never zero as in the gas operated guns, but still so low that it can be minimized. Different guns will give different figures, but they will be in the general area of 15 f.p.s. for .22 caliber rimfire.

An auto-loading weapon will have the same measured recoil, other factors being equal, but the kick felt by the shooter is frequently less. The spring mechanism that is used to load the next cartridge absorbs some of the shock that would be felt by the shooters shoulder or hand. This was explained in more detail in the recoil section where we discussed the law of conservation of energy and how it keeps the recoil the same. To avoid confusion, remember that felt kick and recoil are not the same thing. (See Chapter 6.)

TIME OF FLIGHT RULE OF THUMB

A rule of thumb is available to quickly find an approximate time of flight.

Average velocity is found by adding together the muzzle velocity and the velocity at the range and dividing the answer by 2.

Example: a Winchester .308 Super X 150 grain ST. Muzzle velocity 2,820 f.p.s. and velocity at 300 yards is 2,009 f.p.s. 2,820 + 2,009 = 4,829 ÷ by 2 = 2,414.5 f.p.s. for average velocity over the 300 yard range.

If we then divide the range in feet (note the need to convert yards to feet by multiplying by 3) by the average velocity we will have an approximate time of flight for the 300 yards.

Example: using figures from the previous example: 300 yards times 3 = 900 feet. 900 feet ÷ 2,414.5 = .3727 seconds.

BULLET DROPPED AT MUZZLE

It is sometimes stated that if a bullet were dropped from the gun muzzle at the same time a bullet of equal weight were fired horizontally over flat ground, they would strike the earth at the same instant. No, only in a vacuum. The difference would be slight, only a tiny fraction of a second. The air resistance (drag) and the aerodynamic lift created by a slightly nose high trajectory will keep the fired bullet up longer. Of course, the dropped bullet is also influenced by the atmosphere.

School text-books frequently state that they would hit the ground at the same time. This is an effort to demonstrate that the vertical acceleration of an object due to gravity is not related to the horizontal velocity. As just stated, only in a vacuum, not in the real world.

William C. Davis, Jr., writing in the N.R.A.'s *The American Rifleman* magazine for December, 1982, points out that a 150 grain .30-'06 bullet fired horizontally at 2,900 f.p.s. from a 5 foot height, would hit the ground about 1,330 ft. down range after about .600 second flight time. The same bullet, if dropped the 5 feet, would take .557 second to hit the ground. A gain of .043 second. (Mr. Davis has contributed several valuable charts and other facts to this book. It is very much appreciated.)

Several attempts have been made to measure bullet velocity by the "drop" method. That involves obtaining the velocity from the drop of the bullet over a known range. The drop gives a measure of time from which velocity can easily be computed. It sounds good but is both inaccurate and

unnecessary. The amount of drop cannot be determined close enough for the answer to have any accuracy.
WIND DEFLECTION, VELOCITY & DRAG
Wind deflection is not controlled by time of flight but by the loss of velocity during the time of flight. In the transition velocity range above the speed of sound, the drag increases disproportionately higher than the velocity increases. In other words, if the velocity is increased by a small amount, the drag increases by a large amount. This situation is unique to this velocity range. At all other velocities, an increase in velocity will bring about a decrease in wind deflection. Not in the upper transition range. This is the reason for the special .22 match ammunition that is loaded so it will not go fast enough to get into this velocity range. The higher velocity .22 ammo has a flatter trajectory but is deflected more by a cross wind. Air resistance is much increased in this velocity range and a lot of air turbulence and pressure changes occur. The delay is affected out of proportion. Note: this item is also in the wind deflection chapter.
FIRED FROM A MOVING OBJECT
A bullet fired from a moving object, whether a 2,000 m.p.h. fighter plane or a 60 m.p.h. auto, will be directly affected by that movement. A .30-'06 at 2,700 f.p.s. (1,841 m.p.h.) fired **forward** from a jet at 1,841 m.p.h. would have an initial velocity, in relation to the earth, of 3,682 m.p.h. If fired to the **rear**, it would be zero. If that sounds peculiar, think about it carefully for a moment. It can be no other way. What we have to remember is that we are speaking of the bullet's relationship to the earth. In relation to the aircraft, it would be 1,841 m.p.h. in both directions. (For simplicity, we are ignoring a few minor points that would have no effect beyond about a 1/4 of a percent in velocity.)
ARMOR-PIERCING BULLETS
In armor-piercing bullets, steel penetration leans heavily on velocity and bullet weight, as expected. Ordinary bullets such as soft point, tracer, hollow point, etc. will also go through heavy steel plate if the velocity is very high. A .30 caliber bullet that will not even dent a 1/2 inch steel plate at normal velocity, will knock out a plug bigger in diameter than the bullet at much higher velocity. Armor-piercing bullets do just what the name implies, pierce according to striking energy. Regular bullet types at very high feet per second (3,000 f.p.s. to 4,500 f.p.s., depending on cartridge, etc.) will shear and knock out a plug equal or greater than their diameter. Machinists and die makers will notice the action is similar to a punch and die in a metal press. In this case, the speed and energy make up for the lack of support that would have been produced by the missing die.

HIGH VELOCITY USING POWDER AS A PROPELLANT

If standard powder is used, no amount can push a projectile beyond about 11,000 f.p.s. A more practical limit is about 6,500 f.p.s. Some experts believe 4,500 f.p.s. is the highest efficient limit. Regardless of what the actual practical and efficient limit is, a huge amount of the energy is used just to push the gas down the bore. The gases produced have a high molecular weight and the available energy must push both the gas weight *and* the projectile weight. (See internal ballistics section.) The energy used to push the gas is wasted, as it is in all firearms. Adding more powder charge can only be taken so far until the extra charge is just wasted in trying to move the gas. No energy is left to move the projectile any faster. It reaches a point where it is self-defeating.

The more powder the less efficiency. Rifles have efficiency rates in the 30% to 40% range. Shotguns usually can get slightly over 40%. With pressure gradients that are low and such low efficiency percents, there is little future in velocities over about 5,000 f.p.s. from conventional powder firearms. There is also a maximum velocity that this expanding gas cannot exceed in the barrel without creating a shock wave. (See speed of sound.) **A bullet cannot be pushed faster than the expanding gas will move.**

Barrel erosion is one of the disadvantages of high velocity bullets; partly due to friction and partly to the high temperature. There will also be an increase in recoil from increasing the powder, even if it does not increase the velocity. (See Chapter 6 on Recoil.)

HIGH VELOCITY EXPERIMENTS

High velocity guns, sometimes called hypervelocity, are capable of projectile velocities in excess of 28,000 f.p.s. There have been many test projects over the years involving this area of study. All involve complicated equipment and procedures that generate extreme heat and pressure with almost no barrel life. Mostly, they are for test and research only and have no value in the real world.

In the laboratory, electrothermal enhancement is an effort to bypass the two powder problems. (Limit to gas expansion, velocity and pressure.) The powder and primer are replaced with a fluid and a pulsed electrical current of about 100,000 amps. The heated plasma that is formed will push the projectile up to 50% faster with less pressure. The push is continuous for the entire barrel length with no sudden peak.

Electromagnetic guns do not have the practical value of the electrothermal type. There is no gas expansion and no cartridge. Only the projectile and a magnetic field which is formed as current is induced to two parallel copper rails. Velocity in the area of 28,000 f.p.s. has been reported by these *rail guns*.

Who knows what the future will bring? History has shown that the laboratory experiments of today become the common items of tomorrow.

VELOCITY VARIATIONS

There are many reasons for variations in velocity which can occur even between guns of the same make, model and caliber. Reloading manuals don't agree with each other and manufacturer's factory figures can be different. Your pet load may chronograph faster in Uncle Joe's old junk Mauser than in your new custom rifle.

REASONS: Not all **powder** is created equal and the moisture content and temperature can vary. Lot to lot differences can be noticeable and even the position of the powder in a spacious case.

Primers are not all the same from one manufacturer to another and can have differences from lot to lot. The installation in the case may be slightly different and even the hammer or firing pin blow can be different enough to cause a change.

Bullets come in so many different styles, shapes, materials and jacket types and thicknesses that it stands to reason there will be variations. Even the seating depth can affect velocity, and in some instances, by a large amount.

Barrels can have a major effect on velocity. Their condition and cleanliness is a well know factor. Also consider rifling, over-all length, freebore length, throat length and angle, bore size, and many other factors. All can cause changes in velocity.

Case condition and the capacity which controls the loading density and bulk density should be considered. Is there extra room in the case or is the powder compressed? (See cartridge details in Chapter 15.)

Even the **temperature** and **barometric pressure** affect velocity.

MORE MATHEMATICS

Briefly, we should note that the laws of physics dictate that the velocity created by any force on a moveable object for one second is equal to the force in pounds, divided by the mass of the object. (As we have discussed in several other chapters, the mass is the weight divided by the acceleration of gravity.) The velocity will be twice as much if the force acts for two seconds. If the force acts through a given distance instead of a given time, the velocity that will be imparted to the object will be equal to the square root of two times the force multiplied by the distance through which it acts, divided by the mass of the object.

When used with projectiles, this will give a velocity that is too high if the peak pressure is used because the peak pressure does not remain at that amount during the time it acts. An average pressure is required.

NOTES ABOUT VELOCITY

◆ There is a minor controversy on which is more important to the hand-loader testing new loads; accuracy or velocity. Are they not both important? Appropriate and uniform velocity is needed for bullet expansion, penetration, wind deflection, trajectory, and, yes, for accuracy. But without accuracy, none of the list would have any value.

◆ The importance of velocity cannot be overemphasized, but at the same time, it can be carried to extremes. If pushing a bullet to its speed limit causes a drop in accuracy which can be restored by a reduction in velocity of 100 or 200 f.p.s., so what? While it is true that velocity decreases down range and all bullets expand and perform best at certain velocities, it can be over done. The extra f.p.s. will make little difference. Each caliber, gun and hunter has a maximum effective range. Another 200 f.p.s. will not extend it enough to make any difference. Particularly if it causes a loss in accuracy.

◆ Many wildcat cartridges can obtain muzzle velocities of 4,000 f.p.s. Usually in smaller calibers such as .17 Remington, .240 and .247 Weatherby, .220 Swift, .22-250. etc. Necessary ingredients are a light bullet with a good ballistic coefficient, a case with ample capacity, and a long barrel with proper twist.

◆ Both ballistic tables and computer programs show the advantages of higher velocity. This may include flatter trajectory with less drop, better wind handling ability, excellent expansion, more energy, etc. Use caution and thought for the game or problem involved. In some cases, a heavier but slower bullet may be better for the range and game.

◆ High velocity cartridges have a reputation for being undependable and inaccurate. This may be true in some instances, but it does not have to be. Hand-loaders who use extreme care can obtain excellent results. If a person is used to accepting a 5 percent error, for example, the error will be greater. Tighter tolerances are required to maintain consistent performance.

◆ All of the power of movement in a bullet in flight is in the kinetic energy stored within itself in momentum.

CHAPTER 14

BALLISTIC COEFFICIENTS

Many people use ballistic coefficients in computations without understanding their function. This chapter should throw some light on the subject, even for those that are content to remain in the dark.

We will be discussing three subjects that are intermixed together. *Sectional density* and the *form factor* are included in the ballistic coefficient. It is important to mention that two of these items, the form factor and the ballistic coefficient, are both deceptive commodities that the average sportsman will be obtaining from charts rather than from testing on his own. They are both important because they relate directly to a projectiles ability to maintain forward velocity.

BALLISTIC COEFFICIENT

The ballistic coefficient is a measure of how well a bullet can overcome air resistance and maintain flight velocity. It is a number that is arrived at mathematically and equals the sectional density of the projectile divided by its coefficient of form. It can also be described as the ratio of the sectional density to its coefficient of form. It also is a ratio that compares the bullet to a standard bullet (projectile) that has been tested and therefore has known characteristics. The larger the number the more efficient. **At this point, these terms may be confusing, but they will be explained in detail.**

The ballistic coefficient of a projectile is necessary to a trajectory calculation. Bullets sold for use by hand-loaders have the ballistic coefficient in the manufacturer's data. For some unknown reason, this information is seldom available for factory loaded ammunition. Complete trajectory tables are available but ballistic coefficients for the bullets used are not. (Note; this may change in the near future, but it was true at the time this was written.)

A bullet's ability to retain as much muzzle velocity as possible is an important factor in both trajectory and game killing effectiveness. The ballistic coefficient possessed by the bullet will be a most important factor and the higher the number the better. **The three points which govern the ballistic coefficient are weight, diameter and shape.** The first two can be used to create the sectional density.

As will be shown, there is a benefit in knowing this information and it can be determined for all bullets, even if they are hand-made.

SECTIONAL DENSITY:

Sectional density is the expression used to describe the diameter of a bullet as compared to its weight. This is an important factor with the bullet's ability to sustain its velocity. A heavier bullet has greater kinetic energy than a lighter bullet at the same velocity. A larger diameter bullet will have more air pressure build up at the forward end than one of smaller diameter. Therefore, the most proficient bullet will be the heaviest in proportion to its diameter. Of course, the easiest way to make a bullet heavier and keep the same diameter, is too make it longer.

One way to express it in plain terms is that sectional density is the weight that backs-up the bullets diameter. If the bullet holds together, it will have greater penetration and more killing power. Sectional density is one of the bullet's most important characteristics, along with its shape, in holding on to velocity.

On the other side, high sectional density will resist the push down the barrel more than low sectional density. It is just more weight for the expanding gases to accelerate to a required velocity. Therefore, the heavier bullet will have a heavier recoil than the lighter bullet if they are both loaded to the same pressure. (Remember Chapter 6 about recoil?) It will take a larger cartridge and more powder to push the bullet to speed. A bullet of low sectional density can be pushed to a higher velocity and a flatter trajectory for long-range use, but by the time the light bullet reaches the target, it may have little velocity and energy left. *Retained* velocity at impact may be too low for bullet expansion and penetration. (Both covered in detail in Chapters 16-24-25-26.)

MATHEMATICS - SECTIONAL DENSITY

Mathematically, sectional density is the ratio obtained by dividing the bullets weight in pounds by the square of its diameter in inches. The bullets weight in grains can be converted to pounds by dividing by 7,000. The area of a bullet's cross section increases with the square of the diameter. Therefore, we must square the bullet's diameter before it is divided into the weight.

◆ $SD = W / d^2$

Where; SD = sectional density
          w = bullet weight in pounds
          d = bullet diameter in inches

NOTES ON Sectional Density

◆   The bullet with the bigger sectional density will hold velocity better down range but it may have a more arched trajectory compared to the lighter bullet which probably had a higher velocity.

◆   If the caliber (diameter) stays the same, then doubling the weight will double the sectional density.

◆   Sectional density is strongly influenced by the bullets length, but the jacket thickness, the alloy used in the core, the shape (as blunt nose vs. pointed); all are included.

♦ As velocity goes up, sectional density should also go up. Light bullets at high velocity break up easily upon striking game (or other objects). Higher velocity increases the drag and more sectional density helps to push through the air resistance and deliver energy to the target.

COEFFICIENT OF FORM

The coefficient of form (also called form factor and coefficient of reduction) is a mathematical figure (a number or multiplier) of the bullets shape, its smoothness and the shape of the base. In order of importance, shape is first. The advantage going toward the longer and sharper point and the shape from the tip back to the maximum diameter. The base has little importance on slow bullets but above the speed of sound the base becomes more significant. Smoothness has the least effect, due partly to the fact that few bullets are *rough*.

Form factors are always a comparison of one bullets air resistance to that of another. It formulates the effect of different shapes on bullets with the same sectional density. Another way to express it; **the form factor compares the shape of the bullet in question to the shape of a standard bullet used in a particular ballistic table.** As you can imagine, it has limitations.

It must be emphasized that the form factor is for a particular ballistic table and not for use on any and all ballistic tables. In other words, a particular bullet may have one form factor for an Ingalls table and a different one for C-1 charts and still another one for another chart. In other words, there is no such thing as an absolute form factor for any bullet.

Form factors cannot be calculated in the usual way. **No method is known to calculate and describe the shape in numbers suitable for use in formulas.** Firing tests are conducted for this purpose.

SAMPLE FORM FACTORS

Here are a few sample form factors for examples. Few will be exactly as listed, but in the absence of better information, this should be a help.

Very sharp profile	.60
Moderately sharp profile	.70
Moderately sharp profile with a small flattened tip	.85
Moderately blunt profile	1.00
Very blunt profile	1.20

For more accuracy, subtract .06 for a boat-tail. Add .07 for a small exposed lead tip. If the exposed lead tip is large, add .20.

There are several charts that can be used for estimating the form factor by comparison of the bullet's shape to the shapes shown on the chart. The DuPont Co. and others have worked out ballistic charts of this type. If the person using the chart takes the time and effort to do it correctly, the result can be reasonably accurate.

As just stated, the form factor is always for a particular table and based on the retardation of the standard bullet involved. Therefore, any chart of bullet shapes give in this text would probably not match a chart owned by the reader. A sample is given to show what they are like.

SAMPLE OF A BULLET SHAPE CHART FOR FORM FACTORS

.35 | Ogive 0.5 | Ogive 1.0 | Ogive 1.5 | Ogive 2 | Ogive 3 | Ogive 4 | Ogive 6 | Ogive 8 | Ogive 10

Also, there are formulas available, but they will not be included because ordinarily they do not work as well as the charts and other methods.

BALLISTIC COEFFICIENT

It is well known that a bullet's design can affect the way it handles inadvertent tree limbs and also its stopping power in hunting and self defense. The design also helps the aerodynamic aspects of the bullet. These elements include the phenomena described elsewhere in this book as parasitic drag, bow wave, front air compression, and a partial vacuum at the rear of the bullet. The goal is to keep as much of the velocity as possible while using a bullet design that will do what the shooter wants.

There are several discussions of kinetic energy in this book. Check the index and refer to them if needed. Remember two important points. **(1)** Kinetic energy is directly proportional to the mass of the moving object; a bullet in this case. For example, if their velocity is the same, a 100 grain bullet will impact at half the force as a 200 grain bullet. **(2)** Kinetic energy increases as the square of the velocity. For example, double the velocity of the bullet and the impact force will be increased four times. If the velocity is increased by three times, the impact force will increase by nine. Another way of explaining it would be to fire three bullets, all of the same weight, at 1,000 feet per sec., 2,000 feet per sec., and 3,000 feet per sec. The second will hit 4 times as hard as the first and the third 9 times as hard. If this is given some thought, it is impressive.

The ballistic coefficient measures the bullet's ability to conquer air resistance. Mathematically, it is the ratio between the sectional density and the coefficient of form. The higher ballistic coefficient is preferable as the projectile will hold its velocity better. It is available in charts and can also be arrived at by formula. Generally, it is best to use the charts published by the manufactures because a good amount of test firing is required.

For custom bullets or any bullet that is not listed on a chart, the ballistic coefficient can be taken or estimated from a similar bullet. Simply use care to pick a bullet close in caliber, shape and sectional density.

The ballistic coefficient, when it is calculated from its form and nose shape, frequently works out different when it is test fired. We then have an *effective* coefficient that is more useful than the *paper* coefficient. The reason for the inaccuracy in many paper ballistic coefficients are as numerous as there are misjudgments. The use of a different powder, load or barrel can cause a bullet to fly differently and change the coefficient. The path through the air is frequently a very tiny helical and rarely does it travel directly nose first. (See bottom of page 200.) Shooting through screens or heavy paper will show us that bullets can wobble and this poor stability can be anywhere between the muzzle and the impact point.

The best values are always based on test shooting under laboratory conditions. No matter how it looks on paper, it takes very little to change the flight and alter the ballistic coefficient.

When a bullet's point is not changed, the ballistic coefficient will be raised by an increase in sectional density. The increase will usually enable the bullet to do a better job of holding its velocity. Although, in the interest of safe chamber pressure, the bullet with the lower sectional density may have a higher muzzle velocity and therefor a flatter trajectory. Notice the wording is "may have." Ballistics is full of variables that can change the outcome.

The ballistic coefficient contains both the sectional density factor and the coefficient of form (form factor). As mentioned, charts are the best source for this information.

The ballistic coefficient should not change as the velocity decreases down range. That appears logical because the velocity and the coefficient are separate. Unfortunately, when the ballistic coefficient is used in tables and formulas, it is only true if the projectile is exactly the same as the projectile in the table or formula. The difference may be slight, but it will be noticeable. If the curve is plotted on a graph, the two curves will not coincided exactly.

If this has been too confusing, just remember **the ballistic coefficient is the comparative ability of a bullet to push through the air** and hold its velocity. **A comparison to a standard by a multiplier.** Testing requires measuring the velocity drop over a known distance.

AIR RESISTANCE FORMULAS & TABLES

During the early history of ballistic research, many different ideas and methods were tried to compare projectiles by their drag resistance. All attempts tried to reduce it to a simple mathematical declaration. Many great thinkers of their time, including Sir Isaac Newton, worked hard on the problem. He said that the air resistance was proportional to the square of the velocity. This proved false for the velocities involved in almost all firearms. Later the decision was reached that a simple formula was not feasible.

Testing and experiment was determined to be the only answer. A single bullet of a certain size and shape was used as a standard and all the others were compared to it by velocity and deceleration. This standard projectile had a 2 caliber nose radius (ogival head) on a 3 caliber length. It was given a ballistic coefficient of 1.000 and all others were compared to it. It is still normal to use 3 decimal places for the figure. This particular standard was created in 1881 by Russian Colonel Mayevski and a German named Krupp. For all practical purposes, the constant multiplier called a ballistic coefficient, was created at that time.

The Commission d'Experience de Gâvre of France, conducted their own firing tests in the late 1800's. In 1898, they published a report of their findings. They proved that air resistance could not be stated as proportional to the square of the velocity, as stated by Newton. They also said it could not be reduced to an easy formula. They published a table which showed retardations for each velocity. This became familiar as the Gâvre function. It was excepted and sealed the end of the hunt for a single formula. From then on, **tables of actual values for each velocity, as discovered by firing tests, were used.**

Col. J. M. Ingalls of the U. S. Army used Mayevski's research as a basis, and created his ballistic tables which extend up to 3,600 f.p.s. His tables, and a few variations, are well known in the U. S. One group of his tables (No. 1) will compute the time of flight, the ballistic coefficient and the velocity at a specified place in the trajectory if the muzzle velocity and the velocity at some place in the trajectory are known. Another group of Ingalls tables (Table II) will compute the angle of departure for any range from the muzzle velocity and the projectiles ballistic coefficient.

The British computed tables in 1909 that are about the same as Ingall's except they go on up to 4,000 f.p.s. They are known as the Hodsock tables and were based on a spitzer bullet with a flat base.

Tables developed by the U. S. Army Ballistics Research Laboratory are more refined and use more recent data. In the 1930's, they did a huge amount of research into drag functions at the Aberdeen Proving Ground.

They came up with their own tables for military use which were based on the French tables and are known as G-function or G_1. (G for Gâvre) These tables give the connection between velocity and retardation. Knowing the shape and the velocity, the retardation may be found. The military also researched and produced tables known as G_2 for use on projectiles that, instead of the standard shape, had a long pointed nose and a boattail (tapered) base. It has also been known as J function and J tables. The U.S. military have several methods which they use for their special long range problems. G_5 is used for boat-tail bullets which perform differently above Mach 1 than standard bullets.

A group of charts were produced by the engineers at the DuPont Company and published in 1926. Their numbers were for use in the *Ingalls Ballistics Tables* which are used in the U. S. to calculate both velocity and trajectory. They were developed by Col. J. M. Ingalls.

The C_1 coefficient is another similar method based on the G_1 *drag law*. The ammo. makers use C_1 and it is used in almost all factory charts.

There have been many other different charts and tables published over the years. Of course, they are all approximate.

The charts and their proper use are not included here because of the room involved. To the reader not familiar with them, they appear to be an endless amount of pages filled with columns of numbers. Some charts require the use of high mathematics and for others, the math. is easier. For the most part, they enable us to predict, based on the muzzle velocity, what the velocity would be at any other point as the velocity decelerates. This is based on the retardation created by air resistance. Of course **time** is a major factor as is **space**-the space being the space passed over by the projectile as its velocity is slowing. (In this case, it is usually called space, although distance would mean about the same thing.) In the tables, the time and space is the actual time and space that is used as the projectile slows from one velocity to another velocity. This will be for just the one projectile the chart is based on.

A projectile of the same exact shape and diameter but with twice the weight would require multiplying the numbers by two. If the projectile had the same shape but was one-half the diameter, the basic mathematics we have mentioned earlier tells us that the cross sectional area will be one-fourth so it will have one-fourth the retardation. This is the ballistic coefficient of the projectile and it is covered in detail earlier in this chapter.

Sources for charts are available, including *Exterior Ballistics* by McShane, Kelley and Reno, 1953. To follow their book requires a knowledge of higher mathematics, including differential and integral calculus. *Exterior Ballistics of Small Arms Projectiles*, by E. D. Lowry, Winchester-Western Div., is a long book full of ballistic tables.

A home computer can work the problems fast and accurately with either a purchased program or the *home copy* method available through the National Rifle Association. It is a good organization and the trajectory program alone is worth the membership fee.

SHARP TIP PROBLEMS

Bullets with sharp tips normally have an excellent ballistic coefficient. That is the good news. The bad news is that they are difficult to keep sharp and after damage is incurred, the ballistic coefficient and trajectory suffer greatly.

Tips can be damaged in many ways. Dropping them on the floor or ground, striking each other in pockets and other places; tubular magazines cause problems, and even recoil can smash them into the front of the magazine as the firearm slams rearward. A soft lead tip will be deformed rearward by the high G forces of acceleration. Perhaps the heat in the barrel is a slight contributing factor, but most likely, it is not. The same holds true for heat buildup from aerodynamic friction during flight. While it is theoretically possible, any change would be small and meaningless.

Testing for the changes caused by tip damage is easy. If all the cartridges are from the same lot or hand-loaded the same, then it is a simple mater to chronograph undamaged bullets and then deliberately file the tip and check them. The comparison is interesting. Accuracy is not changed very much but the ballistic coefficient can change dramatically. Muzzle velocity can drop as much as 10% and the ballistic coefficient can be cut in half. These are both large changes and show the value of protecting sharp tips. It also shows how the ballistic coefficient can be altered by a simple action. In this example, in the wrong direction.

Note: There is more on tip damage, including Norma factory tests, in Chapter 16.

TESTING

It is common to use two chronographs to check velocity at two locations and then use a computer program to figure the ballistic coefficient based on the velocity loss for the distance involved.

To obtain ballistic coefficients by testing requires accuracy of the highest level. Distances have to be measured with a tape. Some people have tried home testing using a properly measured 1000 yard range and velocity from an accurate chronograph. The time from muzzle to target then becomes an obstacle. The accuracy has to be correct to the thousandth of a second. For example, a .30 caliber M72 match bullet will take 1.688 seconds to go 1000 yards. An error of .007 (7 thousandths) of a second will change the ballistic coefficient by 5 %.

Testing is done more commonly by measuring the difference in velocity at 2 places. Remember, the loss of velocity is what the ballistic coefficient is all about. In the event that 2 chronographs are used, during the middle of the series, the instruments and their respective screens should be switched so any error or discrepancy will be cancelled out. They also should agree with each other as close as possible. One foot per second would be ideal. This can be accomplished by setting the screens one inside the other. That is one at perhaps 12' and the other at 10'. Both are set with the same midpoint and on bars so they (or just one) can be moved without any change. With test firing, they can be adjusted to read very close to each other and then moved apart to their proper location.

The measurements should be taken as far apart as practical. If possible, 100 yards for rifles and 50 yards for handguns. Again, this is the distance apart of the two pairs of screens. The exact distance is not important, but it will have to be measured and known. Remember to fire a large enough quantity of shots that the result has a genuine meaning.

The temperature, relative humidity and barometric pressure are considered in every serious effort. The air resistance and cumulative drag are influenced by these atmospheric conditions which create an *apparent* change in the ballistic coefficient. Apparent because the projectile only behaves differently because of outside changes. The actual ballistic coefficient remains the same if based on standard atmospheric conditions. (Altitude at sea level, temperature 59 degrees F., relative humidity 78% and a barometric pressure of 29.58" Hg.) An apparent higher ballistic coefficient results when the pressure decreases and the humidity, temperature and altitude increase. The ballistic coefficient will appear lower if the above conditions are reversed. Most factory figures are converted and reported at standard conditions.

As you know from other sections of this book, the drag on a projectile is a function of air density. To correct for non-standard air density, we have to find the figure for the shooting range and the standard for the altitude. **The correction factor is the ratio received by dividing the standard pressure by the shooting range pressure.** The number obtained is used as a multiplier.

Do not use the pressure reading from TV or radio weather reports unless you and they are at sea level. The pressure given is adjusted for altitude. At sea level the static pressure of the air is 2,116 p.s.f. (14.7 p.s.i.). At 40,000 feet, this static pressure decreases to approximately 19% of the sea level value. This gradual change is factored into the weather report pressure readings.

A barometer can be properly set by taking it physically to an airport weather bureau or Federal Aviation Flight Service Station. Both will have an accurate barometer and know their exact altitude above sea level. Both places will provide *ICAO standard atmosphere*. This will yield a smaller ballistic coefficient than the *Standard Metro* atmosphere used in factory calculations. The difference will be very small at less than 2%.

To correct for non-standard temperature, we must add 459.4 to each temperature to obtain a reading on the Rankin scale. (This is explained in detail. See index.) After the additions, the multiplier is obtained by dividing the standard temperature into the shooting range temperature.

The humidity correction is difficult to perform and it makes such a tiny change in the ballistic coefficient, that it will be skipped.

Example: Standard conditions were listed a few paragraphs ago as 59 degrees F., 29.53" and 78%. For our example we will give the shooting range conditions as 90 degrees F., 29.20" pressure and 88% relative humidity. For our example, the original ballistic coefficient can be .342

Barometric pressure correction factor is 29.53 / 29.20 = 1.011

Temperature correction factor is 90 + 459.4 / 59 + 459.4 = 1.059

Temperature factor * Baro. pressure factor * Original ballistic coefficient = Corrected ballistic coefficient   (1.059 * 1.011 * .342 = .366)

While we would like to think that this will give us an accurate figure, it will not. It will be closer than if we did nothing, but it will still be just an approximation. Tables are available that list multipliers made from the ratio between the actual density at the time and place involved and the standard. Corrections that complex are not normally necessary. In our modern times, when that type of accuracy is required, it is usually figured on a computer with a ballistics program.

VARIABLES

The temperature, relative humidity and barometric pressure have been mentioned as factors influencing testing for ballistic coefficients. Also having an effect are the firearm and the amount of twist in the barrel and the velocity. There is a lot of discussion about gyroscopic stability elsewhere in this book, so refer to it as needed, because that is another problem. Any bullet that is unstable, wobbles, yaws or is moving in any way other than dead straight and point-on, will loose velocity faster and have a lower ballistic coefficient. And almost all bullets will have at least one of these or a varying amount of several. Bullets that have traveled far down range will be more stabilized and give higher readings. All of these variables will have an influence on the test results and they are some of the reasons why it is difficult to match factory figures. It is common to even obtain different results from shot to shot. The bullets shape and ballistic coefficient should be the same, and may be, but all of these things cause deceleration of velocity and change the final figures.

BALLISTICS OF SPHERES

The ballistic coefficient of round balls is figured the same as other projectiles. The shape always being spherical, the sectional density, consisting of weight and area, is the main ingredient. It will be low by comparison to other shapes. The usual formula applies.

The sectional density of a sphere varies directly as its diameter. Therefore, if we double the diameter we will also double the sectional density. This, in turn, will halve the velocity loss. Remember that sectional density is the projectiles weight in proportion to the square of its diameter.

To compare a round ball's velocity loss to other bullets, we will use a normal bullet for a standard. One with a sharp point and an ogival head with a short 2 caliber radius. If we give it a velocity loss rate of **1.0**, then a bullet with a longer sharp point would be about **0.6** and **7.0** for a round ball. Notice the decimal point position. A big difference because the velocity will drop 7 times as fast as a standard bullet. In this example, we are assuming all projectiles have the same diameter, but not the same weights.

BASIC MATH. FOR SPHERES

If the bore diameter is doubled, the bore area increases 4 times (2^2). A ball projectile for the bigger bore would be 8 times (2^3) heavier than the smaller. If the density and other factors are equal, the ratio of the bore area to the ball weight increases as $2^3 / 2^2$.

NOTES

◆ Some bullets are deformed in the barrel by pressure and acceleration. This is rare with many modern bullets, but lead bullets are not immune to this problem. This changes the shape of the bullet and alters its aerodynamic capabilities, form factor and ballistic coefficient. Bullet retrieval methods are explained in this book. (See index.) They may help determine if this is happening.

◆ Yaw is a normal occurrence near the muzzle. This will increase the aerodynamic drag compared to a bullet moving properly point-on and, in turn, lower the ballistic coefficient. This is one reason for not placing chronograph screens too close to the muzzle. A proper rifling twist for stability is necessary, as discussed in Chapter 12 on rifling and twist.

◆ A high ballistic coefficient is vital in reducing wind deflection. The time lag discussed is a result of atmospheric drag and the ballistic coefficient measures how well a bullet performs against drag. The result is that a high ballistic coefficient will have less wind deflection, all else equal.

◆ An increase in bullet weight will also increase the ballistic coefficient but it will also lower muzzle velocity, if all else stays the same.

♦ There are always compromises. A game bullet must be effective on game, even if it gives up some trajectory qualities in favor of stopping power.
♦ The ballistic coefficient is as important as velocity and easier to upgrade. If the average cartridge case is lengthened by .200", the velocity will only increase 2% or 3%. Instead, add the .200" to the length of the bullets head and the ballistic coefficient jumps by 35% to 40%.
♦ A bullet with a high ballistic coefficient will spend less flight time over a given distance, all other factors being equal, so it will have less wind deflection. The lag time from drag will be less. A higher ballistic coefficient brings a flatter trajectory and more velocity and kinetic energy delivered to the target. A higher ballistic coefficient will have a lower form factor. A higher bullet weight will increase the ballistic coefficient.
♦ The ballistic coefficient is directly proportional to the sectional density if the bullets have the same shape.
♦ For the utmost accuracy, it would be necessary to create and use a ballistic coefficient that changed with velocity changes.
♦ The ballistic coefficient is a simple multiplier.
♦ A bullet with a lower ballistic coefficient, as compared to a higher one, will have a slightly more arched trajectory, a lower velocity and lower energy delivered downrange. The energy loss will be, in most cases, more pronounced than the other two.

THREE MAIN POINTS (from above)
1. Decreasing the form factor (streamlining) will increase the ballistic coefficient for a given bullet weight and caliber.
2. The higher the ballistic coefficient for a given weight, the flatter the trajectory, the less wind deflection and the higher the velocity and kinetic energy at a specified range.
3. Increasing the bullet weight will increase the ballistic coefficient for a given caliber and form factor (shape).

MATHEMATICS
Air resistance is based mostly on the cross-sectional area of a bullet which in turn varies with the square of the diameter. Mass is one of the main things which will defeat air resistance and it varies with the cube of the diameter. For readers not paying attention, they were both different. The first was based on the square and the latter on the cube. Therefore, for bullets with the same comparative proportions (not properties), their ballistic coefficients will vary with the cube of the diameter divided by the square of the diameter. This mathematically brings us back to the same diameter we started with. So, a .25 caliber bullet with the same shape as a .50 caliber bullet will have a ballistic coefficient of half as much. So, with bullets of different calibers but of similar shapes fired at similar velocities,

the ballistic coefficient will be proportional to the weight divided by the square of the diameter. This will be in ratio to their diameters. (Confused? Read on and it will become easier. If necessary, reread this paragraph.)

There are two basic formulas for determining ballistic coefficients under standard conditions. The second formula is the more accurate and the one used the most.

$$C = SD / i \quad \text{also} \quad C = w / i\, d^2$$

Where: C = ballistic coefficient

   i = form factor from bullets shape

   SD = sectional density

   w = bullet weight in pounds

   d = diameter

Example of 2nd formula: bullet weight 200 grains ÷ 7000 = .02857 grains, diameter .308 inches, form factor .75.

$.02857 / (.75) (.308)^2 = .40155$ **or .40**

The form factor (i) can be stated as the ratio of the drag coefficient of a standard bullet to the drag coefficient of the bullet in question. With all bullets being different in shape, size and weight, they will all have different numbers than the standard. (No one said this was going to be easy.)

NUMERICAL INTEGRATION

Numerical integration was a method in use in the early part of the 20th. century and it was still used through WW2 in some locations. It consists of 2 volumes of tables. They are titled, *Exterior Ballistic Tables based on Numerical Integration*. The first volume gives the elements of trajectories tabulated for conditions at the summit. Volume II gives the elements tabulated for conditions at the gun. These are large books both in basic size and in number of pages. To the reader not familiar with them, they appear to be an endless amount of pages filled with columns of numbers. They require knowledge of advanced mathematics and are seldom seen today. It is a complicated mathematical technique which works out the trajectory based on detailed information on drag coefficients and velocity for the bullet in question. They are used for time, distance and horizontal & vertical velocity component. They were devised and printed by the Ballistic Section of the Ordnance Department, U. S. Army.

TABLE OF BALLISTIC COEFFICIENTS
OF FACTORY-LOADED BULLETS

Brand: R=Remington, W=Winchester, F=Federal
Type: P=Pointed, S=Semi-Pointed, R=Round Nose, F=Flat-Nose,
SWC =Semi-Wadcutter, HP=Hollow-Point, SP=Soft-Point,
SX=Special Expanding, FJ=Full-Jacket, BT=Boat-tail, LD=Lead.

Caliber	Brand	Bullet weight	Bullet type	B.C.(C1)
.17 Rem.	R	25	P-HP	.151
.22 Hornet	R/W	45	S-SP	.13
.222 Rem.	R/W/F	50	P-SP	.175
.223 Rem.	R/W/F	55	P-SP	.201
.243 Win.	R/W	80	P-SP	.255
6mm Rem.	R/W/F	100	P-SP	.356
.250 Sav.	W	87	P-SP	.263
.257 Rob.	W	100	S-SX	.254
.257 Rob.	R/W	117	R-SP	.24
.25-'06	R/W	120	P-SP	.362
6.5 R Mag.	R	120	P-SP	.324
.264 W. Mag.	W	100	P-SP	.254
.264 W. Mag.	R/W	140	P-SP	.385
.270 Win.	R/W	100	P-SP	.251
.270 Win.	R/W/F	130	P-SX	.372
.270 Win.	R/F	150	S-SP	.261
.270 Win.	W	150	P-SX	.344
7mm Maus.	F	139	P-SP	.331
7mm Maus.	R	140	P-SP	.39
7mm Maus.	W/F	175	R-SP	.273
7mm-08	R	140	P-SP	.39
7mm Exp.	R	150	P-SP	.346
7mm R. Mag.	R/W/F	150	P-SP	.346
7mm R. Mag.	R/W/F	175	P-SP	.427
.30 Carb.	R/W/F	110	R-SP	.166
.30-30	W/F	150	F-SP	.218
.30-30	R/W/F	170	F-SP	.254
.308 Win.	W	110	P-SP	.186
.308 Win.	R/W/F	150	P-SP	.314
.308 Win.	R/W	180	S-SP	.248

Caliber	Brand	Bullet weight	Bullet type	B.C.(C1)
.308 Win.	W	200	S-SX	.345
.30-'06	R/W	125	P-SP	.268
.30-'06	R	165	P-SP	.338
.30-'06	F	165	P-BT	.47
.30-'06	F	200	P-BT	.585
.30-06	R/W	220	R-SP	.294
8mm Maus.	R/W	170	R-SP	.205
8mm Maus.	R	185	P-SP	.3
8mm R Mag.	R	220	P-SP	.366
.338 W Mag	W	200	P-SP	.308
.338 W Mag	W	225	P-SP	.435
.338 W Mag	W	250	S-SX	.329
.375 Win.	W	200	F-SP	.215
.375 H&H	R/W	270	S-SP	.326
.375 H&H	R/W	300	R-FJ	.234
.444 Mar.	R	240	F-SP	.146
.444 Mar.	R	265	F-SP	.193
.45-70	R/W	405	F-SP	.281
.458 W Mag.	R/W	500	R-FJ	.345
.221 Fireball	R	50	P-SP	.163
.32 S&W L	RW	98	R-LD	.12
.38 Spl.	R/W/F	158	R-LD	.149
.38 Spl.	R/W	158	SWC	.146
.357 Mag.	R/W	110	F-HP	.099
.357 Mag.	R/W/F	125	F-HP	.124
.357 Mag.	R/W/F	158	F-SP	.145
.41 Mag.	R/W	210	F-SP	.159
.44 Mag.	R/F	180	F-HP	.123
.44 Mag.	R	240	SWC	.143
.44 Mag.	R/F	240	F-HP	.172
.45 ACP	R/F	185	F-HP	.148
.45 ACP	R/W/F	230	R-FJ	.158
.45 Colt	R/w	255	F-LD	.142

This 2 page chart is printed with the permission of the NRA's *The American Rifleman* magazine and Mr. William. C. Davis, Jr. Both have helped with their generous permission at several places in this book and it is a pleasure to thank them once again.

CHAPTER 15

CARTRIDGE DETAILS

We have progressed a long way from the earliest firearms of the 1300's when man used a burning stick or a fiery cinder to ignite a *hand cannon*. A few of the more advanced methods of ignition, the wheel-lock and flint-lock, came later. Different types of percussion caps and primers were used in the early to mid 1800's. LeFaucheax of Paris was credited with the first self-contained cartridge with his pin-fire design in 1836.

One of the most important ideas in firearms was the addition of the priming material, powder and bullet all together in one package called a cartridge. Smith & Wesson of Springfield, Massachusetts obtained U. S. Patent No. 11,496 for this idea on Aug. 8, 1854. The idea was rather crude until S & W refined it with Patent No. 14,147 dated Jan. 22, 1856. This was known as the Volcanic cartridge and as a money maker, it was a flop. It was also a giant step down the proper road.

Beginning in November of 1857, S & W began selling their No. 1 tip-up revolvers with their new invention, rimfire cartridges. Although not granted U.S. Patent no 27,933 until April 17, 1860, for all purposes, the era of modern cartridges began in 1857. Many improvements have been made over the years, but that was the first time shooters had the bullet, powder and primer all together in a single and practical package. This invention enabled the repeater to at last become a reality.

Designed by Morse and Martin, centerfire cartridges soon followed. That invention led to the reloadable case with renewable primers. The modern cartridge and firearm is a result of a gradual development over many centuries.

A cartridge case is more important in the operation of a firearm than frequently realized. It has more to do than just be a container to hold together the powder and the projectile. It has to hold the gas under intense heat and pressure and seal the chamber so the only escape is out the bore.

This chapter will not attempt to discuss the many aspects of cartridge work that are included in hand-loading. We will remain with ballistic concerns and remind the readers interested in hand-loading to purchase one or more good reloading manuals.

RIMFIRE VS. CENTER-FIRE

Modern cartridges are divided into two groups, rimfire (R.F.) and center-fire (C.F.). Both are self-contained with a primer, projectile and the powder together in one case. The primer is ignited by an impact from the firing pin. This flash creates heat and pressure which ignites the powder charge. The main difference between the two types is the primer location.

In a rimfire cartridge, the primer is located in a ring around the outside of the base. It is not practical to reload a rimfire case and this is one of the main reasons the larger sizes disappeared from the scene. Today, most are in .22 caliber, but many fine older guns, such as Smith & Wesson revolvers and Remington rolling block rifles, were made in larger rimfire calibers. The R.F. case is inherently weaker than the C.F. type and it cannot handle as much pressure.

Center-fire cartridges are easily reloaded. The primer is located in a cupped pocket in the center of the case head. This cup is easily changed during the hand-loading process. A small hole, called a flash hole, permits the pressure to force the heat onto the powder and ignite it.

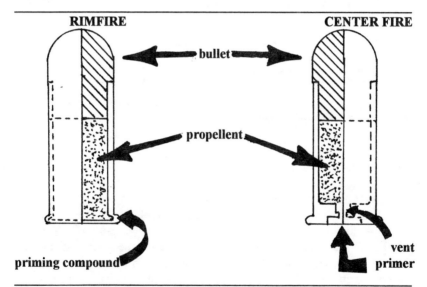

FLASH TUBE DESIGN

The charge is ignited at the back and the rapidly expanding gas frequently pushes the unburnt powder down the bore behind the bullet. The ignition is completed outside the cartridge in the barrel. In an ideal situation, all of the burning will take place inside of the case.

Engineers and experimenters have tried flash tubes since back in the early 1920's, or perhaps earlier. The idea is for the primer flash to move down the tube to the forward end and ignite the powder just behind the base of the bullet. The flash or burn will then move away from the cartridge opening, burning the powder more effectively and with less damage to the rifling and adjoining area in front of the cartridge. Also, the expanding gas does not have to move both the bullet and the powder behind the bullet.

It sounds good in theory and tests have shown an increase in velocity with a flatter trajectory. So why is it not used? It is neither cost effective or practical to manufacture. The initial cost to produce would be much higher and the cases cannot be easily reprimed and reloaded. There are also headaches involved in powder choices. Some experimenters have gone so far as to make a steel base that threads to a brass cartridge. The flash tube was threaded to the base with 48 threads to the inch. It is easy to admire the machining skill that was needed for this project, but it was not without problems. In some of the tests, the flash tube would come loose and blow out the barrel with the expanding gases.

FLASHTUBE DESIGN

CARTRIDGE DESIGN VARIABLES

The case design is an important and yet sometimes overlooked issue. It not only can influence the bullets flight but also the wear and condition of the gun. The shooter has some control over this item with his choice of gun and cartridges. Nevertheless, the average gun owner's options are limited. Hand-loaders and experimenters are not the average gun owner and they are usually intelligent and inventive. They frequently find a way to change something if they believe it can be improved.

What appears to be a straight case will often have a slight taper to ease extraction. Modern firearms have strong extractors but a slight taper is still helpful. But, a straight wall grips the chamber better as it expands under ignition pressure. The grip helps to hold the cartridge from exerting its full force on the breech or bolt face. As you can imagine, the grip involved is small and the reduction is also small; but not enough to make any difference in a well built and undamaged firearm. Many cases will compromise to help extraction and still ease the pressure toward the rear.

Bullets that are short and light weight and moved at high velocity are frequently fired from cases with long necks. Used for powder space in this design, the long necks were used in the past to deeply seat long bullets.

Frequently, a long tapered shoulder will cut into bore life. Large cases necked down can cause the same problem. Some cartridges of this type permit the ignition to push out some of the powder with the bullet. This can happen with any case type, but it is not good. This causes *combustion erosion*. The burning powder damages the starting (lead) section of the rifling and can cause erratic velocities and pressures. The life and accuracy of the barrel are shortened.

A rounded convex shoulder is used to control a slow burning powder and obtain complete burning. An unusually big powder charge is also better controlled in a cartridge with a convex shoulder. The heat is directed back more as with the sharp shoulder.

This is debatable, but many people believe that sharp shoulders do a better job of keeping the powder and ignition in the cartridge case. The heat is directed back toward the powder to improve efficiency.

Note: Hand-loaders and experimenters should use much caution when changing flash hole size or primer types. Some people believe these changes do not increase pressure, but they can. Dangerous excessive pressure can blow up a gun in the shooters face.

Modern rifle cartridges have been greatly improved over earlier versions. Long sloping shoulders are now steeper and sharper. This 25 to 35 degree shoulder helps to hold the powder in the case longer. This, in turn, improves combustion and helps to reduce brass flow which is a major cause of shortened case life. This altered angle also permits a slightly larger powder load. Although, as we discuss in this book under *expansion ratio* and *pressure*, the amount of powder can be excessive for the gun's bore size.

A series of tests were described in *The American Rifleman* during the summer of 1946 and mentioned again in 1981. Three cases were tested, all of the same capacity and all necked to take identical .22 caliber bullets. One case had a conventional body taper and a long 14 degree shoulder. Another had a very long body taper and a 35 degree shoulder. The third had a little body taper and a concave-radius shoulder.

Velocities and pressures were measured by both electronic-transducer and copper-crusher methods. **It was reported that the performance was almost identical, within the expected errors.** This very carefully controlled laboratory experiment found no difference in ballistic performance.

So is all of this deliberation on cartridge design a waste of time? Of course not. But it does show that some theories, no matter how well meaning and well founded, are not as useful as expected.

VOLUME

Cartridge volume is the primary element in projectile muzzle velocity and energy. More volume, more push. It isn't that simple as there are other points to consider, but the amount of powder available is important. The main difference between a .38 special and a .357 Magnum is about one tenth of an inch in case length and the additional volume. (Regardless of the designation, the calibers are equivalent.)

A simple way to determine volume is to weigh an empty case on the powder scales and write down the amount. Then fill the case with water up to the line where the bullets base would be, and weigh it again. The difference will be the volume as expressed in weight of water. Where the bullet base should set can be measured with a depth micrometer on the bullet and then used at the same setting to obtain water depth. Or, file a small lengthwise groove in the neck so the water can be forced out as the bullet is seated to proper position.

The weight can be converted to cubic inch capacity by dividing the weight of the water by 252.4 or if the answer is wanted in cubic centimeters, divide by 15.4.

◆ Let us assume the cartridge case capacity is known and the bullet seating is not accounted for; or in other words, the capacity of the case is figured by filling with water to the very top. We can find the weight of the water displaced by seating the bullet and subtract it from the full capacity to obtain powder space capacity with this formula.

$$W = S * D^2 * 198$$

Where: W = weight of displaced water in grains
S = seating depth in inches
D = diameter of bullet in inches

◆ If instead, the actual volume of the powder space is known, the density is numerically equal to the charge weight in grams divided by the volume in cubic centimeters. This is based on one cubic centimeter of water weighing one gram.

◆ The formula for volume of a straight cylinder is

$$v = \pi r^2 h$$

Where: v = volume
h = height
r = radius

◆ The volume of a sphere is $v = 4/3 \pi r^3$

LOADING DENSITY & BULK DENSITY

This is one of the most frequently misunderstood terms in ballistics. Probably 80% of the dictionaries and encyclopedias that deal exclusively with firearm words and phrases have given false information. Therefore, it is no wonder that hardly anyone understands it correctly. The result could be dangerous for hand-loaders, so it is not something to take lightly; although it can make a nice test. Ask any person who claims to be an expert on firearm ballistics to explain loading density. If they say it is the volume of the charge to the volume of the chamber (percentage of volume), they will be giving a common answer, but also a wrong answer.

The **loading density** refers to the ratio between the charge weight and the water weight that will fill the powder space in the case. If the case is filled with powder to the base of the bullet or has an air space, the actual loading density may be identical. Loading density is a dimensionless ratio with no concern with whether the space available is full or half full or whatever. It is the ratio of the weight of the charge to the weight of water the case could hold up to the base of the bullet. This is related to chamber pressure. The more loading density, the more efficient the powder burns. If that is confusing, keep reading. It will be explained in more detail.

Bulk density, as it relates to powder, is a ratio between the powder weight a case can hold to capacity and the water weight that could be in the same space. Notice that for **loading density**, the powder term was charge weight and for **bulk density** the term is capacity (space). A small but very essential difference.

Examples: (1) We use a powder with a bulk density of .60 and a case that would hold 100 grains of water, and fill the case full with 60 grains of that specific powder. **(2)** A powder with a bulk density of 1.0 would fill any case equal to water; that is, say, 75 grains of water or 75 grains of powder. **(3)** If a case held 62 grains of water and we used a powder with a bulk density of .80, the full capacity would be 62 * .80 = 49.6 grains of powder. The figures used in the examples are all based on no compression or vibrating techniques to settle the powder. As shown, **each example is filled to the limit but only the second illustration has a 1.0 loading density. The first example would have a loading density of .60 and the third a loading density of .80**.

The water capacity is an inverse function of case weight. Brass has a specific gravity of 8.44 and water 1.00. Dividing the specific gravity of water by the specific gravity of brass gives almost 12 % (11.84%). That is the weight of the water displaced by adding more brass to the case. In handloading, for example, if a cartridge case of one type is being substituted for another, the loading density should remain the same. If the case is smaller, the powder charge weight should be reduced by about 12% of the difference in case weights to keep a similar chamber pressure. Another way of wording it; the numerator (top number) of the ratio must be dropped in the same proportion as the denominator (bottom number).

CASE VARIATIONS

Even with what appears to be the exact same cartridge, to paraphrase, not all cases are created equal. From the same manufacturer they can vary in volume and weight. There is always a difference between military and commercial. The weight of the cases can vary by a large amount because of the brass thickness. As expected, the heavier the case the thicker the brass and the longer the case life. Military cases are heavier and thicker and last longer than commercial cases.

Because of brass expansion, case volume, not weight, will increase some after firing, even in a tight minimum size chamber. The difference will be large enough to be detected and measured by the water method.

Case capacity governs the powder amount needed with the heavier case having the smallest inside capacity and calling for less powder; all other things equal. The lighter case would be the reverse and need more powder for a large inside capacity.

If extreme accuracy is desired, hand-loaders will want to vary the powder charge to suit the case. Also notice that velocity variations caused by case volume differences will be somewhat larger with reduced loads.

POWDER COMPRESSING

It is common practice with some cartridges to compress the powder and compact it into the case when seating the bullet. This is normal and safe if done correctly and moderately. The ballistic effect on pressure, velocity and trajectory is almost zero.

CASE BULGES

It has been said the bulge in a case is caused by gravity pulling down the low side. This is totally beyond reason. The accelerating tendency of bodies toward the earth by gravity is almost nothing at 32.17 ft. per. second, per second compared to the 10,000 p.s.i. to 65,000 p.s.i. required to deform the brass in a normal case. The bulge was caused by a thin wall area where the elastic limit was exceeded and a permanent deformation was left after the pressure lowered. In plain terms, an area that was weaker than the rest of the case compared with the pressure involved.

ACCURACY

Hand-loaded ammunition can be adversely affected by bullets that are slightly tilted in the case; in other words, seated poorly. Testing shows a difference of about a 1/2 minute of angle between properly seated straight bullets and those seated at a slight angle to the case. For most shooting, this is not enough difference to be of any concern, however for long range target work it can be the difference between a win or a loss. An indicator reading of run out between the bullet (at just back of the ogive) and the case as small as .005" can open up a group.

For the utmost in long range accuracy, every variable must be eliminated. Cases should be checked for uniformity in all areas including wall thickness, powder capacity, head square with case axis, neck wall concentricity, outside diameter of case in relation to inside chamber diameter, case length, etc. etc. Any case that does not meet strict requirements, even if new, should not be used during a match. New cases should be fire-formed at reduced loads and lubricated so they can slip back in the chamber easily. This will help to prevent the customary thinning and stretching of the case. The case should then be carefully checked again.

The key word is uniformity; no variations of any kind in gun, technique, cartridges, etc. Match winning requires each shot be in or about the same as the previous shot and the one before that, and so on. It is much harder to do than read or write about.

CONCENTRICITY

Concentricity is important in cartridge performance. The problem is, as the word is used in technical jargon, it is very difficult to determine concentricity with accuracy. All round surfaces need a common center. Runout is not the same, although important. A check that uses V-blocks will never give the needed information. Instead of rotating the object on its outer surface, it must be held, even if on the outer surface, so that its true axis centerline will be the center of rotation. If held on the outer surface, it must be held in a manor that will permit shifting and adjustment; as in a 4 jaw chuck in a metal cutting lathe or an inspection fixture. In either case, the center point must be discernible. Holding between centers in a lathe or fixture is acceptable only if the centers are, in fact, on the true axis. As you can see, it is a tough technical subject that is occasionally messed up by even some of the best machine shop inspectors

BRASS CASES

Brass cases are manufactured by hammering and forcing brass material through several forming dies. This requires high pressure and generates heat. Not only is the case forced into the proper shape and size, but the *work hardening* tightens the grain structure and hardens the brass in the critical head area.

Odd and rare sizes of brass cartridges can be made from bar stock by turning and boring on a metal cutting lathe. The biggest problem is that even when the right raw material is used, the hardness is not correct. Usually some annealing will be required to prevent cracks near the mouth and sometimes at the shoulder. The head hardness will also require some careful work.

The bad news is brass is generally harder to anneal and harden with accuracy, when compared to steel. The good news is custom cartridges of this type are usually made for rare old guns that use black powder or equivalent low pressure, so the stress problems are not as critical.

ALUMINUM CASES

Aluminum is used in some cartridge cases, but it isn't a new idea. It was first tried in about 1895 for the .30-40 Krag rifle. Aluminum cases are weak and tend to split so they are not reloadable.

CHAPTER 16

BULLET DESIGN & PERFORMANCE

The importance of the bullet to the performance of a firearm cannot be emphasized enough. After all, the sole purpose of a gun is to propel the bullet toward the target. The design of the bullet has an effect on performance from the moment of ignition, through the trajectory and onto the target itself. At that point, it must perform by either punching a hole in a piece of paper or by proper penetration and expansion in an animal.

Bullets come in many different styles shapes and materials. Some are solid lead. Many are assemblies with a lead or steel core and a covering called a jacket. The jacket may be either gilding metal, gilding-metal clad steel, or copper plated steel. Some military caliber .30 and 7.62 mm frangible bullets are molded of powdered lead and a friable plastic which pulverizes into dust upon impact with the target. Ball cartridges are military in origin and consist of a general purpose combat cartridge for use against personnel and unarmored targets. The bullet normally consists of a metal jacket and a lead slug. The .50 caliber ball bullet and 7.62 mm, Ball M59 bullet contain soft steel cores.

Additional bullet information is scattered throughout this book in other areas where the material applies. The three chapters near the end on terminal ballistics have information about penetration and expansion. (Chapters 24-25-26) The chapters on gyroscopic stability, rifling, ballistic coefficients, wind drift, ricochet and brush deflection all discuss bullets in regard to their subject matter.

LEAD BULLETS:

The use of home made cast bullets by hand-loaders has been well researched and tested. The results can be excellent from all ballistic points if the hand-loader's guides and factory literature are followed closely. Care must be taken because a pit or mark on the bullet (the ogive area in particular) can change the trajectory and cause poor gyroscopic stability. A change in seating depth, sizing and design; all play an important role.

Hand-loading is not covered in this book, but by following the hand-loading rules in other books and the use of much care, the ballistics of cast bullets can be very good. All the rules of ballistics that apply to factory made projectiles also apply to hand cast bullets.

Hard cast lead bullets (antimony added for hardness) can be fired in firearms without major leading problems Some shooters claim more leading problems at lower velocities of about 1,000 to 1,100 f.p.s. While heavy leading reduces accuracy, it can also increase pressure.

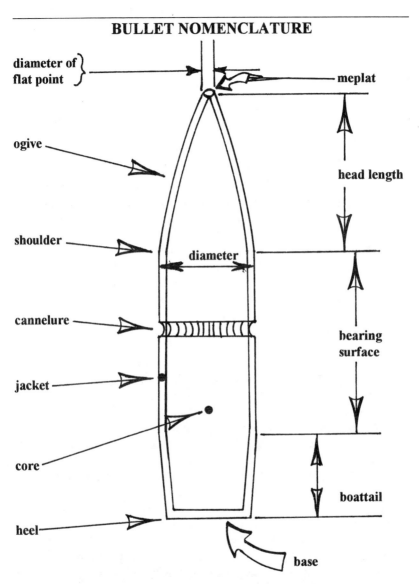

THE LENGTH OF THE HEAD DIVIDED BY THE BULLET DIAMETER IS THE RATIO OF THE HEAD. THIS IS MORE MODERN THAN "RADIUS OF THE OGIVE." THE OGIVE CAN BE STRAIGHT, PARABOLIC OR CIRCULAR AND THE DIFFERENCE IN DRAG WILL BE VERY SMALL.

One advantage to ordinary lead bullets is the chance of reduced bore wear in old rifles. Many older barrels are made of steel that is not as good or as hard as the ones found on most weapons made since around the 1920's up to the present. (Note Chapter 7 on Metallurgy.) There are exceptions, but generally, modern barrels are better suited for jacketed bullets than are older barrels. But at the low velocities, the wear from modern jacketed bullets in even a poor barrel, so long as safety is not compromised, should not be excessive. Fouling is another story. Also, lead may pick up the rifling better and as a result, stabilize better for increased accuracy. Of course, it could be too soft and rip through the rifling without picking it up at all. You may notice that this paragraph demonstrates what is written many times in this book. If you have an old gun and are wondering about this, test it for yourself. There are so many variables that no one answer will work for all situations.

Lead bullets, in most testing, have came up slightly short on velocity when compared to jacketed bullets; not much; usually in the range of about 2 percent. One of the problems with a test of this type is using ammo that is exactly the same except for the bullet jacket and firing enough shots for the answer to be meaningful. This information is not as accurate as desirable, but until more conclusive data becomes available, the 2 percent difference will have to do.

VARMINT BULLETS:

Varmint hunting is a specialized skill that requires bullets of a very special type. The animals hunted are small and usually far away, which demands a high level of accuracy. The hunter also must remember that only a hit in their small vital area will give a one shot kill. The lung, heart and head of a varmint at 300 yards is a challenging target.

Also, these small animals have a thin skin and fragile bones. Very little resistance is offered by their anatomy and a normal bullet will not expand properly, if at all. Special varmint bullets are available that have thin jackets with exposed soft lead and sometimes a large hollow point. Readers that have studied this book will quickly realize their accuracy will not be as good as some other designs, match bullets in particular. Also, it is important to keep velocity in mind. A varmint bullet that is designed to expand in soft tissue at 2,400 f.p.s., for example, may not do well as the velocity deteriorates down range. Other bullets are designed to expand nicely at slower velocities, but there is a catch. (Isn't there always?) Many of the bullets that expand well in soft tissue at moderate velocity are super sensitive, as they have to be. They can come apart in the barrel or in flight between the muzzle and the game. Have you ever wondered why the bullet that should have been a direct hit seemed to just disappear?

The bullet manufacturers have some excellent products on the market. Study their literature and experiment with several that seem suitable for your situation. With patience, you will find one that is perfect.

With a bullet of this design, some accuracy will be sacrificed for expansion and killing power. Which is most important and how far to go in either direction at the expense of the other? Fortunately, the choice is available. Each hunter will have to decide for himself. (Chapters 24 & 25 on terminal ballistics will be helpful.)

CONICAL BULLETS FOR MUZZLE LOADERS

A few people that shoot muzzle loaders have the false idea that conical bullets are a modern invention and that patched round balls were better both ballistically and at killing game. Any reader of this book will realize that neither is correct.

Conical bullets were in use before cartridges and date back to the late 1820's or early 1830's. By the late 1840's, Captain C. E. Minie of the French Army had invented the conical bullet which bears his name. It not only had better ballistic performance but was quick to load. It also required about a third faster twist in the rifling to fly properly. But most conical bullets were used by the military instead of hunters. Then, the same as today, money for new weapons was easier to obtain by the military than by the average hunter.

Conical bullets are heavier than a patched ball in the same caliber. Therefore, they can maintain better velocity and energy for a longer range.

TRACER BULLETS-MILITARY

Tracer bullets are occasionally sold as military surplus and frequently owned and fired by civilians. Their purpose is to leave a trail of smoke or flame to permit a visible observation of the bullet's in-flight trajectory and the point of impact. It is used primarily to observe the line of fire, but it can also be used to ignite flammable materials and for signaling purposes. The tracer element consists of a compressed, flammable, pyrotechnic composition (a chemical filler) in the base of the bullet. This composition is ignited by the propellant when the cartridge is fired. In flight, the bullet emits a bright visible flame. Burnout occurs at a range between 400 and 1,600 yards, depending on the caliber and elevation.

Tracers should not be used for any purpose other than as intended. They can start fires and for that reason, they are against the law in many places. Bore damage is very common and the ballistic performance is poor. The weight of the bullet and the center of gravity change as the filler is consumed by the flame. Gyroscopic stability is hard to maintain as a result. While they are very dramatic, especially at night, they are dangerous and should be avoided.

ARMOR-PIERCING-MILITARY

The armor-piercing cartridge is military and intended for use in machine-guns or rifles against personnel and light armored and unarmored targets, concrete shelters, and similar bullet-resisting targets. The bullet consists of a metal jacket and a hardened steel-alloy core. In addition, it may have a base filler and/or a point filler of lead. The military also have one called an armor-piercing-incendiary that has an incendiary mixture as a point filler. Upon impact with the target, the incendiary mixture bursts into flame and ignites any flammable material.

OTHER MILITARY

There are many other types of military cartridges and bullets which are too unusual to take up space here. They include the armor-piercing-incendiary tracer, the duplex, high-pressure test, dummy & dummy inert-load, line throwing cartridges and various training and match bullets and cartridges.

BOATTAIL BASE

The base of a boattail bullet is tapered inward as the sketch shows. It is also called *taper heel* because heel is a word used for the edge where the side and base join. As explained in more detail in Chapter 13, there is a partial vacuum behind a bullet. This partial vacuum creates a *base drag* that helps to slow the velocity. The vacuum is less behind a boattail bullet than a blunt end. If the slope is to sharp or sudden (not long enough with a steep angle), the air flow behind the bullet will be turbulent and separate. Then the advantage of the boattail will be lost.

The base drag effect is strongest at subsonic speeds. At supersonic speeds there will be a shock wave created at the boattail's shoulder which, combined with a low pressure expansion, will kill some of the good effects of a boattail base. Above Mach 1.2, the advantage will be small. There will be a slight lowering of drag from side friction due to the shorter bearing surface but little from the drag at the rear. A boattail is fine at supersonic speeds, but it must be kept small. It is best at a boattail base diameter of 0.4 of the caliber, or less.

For short range shooting, up to 200 yards, the boattail benefit is not very noticeable. As the range increases, the benefit increases. At 1000 yards, the velocity difference can be improved by as much as 20% with up to a 3 or 4 foot advantage in wind deflection in a modest 10 m.p.h. cross wind.

Even at supersonic velocities, the drag reduction is noticeable, although there has been some opinions expressed that the benefit is near zero. A comparison of two bullets that are almost the same except that one has a boattail base and the other a flat-base can easily prove the point.

For evidence we can use the .55 grain .224 dia. bullet with a 7 1/2 degree boattail length of .134". The flat-base is a 5.56 mm M193 military ball. Both have the same shape, point configuration and weight and a muzzle velocity of 3,000 f.p.s.

Range in yards	Velocity in f.p.s.	
	Boattail	Flat-base
0	3,000	3,000
100	2,641	2,633
200	2,308	2,267
300	1,999	1,921
400	1,702	1,659
500	1,441	1,309
1,000	857	715

BOATTAIL BULLET DESIGN

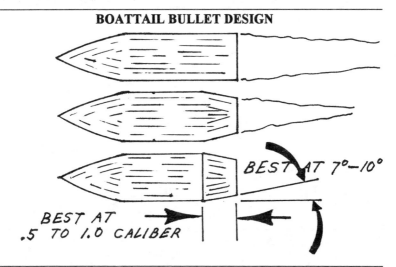

So far, we have been discussing rifle bullets. As mentioned, at ranges below 200 yards, the benefit is small. Handguns, even with scopes and with the best shooters, are not used at ranges long enough to show much advantage. For example, over a 50 yard range, the .45 ACP will take .0029 seconds longer in air than in a vacuum. Cutting the drag by 20 percent would make little difference. The 810 f.p.s. that it started with will still be about 782 f.p.s. Other bullets will create other statistics, but the idea will be the same.

Boattail bullets do not wear a barrel more than any other design. That statement has been around for years and has never been true. It is based on the theory that hot gas can enter the area of the boattail and damage the bore. It is sometimes called the *jet effect* and it is supposed to wash away the surface of the bore and damage the rifling grooves. For years it was believed that any bullet that is undersized in relationship to the bore diameter will have that effect. With the extreme acceleration of the bullet, the gas is not there long enough to affect that area more than another area. If it did have an adverse effect, the steel and heat treating should be suspected of being poor quality. (**Note**: this is not the same as the results from powder burning in the bore as discussed at the top of page 177.)

OGIVE DESIGN THEORY:

The ogive is the curve between the bearing surface and the point. The radius of the curve is usually given in respect to the caliber which ties the picture together in the proper relationship. The ogival radius has significance only as it relates to the bullet diameter. The center point of the radius also is important as it controls the length and helps define the shape. If the bullet has a tangent ogive, the center point of the radius is at right angles to the cylindrical bearing surface and opposite the intersection between that surface and the curved ogive.

The secant ogive has the center point of the radius lower and opposite a point on the cylindrical bearing surface. This gives a good ballistic coefficient for aerodynamic efficiency and usually retains a decent bearing area. (See formula for ogive length on bottom of page 195)

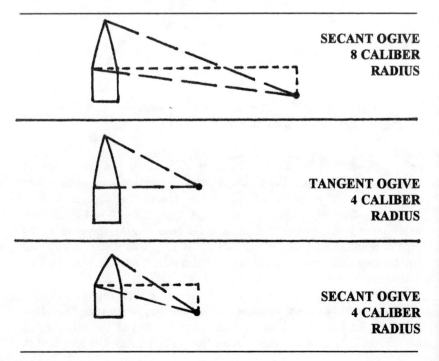

SECANT OGIVE
8 CALIBER
RADIUS

TANGENT OGIVE
4 CALIBER
RADIUS

SECANT OGIVE
4 CALIBER
RADIUS

MEPLAT:
The point of a bullet, the meplat, is generally a small rounded tip. Even a long nosed spitzer shape will have a meplat with a diameter of up to 1/6 caliber, never larger. The drag will be a slight bit less than a sharp point and they are easier to make and handle. At first, it may appear a sharp point would have less drag, but it has to do with air flow, pressure and shock waves. This is discussed in Chapter 13.

BEARING AREA:
The main body or bearing area that rides on the bore must be long enough to hold the bullet in proper alignment. An excessively short bearing area will create unreliable performance. The bullet can yaw while in the bore and exit the muzzle in a yawed or other than point on position. This is not as unusual as it may appear. (Remember the tremendous force that is pressed against the base in milliseconds. This may or may not be perfectly equal across the entire base.) A long bearing area will also give an accuracy boost when shooting from a worn barrel and help to seal the gas pressure for better ballistics. Too short an area can also be stripped by the rifling when the bullet is suddenly forced to turn with the grooves. The length can be over done, though, and increase frictional drag in the bore.

CONICAL   CYLINDRO-OGIVAL   CYLINDRO-CONCOIDAL
NOTE: The conical has no bearing surface and hence, no stability.

SPITZER SHAPE
A spitzer bullet has a nose that is pointed and followed by an ogive radius of at least 7 or more calibers. In other words, some length behind the tip before the main body. It is obvious that the streamlining effect will give it a better ballistic efficiency. The primary advantage is in the aerodynamic shape which reduces air resistance. It helps the velocity hold up well at long range.

Early spitzer designs were poor at expansion. The choice was usually a blunt nose with a large amount of exposed lead. Modern design has changed the spitzer so its benefits can be used along with reliable expansion. Also, there is no evidence that this design is deflected more than other types in brush. (This controversial subject is covered in Chapter 10.)

A study of the ballistic tables will show that with the .30-'06, for example, two 180 grain bullets, one a spitzer and the other a short round-nose, both leave the muzzle at the same velocity and just a few foot pounds difference in energy. At 300 yards, the spitzer has an advantage of 300 f.p.s. in velocity and 475 foot pounds of energy. An impressive difference. The addition of a boattail base will change the figures and all manufactures have products that are similar yet different. Nevertheless, the spitzer, or a design that leans toward that direction, has good merit.

There is a controversy over which is the more accurate, a sharp-pointed bullet or a blunt-point. Most testing gives an edge to the sharp point. If everything else is equal and/or correct; that is twist rate, gyroscopic stability, etc., then theoretically, there would be little or no difference. Of course, that is impossible, so most people prefer the sharp point.

TIP DAMAGE:

Testing has shown that at ranges below 100 yards, minor damage to the tip of the bullet has little effect on performance. At ranges over 100 yards or for critical target work, performance is bound to suffer. For hunting at short range, it doesn't appear to cause any problems in trajectory or velocity.

The Norma factory in Sweden damaged bullet tips by filing the point flat. Different amounts of material were removed at an angle from the tip and ogive. Also, some were filed at an angle on the boattail base. None of these bullets performed badly in controlled tests at 100 yards. The damage was minor and in no case excessive, but it was enough that the results were surprising.

Long range testing by other laboratories have shown keyholing, flyers, and other expected results of instability. But for short ranges, up to about 100 yards, the performance should be satisfactory for hunting. Naturally, contest target work would require the best and no chance should be taken. The same for ranges beyond about 150 yards.

VELOCITY & WEIGHT

In the constant quest for higher velocities, it is well known that a large bore gives a large cross sectional area for the expanding gas to push against. That bullet weight also tends to go up in the larger bores defeats the strategy. A bullet could be made with a large cross section and light weight for excellent high muzzle velocity. The bullet would also have terrible problems in trajectory and energy. The internal ballistics would be good and the external ballistics would be bad. As usual, a compromise is needed.

As the velocity increases, the efficiency of delivering energy goes down. Light weight bullets require more velocity to deliver the same energy to the target, all other things being equal. The velocity to deliver a light bullet at the energy of a heavy bullet is beyond our present technology. There is a strong place for light fast bullets, but most hunters prefer a slow heavier bullet when a lot of stopping power is needed.

Mushroom effect is better at the higher speeds because of the energy involved. Lower speed bullets of proper design can also expand well, and usually after deeper penetration. The energy delivered by a fast velocity light bullet can be high, but the actual stopping effect may be small. Large game such as elk and bear may not be impressed by it. A heavy slower bullet with good controlled expansion may do better.

◆ As mentioned earlier, Newton's 2nd. law says that force is equal to mass times acceleration.

$F = W / g * a$

Where: F = force
w = weight of bullet
g = acceleration of gravity at 32.17

◆ And force is pressure times area.

$F = P * A$

Where: P = pressure
A = area

◆ We can put the two equations together and show the connection between acceleration and pressure.

$a = g P A / w$

Most bullet designs perform best in a narrow velocity range. Too fast or too slow and penetration and expansion will not be as expected. This is a problem for long range work where velocity will decay at the longer distances. Hand-loaders can also experience problems at short to normal range if the bullet choice does not match the expected velocity.

SYMMETRY:

Symmetry or the act of being symmetrical is when an object, a bullet in this case, can be divided into similar or equal parts by a plane passing through the center. Symmetry is vital for proper gyroscopic stability. The center of the mass needs to be on the bullet's axis or the result will be a helical flight path. Sometimes, blunt and round points stabilize better. Not because a blunt point or round point is more stabile or that the long pointed bullet requires more rifling twist to stabilize. It is simply hard to make long pointed bullets that are perfectly symmetrical and with a jacket that is precisely the same thickness at all points. (See Chapters 8 & 9 for more details.)

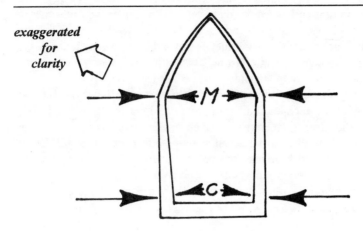

Jacket walls are thicker at the base and get thinner toward the nose.

The Jacket wall at (**M**) is usually (average) about .024" thick, but can be from .010" to .037". At the base (**C**) it will be about .004" thicker, although they may vary from -.003" to +.015", depending on the make and the type.

EXPERIMENTS IN BULLET DESIGN:

Many unsuccessful ideas in bullet design have been tried over the years. Experimenters always repeat some of them because they were not aware of the earlier failures. There are also old ideas where someone solves the problems so they will work. Most of the following ideas had a good solid theory behind them, but they were doomed to failure for one reason or another.

◆ Streamlining the rear of a bullet with a long taper to a point does not serve a useful purpose. At the muzzle, the hot gas tends to turn or upset the bullet and cause poor stability and accuracy. The Army tried this type with a sabot to hold in the gas and help with alignment in the bore. It was not successful. (A sabot is described and pictured in Chapter 21.)

◆ There have been frequent attempts to design and make a successful bullet that has the powder and primer included but without a case. No *Rocket Ball*, as they have been called, has worked very well. Most have been weak and with a tendency to misfire.

The early Voss system was patented in Europe in 1834. Several other European inventors patented ideas along this line in the 1840's and 1850's. The first U.S. Patent was by Smith & Wesson in 1856 for use in the

Volcanic line of firearms. Some people call the items *Volcanic* instead of *Rocket Ball*. Either name is acceptable.

The idea died out for awhile and was briefly revived again by the Germans during WW 2 with a 7.92 mm with no success. The U.S. Military experimented during WW 2 and also more recently with 5.56 mm and 7.62 mm caseless cartridges.

Smith & Wesson has tried again with a 9mm design that is electrically fired. Extensive testing in the mid 1960's showed much promise.

If you haven't guessed by now, the expense of the case is a high percentage of the cost of a cartridge. This is a major reason why it would be nice to do without it.

◆ An unusual bullet design of the early 1900's carried bore lubrication to the extreme. A self lubricated bullet was designed by Smith & Wesson and listed by Union Metallic Cartridge Co.

The center of the bullet was hollow. The base of the cavity had a lead plug. The nose end had four small openings into the ogive area. The center was filled with a lubricant which was forced out the holes as the expanding gas thrust the base plug up into the cavity. The bore was lubricated by this action, or supposedly it was. This unusual bullet was not on the market for long.

◆ In the early 1960's, Malter Arms Co. of New York City made sub-caliber bullets inside a discarding Husk. The Husk, as it was called, was the same as a sabot. They were made in both .270 and .308 diameter. That is, the sabot was either .270 or .308 and made for that size bore. The bullets were smaller at .224 in 82 grain and .243 in 100 grain. They were advertised for a short time and then they also disappeared.

◆ It is frequently suggested that a bullet should be tubular (an open hollow center from the front to the rear) so the air resistance and the void at the rear can be reduced. Various types of plugs or caps at the rear are used to hold the pressure behind the bullet and seal in the thrust while in the barrel. The caps or plugs then drop off like a sabot as the bullet passes the muzzle.

In 1893, the Frankford Arsenal conducted extensive tests on bullets of this design. The bullet was unsuccessful for two main reasons; both ballistic in nature. First, remember that weight is an important part of bullet performance and the tubular design reduces the weight too much. Also, the flow of pressure waves build up inside the tube and block or plug it. The final effect is the same as a solid bullet only of less weight and without a good streamlined nose. (See section on high speed airflow, Chapter 13.)

An interesting side note is the name of one of the officers involved in the test calculations; J. M. Ingalls, who later achieved fame with the

Ingalls Tables. They form the basis for many calculations in ballistics and are mentioned in other places in this book.

Another interesting point on the same subject: Since the very extensive testing on the tubular design by the U.S. Army, many other experimenters and companies have done more experiments on the same idea. A series of very complete tests was made just before the start of WW 2 and then stopped by both the war and a lack of success. Possibly these later engineers were not aware of the 1893 tests.

◆ In the early 1900's, Hoxie Ammunition Co. of Chicago made bullets with an unusual feature. The hollow point of the jacketed bullet had a ball fastened in the opening. The opening tapered down smaller behind the ball so that at impact, the ball would ram back into the opening and force expansion.

While the basic idea had merit, the product was produced for just a short time. Today this type of ammunition would be illegal because it was too good at what it was designed to do.

◆ The use of darts or very small arrows as projectiles in guns is as old as the earliest firearms. It was a reasonable first step and they were used for many years, until the introduction of round balls or at the start, stones.

Our language is not all English, but a mixture of many. American English has borrowed words from the American Indian, or Native American if you prefer, (raccoon, opossum and totem, to name just three). Also from Dutch, Spanish, German and French. In a smaller amount from Yiddish, Chinese and Italian. The french word for arrow is *flèche* and in American English we call darts and small arrows *flechettes*.

The use of flechettes as projectiles in guns has not completely died. Occasionally people still experiment with this idea because of the aerodynamic and ballistic advantages. They have excellent ballistic characteristics in the air but are difficult to launch.

NEW IDEAS
◆ While most of the above experiments in bullet design have not been very successful, some new ideas have worked great. Noseler Inc. manufactures a hunting bullet with a partition extending through the core about 1/3rd. the length forward from the base. At impact, the forward portion expands normally while the rear area remains solid.

They also make a bullet with a plastic nose insert which, because of its unique design, has both a high ballistic coefficient and sudden and violent expansion. This gives excellent expansion when the velocity has deteriorated down range. Of course as expected, rapid expansion reduces penetration depth.

Other manufactures are also making new and better products as the understanding of bullet technology increases.

♦ Bullets with grooves crimped around their body have lower accuracy because the bullet is deformed in an area that disrupts the flow of air and disturbs the aerodynamic stability. Even if the groove is perfectly even and uniform around its full length, which it rarely will be, it will still lower accuracy. It will also reduce the life of the case from over working the metal. This can cause splits at the mouth that are unnecessary. Factory ammo that is crimped is primarily for use in tubular magazines or by companies making a general purpose cartridge for any type rifle.

NOTES ON BULLETS:

♦ A blunt nose bullet should not be used for any distance in a wind, if it can be avoided. The wind deflection will be very poor because of its aerodynamic shape, or lack of it.

♦ For accuracy, many experts consider the base of a bullet more crucial than the front. This is controversial and unnecessary. For accuracy, all aspects of bullet design are significant and which is more so is not important.

♦ A bullet with a long point will hold its velocity and energy over a greater distance.

♦ Separation of the core and jacket is a weakness of some bullet designs and high velocity can be a factor.

♦ A soft point will break up quicker and not be as likely to ricochet. (The pieces may ricochet, but not as a solid piece.)

♦ A very sharp point is not as good (or as easy to handle and protect) as a bullet with a small flat area called a meplat.

♦ Hand-loaded ammo that is carefully and properly loaded will have better accuracy than factory ammunition. Always. (This is not an insult to manufacturers, but all production work has to have tolerances. Hand-loaders can strive for perfection.)

♦ Bullets keyhole at long range from poor stability. One major cause is when the bullet is too heavy for the barrel twist. A lighter bullet (shorter) or a faster twist is required to stabilize the projectile. (See Chapters 8-9-12.)

♦ Controlled expansion ammunition has been used for years for self defense and hunting. A few uninformed people have began to attack them as being offensive and disgusting. Most hunters, police officers and self-defense people want a bullet that stops or kills either instantly or as quickly as possible. After all, that is their one and only purpose.

♦ Ogive length is equal to the **diameter of the bullet multiplied by 1/2 the square root of 4 N minus 1**, where N is the caliber radius. Remember to subtract 1 from the product of 4 times N before finding the square root, then divide by 2 and multiply by the diameter of the bullet. This calculates to a sharp needle point but there will probably be a small flat or rounded section.

CHAPTER 17

TRAJECTORY

Hunters frequently miss a once in a lifetime trophy because of poor sight adjustment, little knowledge of drop, temperature effects, up and down hill slope shooting and perhaps a dozen other problems. All are easy to learn if they understand a few important facts. Target shooters, while they always know the basics, need to know the advanced points as well. **As in all things, knowledge is power.**

The phenomena of a trajectory can only be considered properly by recognizing that a projectile in flight is a rigid body given rapid rotation and undergoing precession and nutation as it travels a path of double curvature. The motion is determined by the continuous action of various forces, including the resistance of the air and gravitation. These last two sentences will not be confusing if the earlier chapters have been read. All of those terms have been thoroughly explained.

The old-fashioned spherical projectile presented a much simpler problem. When the accidental rotation was small, such problems as drift could not occur and yaw was impossible. Dispersion with spherical projectiles is due chiefly to irregularities in jump of the gun and in muzzle velocity. (It is assumed that wind and atmospheric conditions are unvaried during firing. Rare, of course, but a necessary statement, nevertheless.)

With modern elongated projectiles, the angular variation of the axis to air resistance is a significant factor. Even the permanent dynamics of projectiles, such as the position of the center of mass and the longitudinal and transverse moments of inertia, cannot be ignored.

BASICS

This is a frequently misunderstood part of firearm shooting, so it would be best to begin at a basic point and then work our way up to the top. It is important to recognize that while a rifle will be used as an example, handguns and shotguns follow the same rules. There are a few minor differences which will be discussed in their own chapters.

The curved path of a projectile is called a **trajectory**. That information is basic and everyone knows it, but a trajectory is infinitely more complex than most people realize.

Every rifle will hit the intended place on the target or game at only two ranges or distances. For now, let us disregard the close one and discuss the one farther down range. What is that? You say you didn't know there were two? Good! Then you are certain to learn something.

The line of *sight* is the straight line from the eye of the shooter to the target. It goes over the sights, or through them, and is the path the shooter may *think* the projectile is taking. The bullet travels upward above the line of *sight* and then drops back to the line of *sight* at the target or impact point. A line projected through the bore of the gun is the line of *site* or bore line and the bullet will begin to drop below this line immediately after leaving the muzzle. **Note #1:** A casual reader may not have noticed the spelling difference between *sight* and *site*. **Note #2:** *Site* is defined by *Funk & Wagnall's Britannia Dictionary* as the degree of inclination from the horizontal of a line joining the target and the muzzle of a gun. Also called angle of site.

Keep in mind that if we buy a new gun, or perhaps a used gun that is new to us, we don't know where the bullet will strike in relationship to the sights. Later, we will explain how to determine this point and adjust it. For now, let us assume we have a rifle that is zeroed-in to hit properly at 200 yards. Of course, the game we want to shoot is never at exactly 200 yards and few of us could accurately estimate the range anyway. A good understanding of a bullets flight path (trajectory) can help us to choose the best cartridge for the gun and then the best range to set the sights.

Air resistance and gravity are studied in detail elsewhere in this book, but for now, we will keep it simple and just say that they cause the bullet to follow a curved path downward. The pull by gravity is constant and it causes the projectile to start to fall the instant the bullet leaves the muzzle. This gives the fall a constant acceleration. It will not be in a vacuum, as we studied in school, so all projectiles will not fall at the same rate. The fall is not the same because of different densities of the projectiles, different air densities and the varying air resistance which will slow the velocity. Don't believe that because the bullet is spinning from the rifling grooves in the barrel, that it can defy gravity; it cannot.

As can easily be understood, time and the length of flight is important. The longer a projectile is in the air, the longer these details, and others, have to bring about their change.

The trajectory is always curved but all cartridges, calibers and guns are different. Some may have a huge drop and others may have, by comparison, a slight drop. Either through ballistic charts or testing, we can know where the bullet will strike at different ranges. We also need to have the gun sighted in (set to hit) at a certain pre-determined range. In other words, we have to know where in that drop it will be at the range we are using. Also, and almost as important, we have to know at what range the shot we are planning will be. If we know the sights are set to hit at 200 yards, but we have no idea how far away the game is, we may be in trouble. Then again, if we have made a wise choice in cartridges and have sighted in at the best range, perhaps we can be off in our range estimation by 50% and still be successful. Let's see how this can work out.

To hit an object at a distance, we have to point the barrel above that point and let the bullet drop down to it. The *pointing above* is normally done by adjusting the sights properly rather than actually aiming at the top of a target when what we want to hit is the middle. (This was common in earlier times, but as expected, it takes experience and is very difficult.) Aiming above or below is still done in hunting, and it still takes experience. A knowledge of the trajectory for the cartridge you are using will help.

As strange as it may seem, if there were no gravity, not only would the projectile not curve toward the earth, it would stay in a straight path while being slowed by aerodynamic drag. Theoretically, it would just come to a halt on a straight line from the bore. Without air resistance it would continue on in a straight line until ???? (Disregarding the minor influence of yaw, nutation, etc.)

TRAJECTORY DESCRIPTION:

A trajectory is not a straight linear line. Gravity prevents that. Neither is it a true parabolic curve, as some people like to describe it; although it would be in a vacuum with the absence of the drag caused by air resistance. A true parabolic is made by a point that moves in such a way that its distance from a fixed line is always equal to its distance from a fixed point. The projectile makes a steeper curve on descending in the atmosphere than on ascending since the horizontal resistance has affected it longer. The midrange trajectory height is not at 50% of the down-range distance but at about 55%. If it is to be called a parabolic curve, it is a *theoretical* parabolic. The projectile will descend in a steeper path than it took on rising because the horizontal resistance has been in effect longer. (As we just said with different wording.) **Correctly, it is an unsymmetrical line called a ballistic curve.** Still, a parabolic curve is the term normally used to portray the flight path.

TIME:

The time the projectile is in flight is directly related to the amount of drop. A high velocity bullet at 3,000 f.p.s. that takes 2 seconds to reach its target would drop 44.4 feet, as may a slower 1,500 f.p.s. bullet. Only the faster bullet would travel about (not exactly) twice as far in the same 2 seconds. If both were fired at the same target, the faster bullet would reach it quicker and less flight time means less drop.

In a vacuum, where the trajectory would be a true parabolic curve, the mathematics are simple. The vertical drop from the line of bore would vary exactly as the square of the time of flight. For example, consider a one second flight with a 12 foot drop. At double the range and 2 seconds, the drop would be 48 feet. (Remember, we are discussing a theoretical vacuum.)

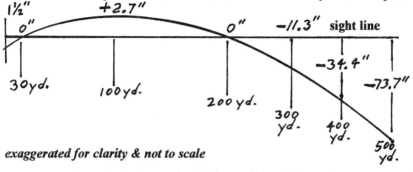

NO ELEVATION

WITH ELEVATION

exaggerated for clarity & not to scale

NOTE # 1: The projectile never goes above the bore line. The drop begins at the muzzle.

NOTE # 2: The projectile does not rise or climb; it cannot. It may appear so in the drawings, but it is only in relation to the sights and not the bore.

It is not likely that anyone would sight and fire a weapon upside-down, but a study of an inverted trajectory may help with an understanding of sights and trajectories.

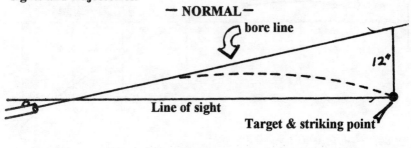

exaggerated for clarity

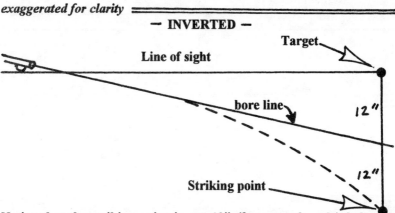

Notice that the striking point is not 12" (for example only) below the target, but 24". Instead of the bore line being 12" above the target, so the bullet can drop toward it, it is now pointed 12" below and the bullet will drop an additional 12".

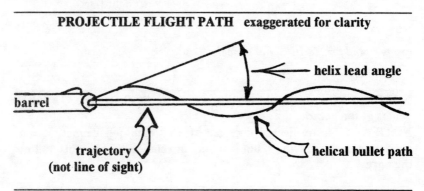

DROP:
It is always good to know how much a projectile will drop below the line of departure at any given range. It is basic and essential to a trajectory.

The constant acceleration rate of gravity is generally considered to be 32.17 feet per second. As explained in more detail elsewhere in this book, the bullet will not fall 32.17 feet in the first second as it starts at 0 and has to accelerate, but it will increase in drop speed at 32.17 feet for each second. Regardless of the drop in the first second, at the start of the next second, the rate of fall will be 32.17 f.p.s. This will increase by 32.17 to 64.34 by the end of the next second. Normally, 16.1 is used for the first second and 48.27 f.p.s. for an average velocity rate. The fall rate would be 64.34 f.p.s. at the end of the next sec. And so on for each second.

The amount of drop can be computed easily by the use of a formula. "When the acceleration is constant, the distance dropped in any amount of seconds is equal to one-half the acceleration times the square of the number of seconds." One-half the acceleration is 16.1. If we use 2 seconds for an example 2 squared (2 * 2) is 4 and 16.1 * 4 = 44.4 . Of course, the acceleration spoken of is vertical drop. The bullets horizontal flight path is not accelerating but decelerating.

◆ Another way of expressing the same formula:

Drop = 1 / 2 $g\ T^2$

Where: T = time of flight

G = gravity at 32.17 (This makes the distance of fall $16.085T^2$)

◆ This formula is basic and gives only the drop in free fall. This will be close enough for short ranges only. For a more accurate drop figure based on the bullet passing through air resistance substitute variable f for constant g. To find the value for f :

$f = g\ [1 - b\ (V - v / V)\ ]$

Where: V = muzzle velocity
  v = velocity at range
  b = .4 for velocities down to the first 2/3 rds. of the muzzle velocity.
  g = gravity at 32.17

This formula is widely quoted and used frequently. It is credited to *The British Textbook of Small Arms* of 1929. **Remember that the age of data is unimportant as long as it is still correct. Facts do not become out of date with age as will an automobile. They can be replaced by new research but not by age alone.**

Drop figures are important in setting sight elevation and it is given in many ammunition manufacture's ballistic tables. Many people today will use a computer to obtain the figures.

ELEVATION ANGLE:

◆ Trigonometry will tell us the angle of elevation with drop in feet and range also in feet. (Not yards because both have to be the same.)

$\text{Tan } \theta = \text{drop} / \text{range}$

The Greek letter theta, or any other for that matter, is normal practice in trigonometry and nothing to get upset over. Check the chapter on mathematics. With a scientific calculator it is just a matter of pushing a button or two. This is just one of the basic formulas of the six functions of an acute angle of a right triangle. It is an adaption of $\text{tan } \theta = $ **opposite side / adjacent side.**

AIR DENSITY & KINEMATIC VISCOSITY:

Some hunters and gun writers say that altitude and temperature have no effect. This is wrong. Others say the effect is minimal or small. This may be, but it will depend on what type of shooting is being done and what accuracy level is acceptable. At 10,000 feet elevation, where some hunting is done, the density ratio is .738 compared to sea level and the pressure ratio drops .31. It is also interesting to notice that if the temperature decrease were standard, the speed of sound would drop 23.1 knots compared to sea level.

The atmosphere which envelops the earth has a recalcitrant effect on ballistics. Kinematic viscosity and air density cause a slight variation in both velocity and the rate of projectile drop caused by gravity. Altitude affects density as the air is thinner at higher altitudes than at sea level and will offer less resistance. Altitude will affect the drop caused by gravity and the velocity loss. The decrease in density with increases in altitude will continue until the air and density disappear completely. (At 100,000 feet, the density ratio is only 0.0132)

Air density also drops during periods of low barometric pressure, normally associated with poor weather and rain.

Air density, to be technical, is the weight of one cubic foot of air. The standard is at 59 degree F. at sea level where the weight or density is .0765 lb. per cubic foot.

Also, the ballistic coefficient changes as the shooter climbs above sea level. At 13,000 ft. altitude, the coefficient is about 1.6 times the sea level number. Technically, as explained in Chapter 14, **it is the *apparent* coefficient that is different. The *actual* coefficient will remain the same.**

The change in performance is not great, but it still can cause a hunt to go sour if things were marginal to begin with. Hunters in the Rocky

Mountains who have rifles zeroed at low altitudes can expect them to shoot higher at the higher elevations. No exact figure can be given because it would depend on variables such as altitude, caliber, ammunition, range, etc. Sighting-in again at the new altitude would be recommended. For a starting point, look for a change in the area of 2".

Kinematic viscosity is the resistance of matter to flow if it is a liquid or yield to stress if it is a solid. It is based on the abstract without reference to force or mass. Simply, for our purpose now, it is the stopping power of air. Wood will have a powerful kinematic viscosity and stop a bullet quickly. Water will have less, but will still stop it after a distance. Air also will stop it after a much longer distance. The kinematic viscosity of wood will vary. For example, soft pine would be less than a hardwood like walnut. Air also will vary but the viscosity of a gas is unusual in that it is generally a function of temperature alone and an increase in temperature will increase the viscosity. The change at altitude is a result of the lower temperature usually found at the higher elevations. (This is similar to the discussion of the speed of sound that was in Chapter 13. There, temperature was also the only factor.)

The coefficient of absolute viscosity is assigned the shorthand notation μ (mu). Since many parts of ballistics and aerodynamics involve consideration of viscosity and density, a more usual form of viscosity measure is the proportion of the coefficient of absolute viscosity and density. This combination is termed the kinematic viscosity and is noted by v (nu).

◆ $$v = \mu / p$$
kinematic viscosity = coefficient of absolute viscosity / density

The kinematic viscosity of air at standard sea level conditions is 0.000158 sq. ft. per sec. At an altitude of 5,000 ft. it has increased to 0.000178 and at 10,000 feet to 0.000202.

Density altitude is an appropriate term for correlating performance in the nonstandard atmosphere. It is the altitude in the standard atmosphere corresponding to a particular value of air density. Involved are both pressure (pressure altitude) and temperature.

Based on altitude alone, the velocity difference at Denver, Colorado, which is about a mile above sea level, would be about 200 f.p.s. faster at 500 yards. (.30-'06 at 2,700 f.p.s. muzzle velocity, 180 grain bullet.) A very high temperature at lower altitude will have a similar effect.

While altitude and temperature play a small part, air resistance is a major force in slowing the projectiles velocity. Bullet shape and type also are important. (Air resistance and bullet shape are both covered in more detail in chapter 13 & 16.) An **example:** the 156 grain .30-'06 bullet with a muzzle velocity of 2700 f.p.s. will meet about 1 1/4 pounds of air resistance.

This will change the bullets velocity 56 times as fast as the change in velocity caused by gravity.

The bullet's shape will affect air resistance. A stream-lined bullet will cut the air cleaner with less resistance and that will let it hold its velocity longer which in turn will cut down its flight time. That will give it a flatter trajectory because of less time for gravity to act on it. But, act it will and at the same rate as any bullet of any shape: less drop because of less time in flight. If, as in our previous example, it is in the air 2 seconds, it will still drop 64.4 feet. Of course, these long 2 second and 3 second times are used to simplify the examples. Most actual shooting would be a small fraction of a second with some long range rifle shooting reaching the longer times.

TEMPERATURE

Air temperature enters into the ballistic picture in three distinct ways. One is with density, another is elasticity of the air and the third is the effect on the cartridge, primer and powder.

The effect of temperature on air resistance is influenced by elasticity acting on the velocity of sound. The elasticity of the air affecting the velocity of longitudinal wave motion in the air is a function of the temperature. Within reasonable bounds, it is not influenced by density changes. For speed of sound and wave propagation, see Chapter 13.

For constant pressure, a given quantity of air changes density with temperature approximately according to Boyle's law. ($pV=k$ where V is the total volume, p is the total pressure, and k is a constant.) The temperature changes may be daily, seasonal, geographical, or irregular variations from weather factors.

Temperature has an effect on air density with cold air being more dense and hot air less dense. As any aircraft pilot knows, the combination of high altitude and hot temperature can be deadly for aircraft performance, but it also affects firearm ballistics. The effects on long range shooting are well known. The change is due not only to air density, but mostly to changes in cartridge pressure. A lower temperature will lower the pressure and the velocity. Shooters that compete at ranges out to 1,000 yards will notice this effect. At this extreme distance, a 10 degree F. change in temperature will need a 1 Minute Of Angle change in elevation. Large temperature differences will cause large elevation changes, all else being equal. (Remember, we are discussing extra long range shooting. There would be a change at a short range, but it can be omitted.)

Cold temperatures retard powder combustion. A major powder manufacturer, DuPont, says that with smokeless powder, a velocity drop will occur below 70 degree F. of 1 1/2 f.p.s. per degree. **Example:** Temperature at 25 degrees F. ; **70 - 25 = 45 * 1.5 = 67.5.** A 67 f.p.s. drop is not enough to make much difference in all but the rarest cases. If someone in the far North is hunting at 41 degrees F. below zero, the drop would be about 167 f.p.s. A full hundred lower than at 25 degrees but still not much for most firearms used in that area. Still, if hunting with a marginal caliber weapon, it should be kept in mind.

Humidity also has a measurable effect with the higher the humidity, the more the air resistance.

Note that the above discussion on density and temperature is also covered from a different perspective in Chapter 14 because of the change that these conditions cause to the ballistic coefficient.

BALLISTIC & TRAJECTORY TABLES:

Ballistic charts are great things to study and compare cartridges, their muzzle velocities and the velocities at different ranges. Energy, trajectory, bullet weights, all of the things that hunters, sportsman and target shooters find interesting. With help from information in this book, the charts can provide a world of valuable data.

A lot of arguments concerning the "best cartridge" can be won or lost with a ballistic chart; but as has been said many times, the proof is in shooting your own gun. Test it on the range and see for yourself.

Velocity figures published by the manufacturers may be a little high because they do the same thing we would all do in their situation. They use the best barrels which are made to minimum dimensions. This will always increase velocity and pressure a small amount over normal. The average gun we purchase will not perform to laboratory standards. This applies to handguns and shotguns as well as rifles.

The ballistic charts are wonderful and valuable toys or tools, depending on your purpose, but don't believe every number. Sort of like your loved ones when you ask them how you look. They won't lie to you, but they may fib a little.

How can a table be used to sight in a rifle? Disregarding wind deflection and numerous other factors discussed in this book, we should now look at how data on a trajectory can help us properly sight in a gun.

A rifle is sighted in to get the point of aim and the point of impact the same at a given range. That range can be anything that the combination of gun and ammunition can reach (energy and velocity considerations not withstanding). Many hunters just pick a distance that is available on their shooting range without further thought. As we shall see, this is an unsuitable method.

Because of gravity and drag and other circumstances, the bullet starts to arc downward after leaving the muzzle. It is wrong to believe it rises on exit for any reason. For all practical purposes, it drops. To compensate for this, the bore is pointed above the intended impact point and the bullet then drops down to meet it. **The rise in the arc is because the barrel is angled upwards, not because the bullet rises or climbs above the bore axis.** To repeat, it does not climb above the bore.

All factory trajectory charts show a distance for a given cartridge that was used as zero in their figures. Rifles will run from 50 yards to 300 yards with 100 yards and 150 yards as the most common. Shotgun slugs are usually given at 100 yards. Many charts will list 2 zero points, one for short range and another for long range. The shooter can take his choice according to his needs. Or, he can pick neither if a third choice suits him better. For a new choice, he will have to work out figures for drop. The information needed is available in this book.

If the vital area we want to hit is about 6" or 8" in diameter, a cartridge zeroed at 200 yards may be within the circle at any range from zero to about 230 yards, depending of course, on the cartridge involved. Beyond that, the gun would have to be aimed higher just to stay in the circle and higher yet to get back to the middle of the circle. If the hunter expected to hunt regularly at 300 yards, the zero should be changed to 300 yards. That would, of course, raise the amount above the sight line for a hit at a shorter range.

With the exception of the slow velocity cartridges such as .30-30 Winchester and .45-70 Govt., most that are zeroed at 200 yards will shoot reasonably close out to almost 300 yards. An average would be about 2" high at 100 yards and 8" low at 300 yards. If only a 100 yard range is available for sighting in, then if set 2" high at 100 yards, they will be satisfactory to 300 yards. Of course this is a generalization. Check a ballistic table for each cartridge in question.

Each cartridge has a maximum range that is usable. The trajectory tables will give the figures that are needed to sight in and use a gun and cartridge combination. The ideal situation is to have the gun sighted-in so that, in the entire distance that we expect to shoot, the bullet will not be above or below the horizontal line by over 3" or 4". In that case, we would not be off target over 3" or 4" on a 6" or 8" vertical line. Also, we should know how high to aim for a longer shot. And just as important, we would realize if it is too long for a likely hit. (Energy, stopping and killing power at range is another important subject. Here we are discussing just trajectory. Also, for target and varmint shooting, those inch figures would be too big.)

Handguns and shotguns with slugs, when used at long range, have the exact same situation as rifles. The only difference is the ranges are shorter and so the highs and lows of bullet drop are not usually as great. If the charts show the bullet drop to be excessive, then we should stop to think if perhaps the range is too long for the gun and caliber.

The vast amount of numbers on the factory charts were figured by a computer. They use ballistic coefficients and muzzle velocity to predict what is expected. These are tested and, at the ranges in most charts, the accuracy is very good. Military tables are accurate to much longer ranges and come from the same information plus time of bullet flight and angles of elevation.

The U.S. Military first measured times of flight for long ranges in 1879 at the Sandy Hook Proving Ground. A new invention by Alexander Graham Bell (1847-1922), the telephone, was used to transmit the sound of the gun firing to a person with a stop watch located near the target. This would move through the wire at the speed of light instead of the speed of sound. (186,330 **miles** per sec. in a vacuum vs. 1,116 **feet** per sec. in standard conditions.) The stop watch was started at the sound of discharge as heard over the phone and stopped at the sound of the bullet hitting the boards used in the target. There were enough shots recorded over the days to arrive at reasonably accurate figures. The ranges were as long as 3,000 yards. The military still uses direct measurement for both long range artillery and small arms up to 1,500 yards.

TRAJECTORY MID-RANGE HEIGHT:

The midrange trajectory height is the highest point the bullet will get above the sight line. Usually at about 55% of the downrange distance, it is valuable information when sighting in and hunting. Note the shorter height of the sight line as compared to the bore line. Ballistics tables can give the height above the bore or sight line. Read the small print to see which method is given.

♦ Mathematically, we can use $h = 4\ T^2$ to find the trajectory height with T being the flight time to the target. The answer will be in feet. This is suitable for most ranges, but for extreme long range or high accuracy use $h = 4.05\ T^2$.

♦ There is a method of calculating the trajectory heights at points other than the midrange. It is an old formula, devised or at least published, by a British officer named Sladen in 1879. It has been well tested and shown to be accurate. It is still in use today.

$$h = 1/2\ g\ t\ (T - t)$$

Where: h = height at any point desired (in feet)
T = time of entire flight (in seconds)
t = time to point where height is desired (in seconds)
g = gravity

Note: This is simple subtraction and multiplication. For $1/2\ g$ use 16.1 (or for more accuracy 16.08) times t. Subtract t from T and take the answer times your other answer.

MID RANGE HEIGHT FALSEHOOD:
   It is a common belief that if the mid-range height and the impact point are the same, the basic trajectory is the same. At first thought, it sounds sensible. If two of the key points are about the same, then they have a similar trajectory. The problem is; those two points can be the only places that are similar and all the rest can be different.

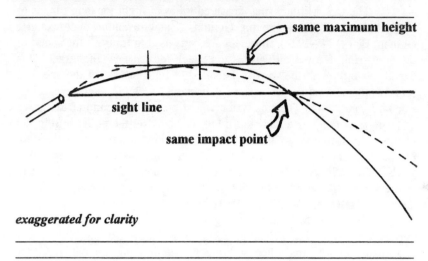

exaggerated for clarity

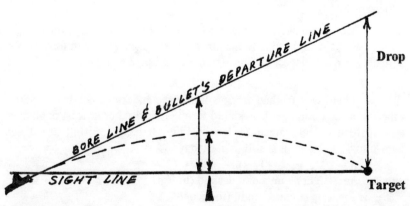

Note: total drop is 4 times the mid range height (MRH) based from the sight line.

exaggerated for clarity

This chart is on the caliber .30 M1 with a 172 grain bullet at a muzzle velocity of 2,700 f.p.s. It is from U. S. Army tests at Aberdeen Proving Ground.

Range in yards	Angle of Departure minutes	Time of flight seconds	Velocity Remaining ft/sec	Energy ft-lbs	Maximum Ordinate feet
0	0.0	0.000	2700	2785	0.0
100	2.3	0.115	2539	2463	0.1
200	4.9	0.237	2384	2170	0.3
300	7.7	0.367	2232	1903	0.6
400	10.8	0.506	2084	1660	1.1
500	14.2	0.655	1940	1438	1.8
600	17.9	0.815	1799	1237	2.8
700	22.0	0.989	1660	1053	4.1
800	26.7	1.177	1526	889	5.8
900	31.9	1.383	1397	745	7.9
1000	37.7	1.608	1275	621	10.5
1100	44.3	1.855	1161	515	13.9
1200	52.0	2.124	1083	448	18.3

The maximum trajectory ordinate is the maximum height above the horizontal range line of the projectile as it follows a ballistic curve between the point of origin (the muzzle) and the point of impact. It is common to use 50% the range for the ordinate but in most cases, due to aerodynamic drag, the actual distance will be closer to 55% of the range.

To calculate a trajectory ordinate, use this formula:

$$y = x / 1000 \ (A_r - A_x)$$

Where: y = height of ordinate in yards (not feet)

A_r = angle of departure of the trajectory in mils.

A_x = angle of departure of the range in mils.

x = horizontal range to ordinate

Note that the mathematics is simple and basic, but this example chart gives departure angles in minutes and the formula calls for mils. Don't forget to convert. Mils is covered in detail in Chapter 18 and it is equal to 3' 22.5" of arc or 3.375 minutes.

SIGHTS:

First, it is important to remember that moving the sights does not change the bullet's trajectory in relationship to the barrel. It simply is getting the sights to point where the bullet hits.

The sights are adjusted to score hits at a set range and the bullet will cross the sight line from below because it leaves the muzzle from below. At very close range, the aim should be a little high because the bullet is traveling low. All trajectories are different but a good crossing distance will be about 15 yards for a .22 R.F. long rifle and for a high velocity bullet it will be out around 45 yards. Inside this short range, very accurate shooting, such as at a snake's head at 10 feet, will require a slight high aim. Try it and make a mental note of how your favorite gun shoots up close. It won't be much, with the 1 1/2" height of the average telescope sight above the bore center line, but the rattler may need a good hit with the first shot. (Details and the math for this crossing distance are on page 212.)

Most sights are mounted above the gun's bore and for mathematical purposes, we usually say that a telescope sight is 1.5" above the center. For metallic sights, the figure 0.9" is frequently used. While these are standard, there will be some variation and it should be changed to suit each gun and sight combination. The height has to be considered in all trajectory computations.

Example: For an example, consider a rifle with the telescope mounted on the side with the same center line as the bore. (Like a top-eject Winchester 94, only lower.) We will consider the center line of each is parallel; laying on the same plane with the sight not elevated to zero the gun at any range. We will use the ballistics for a Federal .308 Winchester cartridge with a 165 grain bullet at 2,700 f.p.s. muzzle velocity. Remember, the sight is at the same height and parallel to the bore.

Checking the ballistic charts, if we aim at a target 200 yards down range, the bullet will strike 10.2" low. To zero at 200 yards, the sight would have to be elevated to raise the strike point 10.2". If the sight is then placed on top in the normal position, the elevation will have to be raised another 1 1/2" (1.5") for a total elevation of 11.7". This was figured at 200 yards. For any other range, multiply the number of yards by 11.7" and divide by 200.

Example at 400 yards. $11.7 * 400 = 4,680$ and divided by $200 = 23.4"$. The bullet drop for 400 yards is given as 46.3" below the bore or 47.8" below the line of sight for the telescope. Subtract the 23.4" from the 47.8" and the hold over figure is 24.4" for this cartridge zeroed at 200 yards and aimed at 400 yards. Using a ballistic table giving bullet drop for the cartridge you are interested in, this can by worked out for any ammo by simply substituting the correct numbers.

SIGHT ADJUSTMENT:

The basic rules for iron sight adjustment: The rear sight is moved in the direction you want the projectile to go. The front sight is moved opposite the direction you want the bullet to go. Rifles, handguns, shotguns; all are adjusted with the same basic principle.

Telescope sights are not all mounted or made the same. If the rear mount is moveable, adjusting it does the same as a rear sight, so it is moved the direction you want the projectile to go. Many scopes are mounted fixed and the adjustment is internal. In that case, either follow the directions on the knobs or move the reticle as a front sight and move it opposite the direction you want the projectile to go.

Make at least 3 or 4 shots before changing the adjustment. You want to establish a group, not what could be one wild shot. The group size has to be on the paper first, and the whole group moved toward the center. Or, perhaps it would be easier to think in terms of moving the sights instead of the group. One thing to keep in mind is the group will not get smaller if each shot is aimed the same. The group center is zeroed in at the chosen range and nothing in sight adjustment will help the wild shots (fliers).

Sight adjustment can be made easier if it is known how much to move the rear sight to obtain the result on the target. This can be computed easily with any gun. It is handy when sights have been changed, a barrel shortened, or a special situation takes place.

◆ $T = R * D / S$

       T = displacement of impact on target
       R = range
       S = Sight radius (distance between front & rear sight)
       D = distance between markings on sight. (Can be measured accurately with calipers.)

This is simple and no example is needed. It can save time and ammunition compared with the usual trial and error technique.

Bracketing is a useful technique when bore sighting or no other method has succeeded in a hit on the target. Then it is necessary to make changes in both directions until a strike can be found. It is not generally good to waste time with small adjustments that are inadequate. Bold changes are in order and even though they are probably too much, it will usually pay off. It is called bracketing and is used in a similar way by artillery. The idea is to find a point on each side and take a midway position and note the error. Split that position with the past one from the other side and continue until the sights are in proper adjustment. The idea is to bring in to center the strike by a reversal of the direction of error back from one side to the other side.

SHORT RANGE SIGHT-IN:

◆   To set new sights or after the purchase of a new or used gun, the sights can be set decently at short range so they will hit close at long range. If the bullet hits properly at the short distance where the bullet and sight line cross, then it also will hit very near (disregarding wind deflection, etc.) at a longer range. For the shooters that want to calculate it, the math is easy. For those that use standard ammunition and don't want to bother with the math., a table is included.

$$CR = R * H / D$$

Where: D = drop
   H = height of sight above bore center line
   R = range
   CR = close range

The sketch shows why this formula and the close sight-in works. It is based on the basic mathematical premise that both the large and small triangles are in the same proportion to each other.

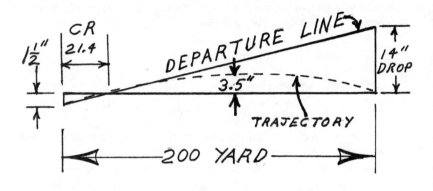

TARGET DISTANCE FOR CLOSE RANGE

Cartridge	Bullet weight grains	Muzzle velocity f.p.s.	Trajectory Range yards	Trajectory Height inches	Targeting distance, yards iron	Targeting distance, yards scope
.22 WMR	40	1910	100	1.7	15	20
.222 Rem.	50	3140	200	2.4	20	28
.243 Win.	80	3550	300	4.7	--	24
.243 Win.	100	2970	200	2.1	22	31
.280 Rem.	150	3050	200	2.5	18	25
.30-30	170	2220	100	1.2	19	--
.30-'06	150	2920	200	2.4	21	29
.30-'06	220	2410	200	3.7	14	20
.300 H&H Mag.	180	2880	200	2.3	21	29
.308 Win.	180	2610	200	3.1	16	23
.375 H&H Mag.	300	2530	200	3.3	15	22

BARREL HEATING:
Barrel heating from firing will change the point of impact. This is little concern to a hunter, as long as his rifle is sighted in for a proper hit with the first shot from a cold barrel. A target shooter should learn what to look for with rapid fire. For accuracy, slower shooting requires the shots be spaced to hold a uniform barrel temperature, perhaps a little warm but not too hot. As stated many times, each firearm and cartridge combination will respond differently. If accuracy is important, experiment with the gun and ammo to be used at the range to be worked. **Going hunting? Remember that the special, once in a lifetime shot, will almost certainly be from a cold barrel.** If you have played around on the range at sighting-in and the barrel has heated, let it cool before the final shots and sight adjustment.

UP OR DOWN SLOPE-HILL:
Trajectory ballistics are said to be *rigid* and hold up through small angles of change. The horizontal standard, as used in ballistic tables, is valid for most shooting conditions. Unfortunately, rigid, meaning not changing or bending, is **only applicable through angles from the horizontal to about 10 degrees up or down.** While this may include many shooting conditions, it does not take in them all.

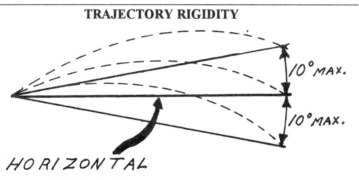

TRAJECTORY RIGIDITY

The main difference between shooting on a flat area and shooting game that is much higher or lower is the visual perception. A common error is aiming too high on an up hill shot or too low on a down hill shot. Although the error may be any number of ways because there are a lot of wrong ideas floating around about this subject. Even among gunners that understand the theory correctly, errors are made in different directions. Every shooter perceives the problem a little differently depending on his visual capacity and mental perceptions. It is hard to estimate the slope distance and slope angle, even if properly trained. As a general rule, errors in estimating range will be toward the long side on hilly or rough terrain (as in up or down slope) and estimated short over open country and water.

SLANT RANGE TRAJECTORIES OVER 10 DEGREES

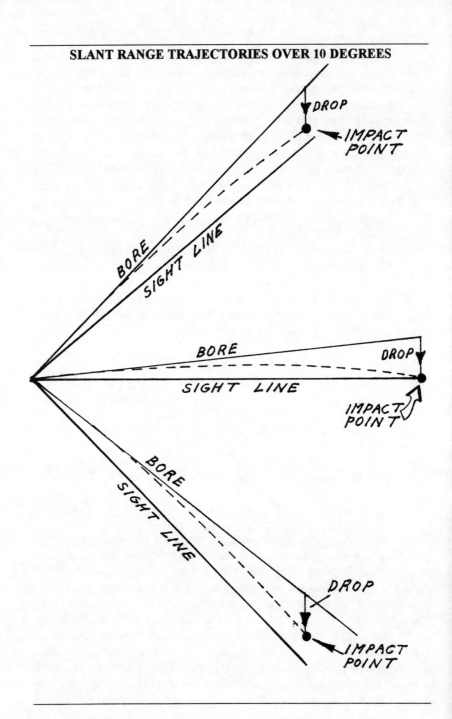

There is a ballistic slope error in an uphill or down hill shot which changes the bullet drop by a noticeable amount. The interesting fact is that it is the same uphill or down-hill. No difference to the bullet, only in the way the person sees it. Although the bullet will always strike higher on a slope shot than on a horizontal shot. Always higher, unless corrected, and it does not matter if the shot is up hill or down.

➔ **Note:** in the following explanation, all the experts agree on the fundamental details, but a few of the small secondary elements are debated. To put that in different words, everyone agrees as to what happens, that is obvious and proven, but there is some argument over why.

One common mistake is believing the velocity is changed by gravity slowing the uphill shot and speeding up the down hill shot. This is generally discounted because the bullet's weight is not sufficient, even in the bigger bores, to accelerate or decelerate it enough to have much effect on either the flight time or the trajectory.

A great deal of theory is based on the angle of the bullets' path in relation to the pull of gravity. The trajectory is flatter because the angle reduces the horizontal distance the bullet travels and the effect that gravity has on the bullet. The same linear distance, if horizontal instead of up or down a slope, would put the projectile in the horizontal gravity pull for a longer distance. In other words, it is the horizontal distance travelled that counts, not the actual linear distance. Remember that gravity works perpendicular to a horizontal line. The idea is that when a projectile is moving at an angle to the horizon, gravitys' force is reduced by the difference between the linear range to the target and the horizontal distance. Of course, the force of gravity remains the same. Nothing here changes the laws of nature, it is simply following the laws as always. Gravity is fixed and the angle of the bullets' flight path to the horizontal is a variable that can change.

Some people with a good background in the subject, believe gravity has such a tiny effect, it is best to disregard it. The changes by other causes, mostly air resistance, are the ones that count. Gen. Julian Hatcher figured that a 150 grain .30 caliber service bullet on an uphill shot would lose speed to air resistance at about 50 times the rate that it would be slowed by gravity. The modern thinking today is that, as just explained, gravity and the horizontal range, are the controlling factors.

The drop at the different angles can be figured. To do so, you must know or estimate the angle and distance. The bullet drop is based on the horizontal distance, not the slant range. Of course, the horizontal distance is always shorter.

As stated, the drop is based on horizontal range yet the slant range is what an instrument or eye will measure. The slope angle can also be measured or estimated, although in hilly or mountainous country, it is sometimes difficult to figure where the horizon would be. The horizon point needs to be known to estimate the slope angle with any accuracy. Instruments with bubble levels are available, but hardly practical to use on a hunting trip. Range estimates are hard to make accurately even on flat ground. The most experienced people with perfect eyesight cannot normally do better than 10% to 15%. At 350 yards, 15% would be an error of 52 1/2 yards. Even at 200 yards, 15% would be a 30 yard error.

After estimating or measuring the amounts, multiply the slant range by .97 for a 15 degree slope angle, .87 for a 30 degree slope angle and .71 for 45 degree slope angle.

Example: Game or a target at a slant range of 310 yards for a 15 degree slope angle, multiply 310 by .97 for a horizontal range of 301 yards. The same 310 yards at 45 degree slope angle would give a horizontal range of only 220 yards. (310 * .71 = 220) Aim to compensate for 220 yards. A high overshoot will happen if aimed for 310 yards.

Always aim as if shooting at the shorter distance.

No hunter is expected to have a bubble level, a protractor, these charts and a calculator with him or her while hunting. The purpose in this section is the same as all other parts of this book. To explain in detail why and how a particular thing happens the way it does. And a lot of hunting shots are missed for lack of understanding this subject. While it is unimportant at the small angles, it can become very significant at the big ones. A good basic knowledge will enable the hunter to estimate or make an educated guess how much to lower his aim. **Lower because, as the text clearly shows, any slant range shot, both up-hill or down-hill, will be higher than if it were a horizontal shot. The range and angle may influence how much higher it will strike, but to be on target, the gun will always have to be aimed lower.**

Depending on the caliber and the sight-in range, slant range shooting will have little effect out to 150 to 200 yards.

Trigonometry for a right triangle.

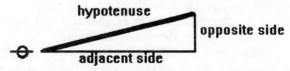

The entire problem is based on trigonometry. The horizontal range is the shorter adjacent side of the right triangle where the slant range would be the hypotenuse. The opposite side is the rise or drop. The multiplier is the cosine of the slope angle.

A complete chart in 5 degree increments is given. While interpolation can be used to find the number for an intermediate angle, if a scientific calculator is handy, just press in the angle and then press COS (cosine) to obtain the multiplier. **Slope up or down is the same.**

Degrees	Multiply slant range by cosine
0 degree	1.00
5	.99
10	.98
15	.97
20	.94
25	.91
30	.87
35	.82
40	.77
45	.71
50	.64
55	.57
60	.50
65	.42
70	.34
75	.26
80	.17
85	.09
90	0

WIND DEFLECTION ON SLOPE SHOTS

It should be noted that wind deflection is based on the slant range and not the horizontal range because that is the actual bullet path for time and effect.

RANGE-FINDING:

All of the trajectory charts in the world will be useless if the shooter doesn't know the distance to his target. Not exactly, of course, but close enough to get the projectile into a vital area. Most errors in long range field shooting are caused by errors in estimating the range. Knowing it is essential.

It is best to have a suitable caliber and a range that is not too excessive because even experienced people with perfect eyesight cannot estimate range closely. 10% to 15% is considered a very close estimate. Errors of 20% are common even by skilled people. A 15% error at 350 yards would be a 52 1/2 yard error. That is enough to promise a wounded animal or a complete miss. Even 15% at 200 yards for a 30 yard error could be a problem if the gun was zeroed for 200 or less and the error was on the long side.

Estimating distance is difficult because a human's eyes are close together. This reduces the three dimensional effect, sometimes called stereoscopic vision. It is more difficult as the distance increases yet because of trajectory, wind drift and other concerns, accuracy is more important. (But harder to obtain.)

The general rule for horizontal shooting is the errors will be toward the long side on hilly or rough terrain, when the game is in the shade or just partly visible, and when the sun is behind the game. The estimate may be short over open country, water and when the sun is behind the hunter. Anything that interferes with vision will make it more difficult. Rain, snow, heat waves in hot weather, shadows from drifting clouds, up or down hills; all of these and more can add to the error. Even high humidity can create problems. Obstructions such as a thin woods and the suns position can change the impression; especially in late afternoon or early morning. On a sunny day when objects appear sharply defined, they seem closer than in weather that restricts vision. Every shooter perceives it differently depending on his visual capacity and mental perceptions.

Unfamiliar terrain adds its own special problems. A hunter from the eastern woods may have trouble in the western plains and hills. Mountains and deep valleys have their own estimating problems.

The most common method of estimating range is by comparing the range in front of him to a known distance in his memory. The football field is the most common because most people are familiar with its look and dimensions. It is not to difficult to guess that the range is, for example, 3 football fields in length.

An approximate estimate of range can be made of any object by pointing at it with the index finger while the arm is stretched toward it. Close one eye and note the finger's position. Then do the same with the other eye. While holding the extended finger still, it will appear to move sideways. Some people obtain good results by rapidly alternating from one eye to the other as they study the setting. The distance to the object is about ten times the distance the finger appeared to have moved on the scene. With a little practice it can work surprisingly well. There are other methods of measuring distance without instruments, but this is by far the quickest and easiest. The reason? The length of the average person's arm is about ten times the distance between their eyes.

Some hunters try to estimate range by picking out a point that is about half way to the target. This distance is estimated and then they mentally multiply it by two, double it in other words, to obtain the range to the target. This works well for some people because shorter ranges are easier to guess.

Another method of range-finding works well if the hunter knows the average dimensions of his game. Experience will help to deduce how the sight or scope picture will look at different ranges for an animal of that size. The problem; most hunters don't hunt enough to become familiar with the sight picture of their game at various ranges. The hunter should become familiar with the details that can be recognized at various ranges for the game he is hunting. Can the details of the fur be made out at 100 yards? What about the eyes at 300 yards? Full size targets in the games silhouette and set up at different distances can be a learning experience. Actual shooting is not required. This can be a big help for hunters who live where most of the animals they see are dogs on leashes.

Regardless of what method is used, it is a good idea to make several estimates and then pick the average of them. Perhaps a long estimate and a short estimate and then use the midpoint.

Hand held range-finders are valuable for hunters. Most operate on the split image principle where a dial is adjusted until two images coincide. Then the distance to the point is read from a dial. None are perfect, and all require that the directions be followed exactly, but some work very well. The largest benefit comes from hunting from a stand or fixed point. Check in advance the distance to various points around the hunting area. If you know the ballistics of your cartridge as well as your own ability, limiting your shooting to that maximum range is a good practice.

The smart pioneers and early settlers expected trouble from Indians and various enemies. From their fort, prairie farm home or western ranch house, they paced off the distance to landmarks within range in all directions. They realized that for a field of fire to be both effective and deadly, it must be at a known distance. In no other way could they protect themselves properly. Pacing off the range or using a range-finder to various landmarks is still a good practice when hunting from a stand or fixed point.

Whatever the method used, it will be more effective if it is done with care and not hurried. Of course, when the game is acting jittery and appears ready to run, it is difficult to have the virtue of patience.

CANTING

Canting is tilting the gun to the right or left as it is fired. If a trajectory could be completely flat with sights that are not higher at the back to raise the muzzle, cant would change nothing. Unfortunately, that is not the case. A cant to the right will swing the impact point to the right and down following a curve or arc; the reverse to the left. The side movement is much greater than the drop. If always done exactly the same for each shot, the sights will be adjusted to compensate and no error is noticed.

The side effect is easy to figure with a scientific calculator. Press in the cant angle and press SIN. The answer is the multiplier.

Examples:

cant angle	sine of cant angle
2 degrees	.035
4 degrees	.070
8 degrees	.139
12 degrees	.208

The multiplier times the drop will tell how much a particular cant will move the impact point. **Example:** a 25" drop at a 12 degree cant. 25 * .208 = 5.2" movement right or left.

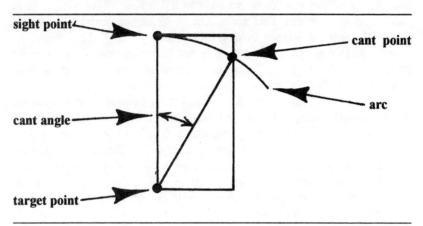

POINT BLANK RANGE

There are three definitions of the expression, *point blank range*.

(1) Many shooters believe point blank range is any unspecified distance that is very close to the muzzle. This is common in pulp fiction and television detective shows when they are written by writers with no knowledge of firearms. In colloquial use (slang), it is used to indicate a very close range with no precise limits. It is interesting, but not correct.

(2) The expression is also used to indicate the first down-range distance in the trajectory that requires no elevation change either up or down to hit point of aim. This first point will be just a few yards range even when the 2nd. primary point is 100 yards or 300 yards. In other words, the point where the projectile crosses the line extending through the sights because of the angular difference with the bore line. Knowledge of this distance for the cartridge in use can be handy at times. (See sketches) From a technical standpoint, this use of the term is correct and a few dictionaries give this definition. But, it is seldom used in this context.

(3) Here is the most common use of the 2 correct meanings of the term. The maximum distance that will result in an accurate hit in the vital area with the sights set on the target correctly. For hunting rifles, if we consider the target to be an abstract circle over the game's vital area, then point blank range would be the range where the bullet would hit in that circle. If the diameter is 8", then the sights would be adjusted for the range that would not let the bullet pass more than 4" above or below the center of the circle. This requires sighting-in at the range that the maximum ordinate is 4" above the line of sight. The midrange trajectory is important in this situation. It takes a study of the ballistic tables to determine this distance.

Many computer ballistic programs will compute the point blank range if we have the information to feed to it. We need the diameter of the vital area which will be different for different game. For varmint hunting it may be just 3" and for big game perhaps 6", 8" or even 10". In addition to determining the target area diameter, we also have to know the ballistic coefficient and muzzle velocity.

A close study of manufacture's ballistic tables will enable most shooters to estimate the distance closely. Remember, the velocity of the ammo from the gun being used may not be as high as the factory figures. Also, the data is based on sea level and the hunt may be in the mountains. These, and other things will make a small difference. As a general rule, it will take at least a 200 f.p.s. difference to make a significant change in trajectory. A study of how much this will affect things and in what direction will permit the shooter to make slight corrections.

Will changing the sight-in range to point-blank range help a hunter bag his game? Probably. It will do a little to compensate for errors in range estimation, sight adjustment and aiming. And we all know that anything that is even the slightest help is worth doing.

.45-70 GOVT. TRAJECTORY:

The .45-70 cartridge was mentioned in the discussion of charts and tables and it owns some interesting facts.

The .45-70 Springfield was introduced in 1873. The U.S. Cavalry and Infantry both used the gun as either a rifle or carbine. It was designed for a breech pressure of 25,000 pounds with black powder cartridges.

As with all military arms of both yesterday and today, a lot of experimenting and testing was performed. Standard testing was conducted at ranges up to 1,000 yards. A 500 grain bullet (adapted in 1884) took a little over 3 1/2 seconds to reach the 1,000 yard point with a muzzle velocity of 1,315 f.p.s. The bullet still had 711 f.p.s. at 1,000 yards. (The 405 grain bullet started 50 f.p.s. faster but could not hold it as well. At 1,000 yards it had only 536 f.p.s.) To hit the bulls-eye at that long range, the 500 grain bullet had a mid-range trajectory of 43.5 feet. In other words, the bullet was

43.5 feet above the line of sight at 550 yards (approximately). The bullet would go through a foot of pine at a range of 400 yards.

Testing by Springfield Armory showed the 500 grain bullet would give an average distance from the bulls-eye center to each hit of 1.3" at 100 yards and 4.2" at 300 yards. At the 1,000 yard range it was 21.6" and this was fired from the shoulder with standard issue service rifles. No bench-rest was used. Not bad for the 1880's or, everything considered, even today.

VACUUM TRAJECTORY:

◆ Here is a formula for trajectory in a vacuum with no air resistance. It is modified for firearms use and given here for academic reasons only. The text books list several other formulas for vacuum trajectory but unless we are planing a space flight, we will not have a practical need for them.

$$R = V^2 \sin 2\theta / g$$

Where: R = range in feet

V = muzzle velocity in f.p.s.

θ = angle of elevation (departure angle)

sin = sine of angle

sine $2\theta = 1$ for a 45 degree angle

◆ Also for **elevation in a vacuum** using the same symbols as before:

$$\sin 2\theta = Rg / V^2$$

NOTES:

◆ The point of a bullet turns in flight so that it remains point first. In other words, if it is fired at an angle of 20 degrees above the horizon, it will not hit at that angle down range. A properly stabilized bullet will still strike point first. There is a complete discussion of this in the section on gyroscopic stability.

◆ The projectile does not rise or climb. It cannot. It may appear so in the drawings but it is only in relation to the sights. Not the bore. The projectile never goes above the bore line. The drop begins at the muzzle. (Disregarding, nutations, gyroscopic unstability and yaw.)

◆ Hunters that shoot game at short, moderate and long range will find it a big help to carry trajectory figures with them. If, for example, a gun is sighted to zero at 300 yards, it is invaluable to know how high to hold over the vital area at 350 or 400 yards. If written in tape with ink that won't run in wet weather, it can be taped to the stock and referred to as required. Be sure to use figures for your ammunition, not just something similar.

◆ In trajectory studies, don't forget the valuable information in other sections of this book. Wind deflection and drift, ballistic coefficients, gyroscopic stability, penetration and expansion. All of these and more have to be considered in making the choices.

CHAPTER 18

MINUTE OF ANGLE and EVALUATION OF TARGET GROUPS

MINUTE OF ANGLE

Military rifle sights are made with minute of angle adjustments called *clicks*. Minute of angle, frequently abbreviated MOA, is used in target shooting and other aspects of the shooting sport. It can be misunderstood, so here is a long explanation.

We know a circle is divided into 360 degrees. Each degree is divided into 60 minutes and each minute into 60 seconds. No, we are not mixing circles and clocks. That is the correct terminology. Minute of angle, as used in firearms and sights, is one minute or 1/60th. of a degree. There are 21,600 in a complete circle (360 * 60 = 21,600), so 1 MOA is also 1/21,600 of a circle.

Most of the time, MOA is just rounded off to 1" at 100 yards, 2" at 200 yards, 3" at 300 yards, etc. This has an error of 4.6%. For readers that want to be very exact, the number is 1.0472 inches. The figure is based on dividing the number of inches in the circumference of a circle with a radius of 100 yards by the number of minutes in a circle. As in:

100 yards * 36" = 3,600" radius * 2 = 7,200" diameter

$$7{,}200\,\pi = 22{,}619.462"$$
$$60 * 360 = 21{,}600 \text{ minutes}$$
$$22{,}619.452 / 21{,}600 = 1.04719" \text{ or } 1.0472"$$

Or for a different view, the circumference of a circle equals 2π * the radius. Therefore, for any range (R), a minute equals **6.2832 * R ÷ 21,600** which is .000291 when divided out

◆ Some people like the easy route and use $\pi / 3 = 1.0472$.

Sight coverage. The advantages of MOA's goes beyond the obvious use in adjusting sights for range. Want to know how much the front sight covers the target? Determine the distance from your eye to the front sight. Determine the width of the sight. If the first distance is, for example, 30", then 30" times .0003 equals .009 per MOA. Divide the width of the sight by the .009 (in this example) and the answer is the area in inches covered by the sight at 100 yards. Double it for 200 yards, etc.

Drop figures can be changed to MOA by dividing the drop by the range in 100 yards. For example; A drop of 8.2" at 200 yards. 8.2 is divided by 2 = 4.1 MOA. This does not consider the height of the sight above the bore center line. For example; a telescope sight mounted 1 1/2" above the bore center line, divide by 2 for 200 yards = .75 MOA for the sight by itself.

Elevation angle computation. Combining areas of knowledge will lead us to learning. Take the fact that total drop is 4 times the mid-range height. Combine with the knowledge that one minute of angle is equal to about 1" for each 100 yards. From this it is easy to compute the approximate angle of elevation. (If necessary, review trajectory basics.)

For example; suppose a cartridge with a 0.7 feet mid-range height. Multiply times 4 for total drop. In this case, 2.8 feet. MOA is in inches, so we multiply times 12 for 33.6 inches. At 300 yards, one minute of angle is about 3". (For more accuracy, 3 * 1.0472 = 3.1416) 33.6 divided by 3 is 11.2 minutes. (For more accuracy, 33.6 divided by 3.1416 = 10.695 min.) This example gives 11.2 (or 10.695) minutes of elevation. As with many things in ballistics, the longer the range the more error. This procedure is close enough for general use at the shorter ranges but gets more inaccurate as the range increases. To be accurate at long range, say beyond 300 yards, a knowledge of the bullets ballistic coefficient is necessary.

1,000 YARD RANGE

Many shooters consider 120 yards a decent range for consistent groups as small as 6" or 8". Some 1,000 yard rifle shooters can do this well, and better, and 1,000 yards is 120 yards plus a half a mile. That is what is called "a pretty fur piece" in the southern Appalachian Mountains, and they are not speaking of mink jackets

MOA FOR DIFFERENT RIFLES

The average hunting rifle should be able to group within 3 MOA. In other words, roughly 3 inches at 100 yards and 6 inches at 200 yards and 9 inches at 300 yards, etc. A good grade center fire target rifle should group within 1 1/2 MOA. Or even 1 MOA if the shooter expects to win in competition. A good varmint rifle should group better than the standard hunting rifle and as good or better than the average target rifle. For consistent hits, 1 1/4 MOA should be expected. Some bench-rest rifles can shoot 1/2 MOA groups at 300 yards and longer.

MILS

The error of 4.6% in using 1" at 100 yards for MOA can be avoided by converting to true linear mils. A mil is one thousandth of the range and in the metric system one mil equals 100 mm at 100 meters, 200 mm at 200 meters and so on. For conversion from the American system in inches, divide the measurement by 3.6 for 100 yards and 7.2 for 200 yards and 11.81 for a 300 meter range. This is based on 100 yards equals 3,600 inches and 3,600 inches divided by 1,000 equals 3.6 inches.

There is a MIL chart on the next page and the subject is covered in more detail on pages 292 and 293.

While this is basic math., these two charts chart will be useful for readers who do not like to work problems.

MIL	mm at 100 meters	inches at 100 yards
.1	10	.36
.2	20	.72
.3	30	1.08
.4	40	1.44
.5	50	1.80
.6	60	2.16
.7	70	2.52
.8	80	2.88
.9	90	3.24
1.00	100	3.60

MOA	mm at 100 meters	inches at 100 yards
.1	3.438	.1047
.2	6.875	.2094
.3	10.313	.3142
.4	13.751	.4186
.5	17.189	.5236
.6	20.626	.6283
.7	24.064	.7330
.8	27.470	.8378
.9	30.939	.9425
1.00	34.377	1.0472

EVALUATION OF TARGET GROUPS

Is it possible to test firearm velocity and target groups without reference to statistical methods and have a clear and precise picture of the situation? Is it also possible to fire a good group on a 1,000 yard range with no knowledge of wind deflection and trajectory? In both cases the answer is no. It would be blind luck to obtain anything except failure.

Figures can be interpreted in several different ways and different conclusions drawn from them. All versions may be *computationally* correct, yet not all, strictly speaking, may be *mathematically* correct. There are proper ways to analyze figures and obtain defensible and reliable results. In other words, there are ways to be sure about the accuracy of your hand-loads and rifle that go beyond just firing a dozen rounds and then saying to your friends, "I think it did better than the last time, don't you?"

MINI GLOSSARY

There is a glossary at the end of this book, but it is important that these words be understood in their proper meaning as they relate to ballistics. And let's be honest with each other. Most of us are not going to look these up, so they are repeated here.

If the words **accuracy** and **precision** are looked up in a standard dictionary, they appear to have similar meanings and be interchangeable. In the field of target shooting, their meanings are different and each can exist without the other.

♦ **Accuracy** is how close a strike is to where it was intended to hit and measurements are to that one point only.

♦ **Precision** is how well the shots grouped and the measure is to each other no matter how close or far from a target center.

♦ **Extreme spread** is a common term in group evaluation. It is frequently called *group diameter*, but that is incorrect. Properly, it is the longest distance between the most widely separated shots. It may be the group diameter or it may not. Sometimes it is called *maximum spread* or *group size*. If, out of 20 shots, 19 are about the same and one is far off, it will be counted and destroy the consistency of all the others.

It is also frequently used with the distance between the lowest velocity and the highest velocity. A large spread of about 150 or 160 f.p.s. will make any average velocity figure almost meaningless. Even at a 100 f.p.s. spread, the confidence level can be low. (These figures may be acceptable industry standards for some factory ammunition, but hand-loaders can and should do much better.)

Extreme spread is very easy to compute as only 2 points are considered. The widest-spaced target hits or highest and lowest velocity.

♦ **Extreme vertical or horizontal dispersion** is the longest vertical or horizontal distance between shots. Note: in this book a lot is written about *deviation*. This is *dispersion* which the dictionary describes as breaking up or scattering in all directions. That is strong language for even the worst target shooter. The people that originated this word usage could have saved us a lot of trouble by using the word *distance*, which it is.

♦ **Mean vertical or horizontal deviation** is the average distance from the center of impact of the group.

$$M_A = \Sigma \, d / n$$

Where: M_A = Arithmetic mean measure of central tendency

d = distance

n = number

Σ = indicates the process of successive addition (cumulation).

◆ **Maximum mean radius** is a term used occasionally in military specifications for ammunition. This is a method of giving the grouping ability of the ammo. After a group has been fired, the center is determined and the distance is measured to each hit. This distance is totaled and then divided by the number of shots in the group. The answer is the mean radius of the group. Also technically described as the average of the radial distances from the center of impact.

◆ **Diagonal** is the line, and more important the length of the diagonal line, in a square drawn around the rectangle created by drawing vertical and horizontal lines around the outside of the hits on a target. That sounds confusing at first, but if the knowledge is important, read it again slowly. It is a good explanation that at first glance appears complicated. The first lines drawn will almost always create a rectangle. Then the 2 short sides are extended and another line drawn to create a square. This increases the length of the diagonal line.

STANDARD DEVIATION
Definition

This term is frequently written as SD (or sd) but this is easily confused with *sectional density* so it will be spelled out in full. Standard deviation is a term used mainly with velocity to show a variation from the average. The term can also be used in many other types of problems. Handloaders will frequently try to achieve as low a velocity standard deviation as possible believing it will give the lowest group size. Their belief is usually correct, but as in most things, there are exceptions. It is primarily a term in mathematics that is used to organize and understand dispersion and central tendency. An *average* may mean nothing at all. We have *quartile* deviation, *mean* deviation and *standard* deviation to chose from. Standard deviation has three advantages. (1) Every item affects its value. (2) It is subject to further algebraic computation. (3) Deviations from the arithmetic mean yield a minimum sum. The only disadvantage is that it gives greater emphasis to extreme values because the deviations are squared. Some users don't consider this a shortcoming.

SAMPLING SIZE REQUIRED

To properly apply a standard deviation figure, certain criteria must be met. For example, the items must possess the same characteristics. Also, a sufficient number of tests must be conducted to assure that the figures are a true representation. In other words, a large sampling is needed. If the test involves target groups, the larger groups are best (in number of shots fired), but they can increase the group size. The shooter may become tired and use less care and bore fouling can increase. Thin barrels, especially, can be changed by heat from repeated use. There is also an increased chance of a flyer. In other words, the more shots fired, the larger the group, yet the statistical accuracy is still better.

When we speak of a large sampling, the number 30 is considered ideal. Less can cause an erroneous answer and after a threshold is crossed on the high side, it becomes repetitious. Of course, the size of the batch has to be considered. Five might be a good sample from a batch of 50 rounds. The ammunition manufacturers generally test about 1% and accept a variation of as much as 125 f.p.s to 140 f.p.s. This is understandable. A highly precise product is expensive to make and they are in business to sell ammo, not shoot it up. Some makers exceed this by a large amount, but they are an exception to normal.

Shots fired at a target that is perpendicular to the gun (technically, perpendicular to the trajectory) will normally group in a circle. From a statistical point of view, it can be said that the number of shots in a group governs the coefficient of variation of the maximum diameter of the circle. In simpler words, the more shots fired, the more likely it is that the result will be accurate.

The proper number of shots to fire can be found by asking yourself three questions. **(1)** What is the amount of acceptable error? **(2)** How big is the variation usually found in this type of test? **(3)** What is the confidence level I want to have with the result? Of course, there is no correct answer to these questions, but an honest response may help you to reach a decision.

Standard deviation can be measured in different ways. One is the actual deviation for the rounds fired. The other common method is the deviation that is postulated for an entire lot based on a small sample. While the first method can be considered precise, the second is, at best, a comparative measure that is conditional on unknown factors. Simple language? Don't bet any money on an answer based on a small sampling.

MATHEMATICS

Standard deviation takes into account every shot fired but gives emphasis to the extreme variations. It is based on the squares of the deviations, and all squares are positive, so the extreme values have importance.

The symbol used to express standard deviation is the lower-case Greek letter, sigma. (σ). There are a dozen different methods in engineering statistics to solve for standard deviation. This is perhaps the best and much easier to work than it appears.

$$\sigma = \sqrt{\frac{1}{N-1} \sum_{n=1}^{n=N} (V_n - \overline{V})^2}$$

Where: $\sum$ = Indicates adding or "the sum of" when measuring the central tendency or average.
 n = number of items such as shots or hits
 V = value of items such as velocity

To work the formula, determine the average velocity of all shots and find the difference between each shot and the average. Each difference is squared and the total added together. This total is then divided by the number of shots minus one. (Don't forget to use one less as the divisor.) The square root of the answer is the standard deviation. While it takes a little time, with a scientific calculator and a sheet of paper and a pencil to keep track of the figures, it is easy. **Many modern chronographs will perform this operation in a fraction of a second as a routine function.**

An illustration would be a cartridge with an average velocity of 2250 f.p.s. and a standard deviation of 15 f.p.s. It is likely, then, that a shot will fall in the range between 2235 f.p.s. and 2265 f.p.s.

Target groups

For standard deviation of a target group, the formula and math are the same as for velocities, but the measurements are more involved than most people are willing to do. First, the center of the group must be accurately found by measuring each bullet hole's top and bottom points from a standard reference line and the averages computed. Each bullet hole must be measured from its center to the center of the group.

Target groups are only an indication of accuracy if interpreted correctly. If just one shot is fired and it hits the center of the target, that is regarded as great accuracy. If 5 shots are fired and the shot that is dead center is the last, the same will be believed. The other shot's poor locations will be blamed on errors corrected by the final shot. Actually, the other shots have to be considered and that is the reason for the insistence on a sufficient number to assure a true representation.

DISPERSION AT RANGE

It is an error to believe that group size will stay in proportion as the range increases. In other words, if a rifle will shoot a 2" group at 200 yards it will **not** shoot a 4" group at 400 yards. When the distance is doubled the group size will almost always be more than double.

With the decreasing velocity and both the side deflection and drop increasing by accelerated motion, the group cannot change at a proportional rate. Under ideal conditions and with both a great gun and a skilled shooter it may be close. For most of us it will be a larger group size.

NOTES

◆ In most groups, the mean vertical and horizontal deviation will be the shortest number of inches. The next shortest will be either the standard deviation or the mean radius. Next will be the extreme spread with the diagonal as the last and the longest.

◆ The dimensions that extend to the outside edges of the group are most responsive to the amount of shots in the groups. The dimensions of mean deviation and mean radius and standard deviation are not as affected by the amount of shots and are more efficient.

◆ Groups fired in bench-rest competition are measured from the center of the holes. This center to center method of measurement gives figures that may appear strange to people not familiar with the sport. For example, a .308 caliber hole that was exactly 1 inch center to center from another .308 caliber hole would have a 1.308 inch distance from the outer edge to the farthest outer edge. It would also be possible, and is common with top shooters, to have a group size smaller than the diameter. In this example, if the center of a .308 diameter bullet hit .125 inch from the center of the other, the hole would be, at its widest, .433inch. The group measurement would be smaller than the diameter at .125 inch.

◆ *Power factor* is not really a ballistic term, in the true sense, but it is a figure used to determine what firearms are acceptable in competition. The National Rifle Association Action Shooting contest requires a minimum of 9mm and a power factor of 120,000 minimum. This is obtained by multiplying the velocity in f.p.s. by the bullet weight in grains.

IPSC rules add one more step. They require the bullet weight multiplied by velocity and the answer is then divided by 1,000. To be classified as in the "major" category, the final number must be at least 175.

CHAPTER 19

MAXIMUM RANGE

This discussion of maximum range is not for the purpose of shooting game at long distances because we are discussing ranges too long for practical hunting. Safety and essential knowledge are our primary goals. The hunter shooting at a deer on a hill, for example, could miss and hit a child in a farm-yard 2 1/2 miles away. (The 7.62 NATO cartridge at 30° elevation has a maximum range of 4,400 yards or 2 1/2 miles.) Every hunter, police officer or anyone else that owns a gun, has a duty and a responsibility to not hit innocent people. This is so obvious it should not be necessary to mention it, but every year a few people are killed from these senseless accidents. That adds fuel to the fires that are stirred up by the anti-gun crowd. Life is precious. Let us always be safe. If a criminal needs to be shot in self defense, so be it. But don't hit a child a mile away.

ELEVATION & RANGE

With no air, maximum range would be at a 45° elevation and the only point to consider would be velocity. In this case, divide the velocity by ten and square the result for an answer in yards. Thus, under vacuum conditions, a projectile at a 2,060 f.p.s. muzzle velocity would go about 14,145 yards. This is a theoretical vacuum range, and remember, it is pulled down by gravity.

This is only for academic knowledge, because none of us will ever be firing a gun in a vacuum, but the basic idea helps us to understand the real world.

Because of air resistance, maximum range will be with the gun barrel elevated to an angle well below 45°. The maximum will normally be between 27° and 35°. It will usually be closer to 31°. As the barrel is raised higher than the optimum angle, say 31° for example, the range will start to get shorter. The most efficiency is obtained at the lower angles. In many cases, as much as 65% of the maximum range can be obtained at an elevation as low as 5°. The higher the barrel is raised beyond a rather low point of about 7° to 8°, the range will increase but at a very low rate. Generally, 8 degrees elevation will deliver about 85% of maximum range. This clearly shows the danger of shooting without a backstop. (A bullet with a 2 1/4 mile range would travel 1.9 miles at 85%. - 3,344 yards or 10,098 feet.) This is based on generalities and all bullets and loads will be different.

There is no satisfactory formula for working out maximum range in an atmosphere environment. Manufacturer's charts and military data are the best source for this information. Properly conducted field testing is within their capability, but it is not practical for the sportsman.

Many other subjects discussed in this book are also pertinent. Air density affects range as does bullet weight, ballistic coefficient, velocity, etc.

A method was given in the second paragraph of this chapter to determine an approximate distance in a theoretical vacuum and most large caliber bullets at low muzzle velocity will travel from 15% to 20% of that distance. The lower the velocity and the higher the ballistic coefficient, the better the bullet will do compared to its no-air range. Light weight hunting bullets at high velocity will attain only about 4% to 10%. Pistol bullets of lower velocity, 10% to 20%.

From the chart and the discussion presented, most other bullets of similar ballistic coefficient, velocity and weight can be interpolated or closely estimated.

Thin, less dense air from high altitude or high temperature will also increase range. Air resistance can never be forgotten. After all, air resistance is the factor that lowers the elevation for maximum range from 45 degrees down to 27 to 35 degrees. At an elevation of 12,000 feet, the air has thinned enough that the maximum range for most bullets is extended 38% as compared to sea level.

HELPFUL CHARTS

The British Textbook of Small Arms has complete figures for a boat-tail Swiss bullet of about .30 caliber, 174 grains with a ballistic coefficient of .406 and a muzzle velocity of 2,600 f.p.s. The maximum range obtained is 4,457 yards at a departure angle of 34° 42'. The book and the figures are from 1929, but it is worth repeating because time does not change things of this nature. The basic points and the laws of physics are still good today. Modern cartridges and bullets will follow the same general pattern.

Elevation in deg.	Range in yards	Flight time in sec.	Final velocity in f.p.s.	Angle of fall in deg
5	2464	7.56	588	11.25
10	3273	12.65	447	24.30
20	4097	20.95	390	47.68
30	4423	27.91	405	62.40
34	4455	30.20	416	66.15
35	4456	31.06	417	67.40
40	4413	34.01	429	71.35

CALIBER and BULLET	BULLET WEIGHT in grains	MUZZLE VELOCITY in f.p.s.	BALLISTIC COEFFICIENT	EXTREME yards	RANGE miles	PERCENT of maximum vacuum range
.22 LR	40	1145	0.128	1500	0.85	11
.22 WMRF	40	2000		1800	1.02	5
.223 Rem.boattailSP	55	3240	0.46	3867	2.19	4
.243 Win. flatbase	100	2060	0.37	4000	2.27	5
.30 M1 carbine	111	1970	0.179	2200	1.25	6
.30 ball M-2	152	2880	0.4	3500	1.99	4.5
.30 boat-tail M-1	172	2600	0.56	5500	3.12	8
.30-'06 flatbase	180	2700		4167	2.37	6
.30-'06 boat-tail	180	2700		5667	3.22	8
.30-40 Krag	220	2000	0.34	4050	2.31	10
.40 Mauser	355	1580		3554	2.02	14
.433 Mauser 11mm	386	1427		3281	1.86	16
.45 Springfield Carb	405	1150		2800	1.59	21
.45 Springfield	500	1315		3500	1.99	20
.380 ACP	95	970	0.08	1089	0.62	12
.38 Special +P	158	890		2133	1.21	27
.357 Mag.	158	1235		2367	1.34	16
.45 ACP	234	820	0.16	1640	0.93	24
.44 Mag.	240	1390		2500	1.42	13

Here is a brief list of some cartridges which have had their maximum range determined by military ordinance and ammunition manufactures. Notice the percent of maximum vacuum range and that some of the distances are very long.

The .45 ACP at 820 f.p.s. and 1,640 yards was based on a standard 4" barrel at a 30 degree elevation. Fired from a longer barrel in a machine gun, the muzzle velocity increased to 920 f.p.s. and the maximum range increased to 1,760 yards at the same 30 degree elevation.

Most cartridges have not been tested for maximum range, but a little careful thought about those that have been, will yield a rough estimate for the others. Military cartridges, such as the .30-'06 and modern NATO and U. S. Military calibers have been well researched. Comparisons with other cartridges can be made for approximate maximum range. The ballistic coefficient, velocity, bullet weight and more must be considered.

.22 R.F. LONG RIFLE ELEVATIONS (standard velocity)

This table gives the angles of elevation and the range out to the extreme. Note the maximum range is at 25 degrees. Any increase in elevation beyond 25 degrees has little change in range to about 33 degrees. Above that, the range decreases.

Range in yards.	Angle of elevation	
	deg.	min.
100	0	16
200	0	34
300	0	56
400	1	29
500	2	10
600	2	56
700	3	51
800	4	56
900	6	13
1000	7	44
1100	9	32
1200	11	41
1300	14	16
1400	19	00
1450	25	00

Research by the U.S. Ordinance Dept. found .22 R.F. long rifle cartridges to have an extreme range from 1,050 yards at 8.8 degrees to 1,252 yards at 17.8 degrees. The same elevation, 17.8 degrees, obtained the maximum range of 1,324 yards. They concluded a maximum range between 1,250 to 1,350 yards for the .22 long rifle cartridge. A tail wind would increase that somewhat.

Researchers have a problem in gathering data on maximum range because it is difficult to be certain where a bullet has landed. Smooth water or wet sand is good, but still there are problems. (This is similar to the testing on vertical shooting as discussed in Chapter 27.)

U.S. ARMY ELEVATION TABLE

7.62 m.m. service ammunition. Note that *mils* is used by the military because a mil is about one-thousandth the range. One mil equals 3.375 minutes of angle. (The mil is explained in detail in Chapter 18.)

Range in meters	Range in yards	M59 Ball mils.	M59 Ball min.	M80 Ball mils.	M80 Ball min.
100	109	0.7	2.4	0.8	2.7
200	219	1.6	5.4	1.6	5.4
300	328	2.6	9.1	2.6	9.1
400	438	3.7	12.5	3.7	12.5
500	547	5.0	16.9	5.0	16.9
600	656	6.5	21.9	6.6	22.3
700	766	8.3	28.0	8.5	28.7
800	875	10.5	35.5	10.8	36.5
900	985	13.1	44.2	13.5	45.6
1000	1094	16.2	54.7	16.7	56.4
1100	1203	19.9	67.2	20.5	69.2
1200	1313	24.1	81.5	24.8	83.8
1300	1422	28.8	97.3	29.5	99.6
1400	1532	34.0	114.8	34.7	.117.2
1500	1641	39.7	134.0	40.3	136.0

The use of the word ball, does not mean the projectile or bullet is in a spherical shape. It is just a military term for a general purpose bullet designed for combat and training requirements.

M-59 Ball is a U. S. cartridge 7.0 gm non-streamlined bullet, lead core, Vo 853m/sec in M14 rifle.

M-80 Ball is a U. S. cartridge 9.65 streamlined bullet, lead core, Vo 854 m/sec in M14 rifle.

BALLISTIC TABLE, 7.62 x 51 Ball M80

Range meters	Velocity m/sec	Energy J	Drop mm	Elevation mils	Vertex mm
0	854	3,519	0	0.00	0
100	778	2,920	79	0.75	20
200	709	2,425	317	1.60	85
300	642	1,988	659	2.63	211
400	578	1,612	1,455	3.77	418
500	518	1,295	2,456	5.09	730

SHOTGUN RANGE

The maximum range for shotgun pellets is limited because a sphere has a poor ballistic coefficient compared to a pointed projectile. *Ordnance Technical Manual 9-1990* gives maximum ranges for various shot fired in full-choke guns. No. 1 buckshot at 600 yards; no. 8 shot at 230 yards; no. 9 shot at 210 yards, etc.

Journée's formula for determining the approximate maximum range for a shotgun is used frequently. The formula simply states that the maximum range in yards is about 2,200 times the shot pellet diameter. While the figures do not agree perfectly with the others, they are close enough for most uses.

These figures are calculated with Journée's formula.

Load	Diameter	Max range in yards
12 ga. rd. ball	.645	1420
16 ga. rd. ball	.610	1340
20 ga. rd. ball	.545	1200
410 ga. rd. ball	.38	850
OO buckshot	.34	748
O "	.32	704
1 "	.30	660
#1 shot	.16	352
2 "	.15	330
3 "	.14	308
4 "	.13	286
5 "	.12	264
6 "	.11	242
7½ "	.095	209
8 "	.09	198
9 "	.08	176

A shotgun pellets maximum range varies with velocity and mass. A tail wind of only 10 m.p.h. can increase the range by about 15%. Thin, less dense air created by higher altitude or high temperature will also increase range. The reason, of course, is less air resistance (drag).

Slugs are discussed in detail in Chapter 21. For maximum range, it is important to remember that while they start out with a good muzzle velocity, they cannot hold it and run out of velocity, energy and range very quickly. They have poor ballistic coefficients and poor sectional density as compared to handgun and rifle bullets.

CHAPTER 20

WIND DEFLECTION

TERMS

Before we get started with this interesting section, we need to clear up a problem in terminology. The sideways movement of a projectile in flight by a cross-wind is correctly called deflection, not drift. Drift is the term for the side-movement caused by spin and gyroscopic precession. Deflection is the proper term for the side movement caused by the wind. You say you have been calling it drift for 40 years and can't stop? Fine. That's okay. Just so you know that it is supposed to be called deflection.

To make it even more confusing, if we check the dictionary, we find that drift is the best word for movement by air. In aircraft navigation, drift is used to describe the movement from the intended track by wind while deviation is used as the difference between the indications of a compass on a particular aircraft and the indications of an unaffected compass at the same point on the earth's surface. In other words, the magnetic effect caused by the aircraft itself.

With the passing of time, this word usage has became standard and proper. If it is confusing and seems silly, so are many other terms in the English and American language. As stated at the beginning of Spin Drift, as long as the listener or reader understands what is meant, either term should be satisfactory. For this book, we will use the words as explained.

Windage is used to describe wind deflection and it can also mean the sideways movement of a sight. Kentucky windage has become the term to use for hold-over to compensate for wind by guess or estimate. Early Kentucky rifles (most made in Pennsylvania) did not have adjustable sights so holding-off was required.

In the early days of firearms and cannons, windage had a very different meaning. With all smooth-bore arms, particularly cannons, windage was the difference between the diameter of the ball and the diameter of the bore; in other words, the clearance or looseness. In cannons, it was usually about 1/40 the diameter. It helped in loading and made the balls useful in different weapons.

Windage has also been used as the term to describe the surge of air (aerodynamic compression wave) caused by a moving projectile.

In this section, we will use windage in today's common usage as it relates to the allowance for wind deflection.

The wind simply blows the bullet off course and a correction angle into the wind is required.

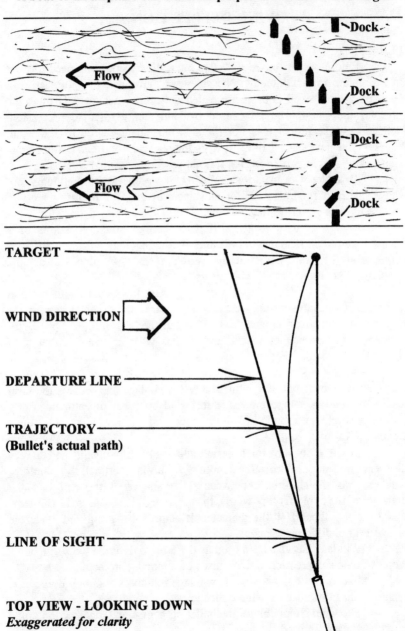

ESTIMATING WIND VELOCITY

Wind deflection is always a problem because it has to be estimated. Range is known in target shooting and for most hunters, easier to judge. Wind speed estimates and figuring the correction angle is called *wind doping*. While it is difficult, it can be made a lot easier by knowing the ballistics involved.

The shooter must estimate the wind speed by how it affects both him and his surroundings. Of course, an instrument known as an anemometer measures the wind velocity, but they are seldom available and usually not practical. They measure the wind at only one location and the wind can vary from place to place.

Flags set along the range can help in target shooting, but they can also be a source of aggravation. Hunters can sometimes get a feel for the wind across a space by the movement of tall grass or leaves and small branches. Experience and intuition can help.

A wind speed of 1 or 2 is calm and can be disregarded. From 3 up to about 6 is a light breeze, noticeable but not a problem to hold a match or a paper. From 7 up to 12 it is getting rather strong and on a cold day it will begin to feel very uncomfortable. If you are a smoker, lighting up will require a shield. At 15 it will get impossible to keep your cowboy hat on your head. At 20 it is at the point where if the wind is from anywhere more than about 15° from the bullets path, an accurate shot at the longer ranges will be extremely difficult and require a high level of skill.

At a wind speed of 25 or more, if you are target shooting for fun only, fine. For any other reason, good luck. Hunting in a high wind? Practice your Indian (Native American) stalking techniques because in a 25 m.p.h. wind, you had better be darn close.

The shooters who claim they can get consistent and accurate results in a high wind deserve a big compliment and are to be commended. They are indeed rare. (Or are full of baloney.)

CROSS WIND ANGLE

In long range shooting, cross wind deflection will cause more problems than elevation errors, even if the cross wind is mild.

The angle of the wind from the bullets intended flight path was given in an earlier paragraph as 15°. Usually a comparison with a clock is applied, as in the old movies where a fighter pilot yells into his microphone, "Bandits at 2 o'clock," or whatever. The shooter will visualize themselves at the 6 o'clock position and the target straight ahead at 12 o'clock. The 90° wind from the right would be 3 o'clock and 90° left would be 9 o'clock. Generally, winds from 2 and 4 on the right and 8 and 10 on the left will have nearly as much effect in deflecting the bullet as the 3 and 9 positions. (See next page.)

WIND DIRECTION *CLOCK POSITIONS* & DEFLECTION EFFECT

A wind speed of 20 at the 3 or 9 position will have an effect of about 18 m.p.h. at the 2, 4, 8 and 10 positions. These will have a cross wind component of 86 percent. The same 20 m.p.h. wind at 1, 5, 7 and 11 o'clock positions will have a 10 m.p.h. component. In other words, half the effect with a cross wind component of 50 percent.

In military terms, winds from 2, 3, 4, 8, 9 and 10, o'clock positions are called *full-value*. Winds from 1, 5, 7, and 11 o'clock are called *half-value*. Winds from 6 and 12 o'clock are called *no-value*. You will notice from the chart (figure 32) that the military wording is not very good for the 2, 4, 8 and 10 o'clock positions.

No wind is exactly steady. It always is a little gusty and can change speed by as much as 50 percent in just seconds. The direction of the wind also changes quickly and can shift back and forth by as much as three hours. (Using the clock method.)

Wind that is close to head on in the 12 o'clock position, or tail on in the 6 o'clock position, are not as favorable as they may appear. While it is true that no wind correction is needed, if the wind is moderate to strong in velocity it can affect the bullets path. The wind direction will constantly fluctuate, sometimes by as much as 30°. The result will be group shots that are bigger than expected or missing a rabbit that should have been hit. Technically, a 12 o'clock head-wind will lower the impact point and a 6 o'clock tail-wind will raise it. However, the amount will be very small.

One way to see wind changes is by watching the heat waves known as the mirage. Best seen by telescope-sight or spotting scope, it is the shimmering of the distant view by the heat. The light rays are bent as they pass through different layers of density. Refraction is a word used for this phenomenon. The deflected light rays must be considered in long-range shooting and can by used to dope the wind changes. Long-range target shooters get very expert in mirage reading. This is a fascinating subject, but it does not come under ballistics, so we will not go into it deeper.

Caution note: There is little benefit in using winds from the weather bureau because the wind can vary considerably from one area to another and from one shot to the next. Just in case, remember they give wind velocity in knots and most of us are expecting statute miles per hour. To convert, multiply nautical miles by 1.15.

EFFECT OF WIND

Of course, the whole point in this discussion is that the wind will push a bullet off course. In some cases, by a huge amount. **Delay time** is the main point to discuss, but other ingredients enter into the recipe.

Time is a major factor. The more time a bullet is in flight between muzzle and impact, the more exposure to the wind and the more it will be blown off course. Normally, a longer range increases time and has the same effect. If other things are equal, the deflection will be directly proportional to the length of time the projectile is in flight. If two bullets have the same shape, density, stability, weight and an equal wind blowing on each, the bullet in the wind the longest will obviously be blown off the most.

Weight is also a factor because momentum is a result of weight times velocity and momentum tends to keep the bullet straight. If one bullet weights 200 grains and another 100, the heavier will be harder to push away from it's path. As mentioned, the higher velocity, besides spending less time in the cross wind, will also have more momentum.

Density is a factor as well. If two bullets are of the same weight, the less dense would be bigger and have more surface for the wind to act upon. In other words, the more density, the less the bullet will be influenced by the wind.

Velocity also has an effect. If we have a light bullet moving at a faster velocity than a heavy bullet, so that the momentum is equal, the lighter bullet will lose speed at a faster rate. The heavier bullet will then be blown off course less. If we raise the velocity of a light bullet so its momentum is higher than a slower-heavier bullet, the lighter bullet will be moved less by the cross-wind. The problem with this situation is that the lighter bullet will not be able to hold velocity, as compared to the heavier bullet which may hold it fairly well. At some point in their trajectories, the momentum of the light bullet will drop to a point where its advantage will be lost. The slower-heavier bullet may then become the one with the best wind resistance.

As much importance as we have put upon weight, the ballistic coefficient and high velocity are the main factors in cutting wind deflection.

Caliber	Bullet type & weight	Muzzle velocity f.p.s.	Defl. with 10 m.p.h. x-wind at 4 o'clock			
			200 yards in inches	300 yards in inches	600 yards in inches	1000 yards in inches
.30-'06	172 grain M1 boattail	2700	2.2	5.8	25.2	69
.30-'06	150 grain spitzer w/ flat base	2700	3	8	36	115
.30-40	220 grain round-nose	2160		15.5	68.5	191
.45-70	500 grain	1200		25	75	220

DELAY (LAG) TIME

Delay time or lag time is the difference in time of flight that a bullet takes to traverse a given portion of its flight versus what that time would be without drag (air resistance). To express it simpler, it is the difference between the bullet's theoretical flight time in a vacuum and the actual flight time. Air resistance is explained in detail several places in this book, especially in Chapter 13. For now, just remember that it slows the projectile's velocity so it does not remain constant. It is decelerating.

The delay time will be almost the same in each section of a long range shot. Not exactly the same, but so close that it is unimportant for our computations. The bullet's velocity slows during a long range shot and that in turn slows the time to cover the same distance. They just about cancel each other out and the delay time in the first 50 yards will be close to the delay time in the last 50 yards.

♦ For delay time, divide the range in feet by the muzzle velocity in feet per second and find the difference between that time and the actual flight time in seconds or fractions of a second.

$$T_v = R / V$$

Where: T_v = vacuum flight time

R = range in feet

V = muzzle velocity in feet per second (f.p.s.)

The vacuum flight time (T_v) is usually expressed as R/V which is the range divided by the muzzle velocity in f.p.s.

For a clarification, we could compare a bullet with an airplane. Say that an airplane started a planned 1 hour flight. If it ran into an unexpected head-wind and the trip took 1 hour and 10 minutes, the 10 minutes could be called its delay or lag time.

Slow bullets have a high delay time. This drops as the velocity is increased to about 750 or 850 f.p.s. Then as the velocity enters the transonic range, the delay time goes up fast through the entrance to supersonic and starts to drop again at the high side of the transonic range. (These terms are explained in detail in velocity and drag, Chapter 13.) From this point, as velocities increase, the delay time drops in a steady rate with no more increases. To express that in simpler terms, as the velocity increases, the delay time drops except for a sudden increase for bullets moving at or anywhere near the speed of sound. Another section explains the complex forces in action in the transonic range.

Delay time is about in proportion to the square of the range. Double the range and the delay time will increase 4 times. Triple the range and the delay time will be about 9 times as long. Notice the use of the word *about*. This is just an approximation.

The lowest delay time will result from using the longest bullet available for the caliber. Why? Remember weight and density, etc.

◆ **Displacement in feet:**

For short range and where the barrel is at a low angle to the horizon, the following formula is used. Known as the Didion's formula, **it is the most common formula used for wind displacement.**

$$D = W_z (T - T_v)$$

Where: D = deflection (displacement) in feet.

W_z = cross wind velocity or wind component if not at 90 degrees. (in f.p.s.)

T = Time of flight in seconds for the range involved.

T_v = Vacuum flight time (flight time if without an atmosphere). Usually stated as R / V which is range in feet divided by muzzle velocity in f.p.s.

◆ **Displacement in feet with high elevation angles:**

For long range firing that involves high angles of elevation in relation to the horizon, this formula can be used. If the reader has math tables or a scientific calculator, it is easy. For moderate range, cosine y can be omitted, although that removes the advantage of this particular formula.

$$D = W_z [T - (X / V \cos. y)]$$

Where: D = bullet wind displacement in feet.

W_z = cross wind velocity or wind component if not at 90° (in f.p.s.)

T = time of flight

V = muzzle velocity in f.p.s.

X = range in feet (Note: feet not yards.)

cos. y = cosine of elevation angle. (See figure 35)

For an example, consider a .338 Win. cartridge at a range of 500 yards times 3 = 1,500 feet. A 10 m.p.h. wind times 1.467 = 14.67 f.p.s. Flight time 0.744 second with a muzzle velocity of 2,700 f.p.s. and for our exercise, a 5° elevation angle. Both a chart of Natural Trigonometric functions and a scientific calculator agree on .996 for a cosine of a 5° angle. 2,700 times .996 = 2,689.2 divided into 1,500 = 0.5578. For T we use the .744 subtracted by 0.5578 = 0.1862 times 14.67 = 2.73 feet. At first this appears too low because we are expecting bigger numbers in inches, but it is in feet and therefore correct. As you can see, the low 5° angle caused the velocity to be multiplied by .996 and changed the answer very little. Higher angles will cause bigger changes. For example, a 30° angle would have used a cosine of 0.866 for a larger change

◆ **Displacement in inches:**
Below is a formula for displacement in inches. It is easier than it appears.
$$(R / V = DT) * W_z = D$$
Where: D = displacement in inches
R = range in feet
V = muzzle velocity in f.p.s.
DT = delay time
W_z = cross wind velocity or wind component in inches per second.
(Note: inches.)

◆ **Displacement in Minute of Angle**
Below: The U.S. Army used this formula for target shooting with the M1906 rifle and M2 ammunition. The answer is given in Minute of Angle. (MOA)
$$WC = W_v * R / 10$$
Where: WC = windage correction in MOA
W_v = cross wind velocity or component
R = range in yards

◆ **Time of flight:**
Below is a formula for computing flight time.
$$W_z = 16.1 * T^2 / T - R / V$$
Where: W_z = time of flight
16.1 = 1/2 gravity
R = range in feet
V = muzzle velocity in f.p.s.

All of these formulas are much easier than they appear. The first is one of the most common, but they all serve a purpose. With a look at the terms that are used in the formulas and the answer that is obtained, the advantage of each can be seen.

This whole wind deflection problem can be confusing. Earlier we mentioned the bullet's time in flight as being a factor, which it is. But it is by no means all the picture. If we drop a bullet to the ground from the muzzle in a 10 m.p.h. wind it will not be blown to the side as much as a bullet fired at a target or game, even if they are both in the air about the same amount of time. Key points are the delay time and the speed of sound.

Several key points can be learned from the formulas that have also been proven by testing.

If the velocity, among other things, is equal, the heaviest bullet is deflected the least. If the pressure level is kept equal, the heavier bullet and the lighter bullet, (other things also equal) will be deflected about the same by the wind. The heavier can buck the wind better but the lighter will have more velocity. (Note: we are discussing only wind deflection at this point. Not killing power for hunting or self defense. That may be a lot different.)

The ballistic coefficient of a bullet is explained in detail elsewhere in this book, so no details will be repeated here. However, it does affect wind deflection as it lowers or increases air resistance and thus influences the time of flight, velocity holding ability, etc. A bullet with a good ballistic coefficient will do better in a wind. The muzzle velocity is important, but the bullet has to be able to maintain a fair amount of it down range.

.22 CALIBER PROBLEMS

An increase in velocity decreases wind deflection, right? Normally. An exception to that rule is a .22 long rifle bullet. The airflow at Mach 1.0 is the reason. Air resistance is increased in this velocity range and a lot of air turbulence and pressure changes occur.

The delay time is affected out of proportion. The effect is in the velocity range of 1,000 to 1,350 f.p.s. but the .22 rim fire is almost the only cartridge in this range. (The .32-20 Win. and the .44-40 Win. and a few others are also included.)

Wind deflection is not controlled by time of flight but by the loss of velocity during the time of flight. During the transition velocity range above the speed of sound, the drag increases disproportionately higher than the velocity increases. In other words, if the velocity is increased by a small amount, the drag increases by a large amount. This situation is unique to this velocity range. At all other velocities, an increase in velocity will bring about a decrease in wind deflection. Not in the upper transition range. This is the reason for the special .22 match ammunition that is loaded so it will not go fast enough to get into this velocity range. The higher velocity .22 ammo. has a flatter trajectory but is deflected more by a cross wind. (Note: for more details, see Chapter 13.)

Many people incorrectly believe that an increase in velocity has no benefit for cross-wind problems. This idea is due in part to when the .40 grain .22 long rifle bullet was introduced in the late 1920's. Its approximate 1,330 f.p.s. performed worse than the standard bullet at 1,150. No one in those days knew of the problems with velocities near the speed of sound.

WIND & PROXIMITY TO SHOOTER

This is a little known point that is both important and intriguing. **A wind close to the shooter will cause more deflection than the same wind at the target: not a little but a lot more.** The reason is the closer to the muzzle the wind is located, the quicker the lateral momentum or deflection will begin. Once it has begun, it will continue to move off more and more. The projectile will have a sideways momentum established. Only a slight wind will be required to accelerate the deflection.

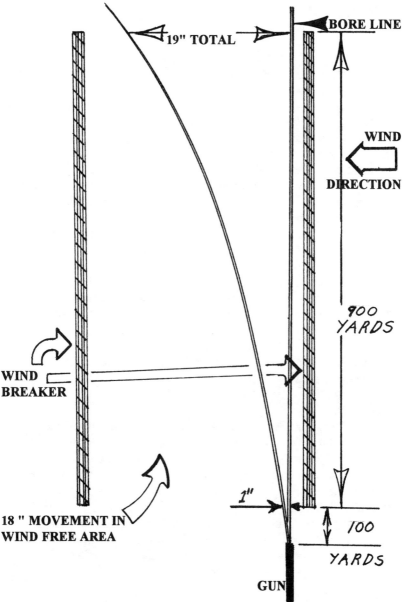

1,000 YARD TOTAL RANGE. *Exaggerated for clarity & not to scale.*

A very strong wind at the target area is not too important if the rest of the range is protected. Of course this will also apply to a hunter in the woods with a clear shot at game in a windy open area.

For an example of this, consider a 1,000 yard range that is shielded from the wind at all but the first 100 yards. A culvert, a long building, a mountain, thick woods, a tunnel; anything that is 900 yards long. Also we will consider a bullet that is blown off or deflected by the wind 1 inch in 100 yards. The wind speed for our example is unimportant, but we will say 17 m.p.h. As the drawing shows, the bullet is blown 1" from its straight path at the end of 100 yards.

That part is easy. Now it gets more tricky. The bullet leaves the muzzle at a cross component of zero. The rules of physics state that to reach 100 yards at a 1 inch side displacement, it must obtain a 2 inch cross component. (Remember, it has to accelerate from zero.) The bullet will enter the shielded area at 1 inch off the zero line but with a sideways movement of 2 inches per 100 yards. (.02 inch per yard.)

The bullet enters the protected area with a side movement of 2 inches per 100 yards already established and Newton's first law of motion says that it will continue its lateral movement without the wind's push. In the 900 yard shielded area it will move over another 18 inches. This 18 inch side-way movement was caused only by the wind in the first 100 yards. If the range was all exposed to the wind with no shield, simple math would show a total deflection in the 1,000 yards of 100 inches.

In the no shield case, the bullet would move sideways 19 inches in just the last 100 yards, but that is not because the wind at impact end is what is important. The bullet had reached that side-way velocity at reaching the 900 yard point.

VELOCITY

Although some gun authorities and sportsman believe otherwise, it is proven that for the least wind deflection, the highest velocity is best. The bullet should be able to hold that initial velocity and a bullet shaped for the least drag and with some weight will do the best. The sectional density is also a consideration.

BULLET CHOICE

Any bullet with a large blunt nose should not be used for long distance in a wind, if it can be avoided. The resistance to wind deflection is very poor because of its lack of a good aerodynamic shape. Bullets and their design is covered in detail in Chapter 16.

HOLD-OFF & SIGHT ADJUSTMENT

Holding-off to compensate for a cross wind requires moving the aim point into the wind and letting the wind push the bullet over toward the targets center or the intended point of impact. This requires skill to do well. Experience and practice will pay off with more accuracy.

Tables have been drawn up that can be used to make estimates of wind drift. Some basic mathematics is still involved in interpolation between ballistic coefficients, cross wind angle, etc. The chart-makers considered ballistic coefficients, range, cross wind and muzzle velocity. The tables are too long and require too many pages to reprint here. The Hornady Manufacturing Co. of Grand Island, Nebraska, prints an excellent reloading handbook with these tables included. It also has their ballistic tables and cartridge and bullet information.

EXAMPLE OF HOLD OVER TO STRIKE OCTAGON

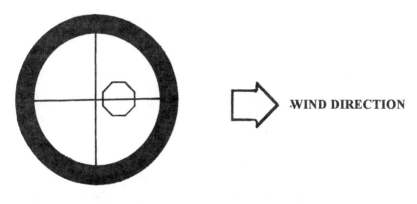

NOTES:

◆ It is probably unnecessary to mention this, but it is important to remember that wind deflection is not just a problem for rifle bullets fired at long range. The short and moderate ranges can be influenced as well. It also affects shot pellets, shotgun slugs and sabots, handgun bullets, etc. Even an arrow or a thrown rock. The wind does not discriminate.

◆ Another "unnecessary to mention" point is that all wind corrections are made into the wind.

◆ A seldom used rule of thumb formula that determines the approximate wind correction in MOA requires a constant for each cartridge. It involves multiplying the range in hundreds of yards by the wind velocity in miles per hour and then divide by a constant for the cartridge. The answer is the MOA wind correction angle.

CHAPTER 21

SHOTGUN -- PELLET & BUCKSHOT BALLISTICS

As we begin this discussion, we must remind ourselves that the physics relating to ballistics are not suspended for shotguns. Their projectiles have to obey the rules of nature the same as a bullet. Steel shot, lead shot, slugs; they all are affected by drag, bore friction and trajectory problems; all the numerous items discussed in other parts of this book. The pattern can be influenced by wind direction and velocity, humidity, temperature, shot size, ammo used, ammo type, load, etc. The gun itself has built in problems with the choke constriction, length of straight portion before choke, angle between straight portion and constriction, etc. A slug is a single projectile fired from a smooth bore or rifled barrel while the pellets are fired in clusters or groups. The slug follows the rules as does a bullet, but with its own special problems. Shot-shell ballistics are controlled by gage, choke, barrel length, shot weight, type of crimp, crimp resistance, altitude, pellet size, wad compressibility, wad sealing ability, case type and shape, powder type, powder amount primer type, etc.

CHOKE

The choke is a constriction that *squeezes* the pellets and controls the pattern as it leaves the muzzle. Perhaps squeeze is not the best word to use, but modern science is not 100% certain how a shotgun choke operates. Maybe 99.99%, but not completely certain in all details. Therefore, for lack of a better word, we will say it is squeezed.

As it leaves the muzzle, the shot is somewhat like a heavy liquid, perhaps a thick syrup. To some extent *fluid*, in other words. While science cannot tell exactly how the choke works, it can tell us how it doesn't work. It is not, as some believe, like a garden hose nozzle. The restriction is slight, by comparison, and it does not increase the velocity.

Most modern chokes are either cut in the inside of the barrel by reamers that are made to the desired dimensions and shape, similar to the method of cutting chambers, or they are swaged.

Rotary swaging is a method of reducing the cross-sectional area of the barrel. This is a process involving squeezing the steel blank on the outside. The shape is formed by compression from high pressure. This increases the hardness and the tensile strength of the metal, so long as fracture does not occur from excessive action. Very close tolerances can be held by what is essentially a rotary forging machine. Formerly, some barrels were left with the outside of the barrel tapered to a smaller diameter at the choke which gave both an odd appearance and a weak wall section. Now they are machined bell shaped on the outer diameter so it will be straight after swaging.

There are different amounts of choke, from full down to a *cylinder* bore which has no constriction and therefore, no choke.

The cylinder or unchoked bore permits the pellets to disperse some at the muzzle and air resistance disperses them more. The result is a large pattern that is open and spread out with no protection from air resistance.

Shot fired through a full choke exits in a straight line with the pellets strung out in a row. The pattern at the target should be small. The pellets in front block the air somewhat from the pellets farther back in the string. Although strung out, they should stay tightly bunched.

Testing by the University College London Ballistics Research Laboratory in London, England, has shown that the trailing pellets in soft-lead patterns have only 50% of the energy at the target as the pellets in the forward bunch. This same laboratory has discovered that some long strings of pellets have some of the pellets so far behind they serve very little purpose. With hard shot loads, they find that some full chokes will create a very short pattern that has higher than normal velocity. They did one test that produced a 10% higher velocity with full choke than with cylinder choke at 20 to 40 yards. (Winchester 7 1/2 trap-loads.)

The full choke puts the most shot pellets in the smallest area, but that is not always desired. Quail that are flushed up and fly close to the hunter are an example where a full choke is usually too much. They would be harder to hit, and if the hunter was lucky enough to hit one, it would be damaged beyond use. The improved cylinder or modified is frequently used in this case. Most flying game are hunted best with improved cylinder or a modified choke. This includes pheasant and ducks at 40 yards, although frequently an improved choke will not put enough shot into the bird.

Most shotgun kills on game such as rabbits, quail, grouse etc. are shot at an average range of about 20 yards. Not very far at all. Close enough that more hits could be counted with improved-cylinder or skeet bore or perhaps in some cases, modified. Many shooters want full choke bores when they would do much better without so much choke. The pattern is too narrow and crammed. Many guns will pattern about the same at 20 yards if chocked between full and modified but will open up 8" to 10" for improved-cylinder and another 3" or 4" for skeet choke. An average at 20 yards for 12 ga. will be around 18" to 22". This may be too tight for the range at which most upland game is taken. Informal testing shows that heavier shot loads can give bigger patterns so a substantial load may be best for any choke. As declared at other places in this book, **test your own gun**. It may not perform as expected and your needs may be different than average.

The 1994 Winchester shot-shell guide recommends full choke for only turkeys. Many shooters like it for high flying ducks, and a few other select uses. Ground game fit into the same pattern with one choke not suitable for all game. Practice, experience and pattern testing of the gun will be the deciding factor. Generally, if a manufacturer of such a high caliber (no pun intended) as Winchester recommends a particular gauge, shot size or choke, it would be wise to pay attention. Many years of research go into their reports.

Choke constrictions have became less over the years, mostly because of changes in ammunition. In other words, a full choke needs less constriction to obtain the desired per cent of pellets in the circle. Constriction may vary with manufacturer, but all are using less and the published charts and tables clearly show it. A .040 inch constriction delivered a 65% to 70% pattern and now it is common to obtain 80% with only .030 inch.

Ammunition changes in shot types, crimps, wads, shot protectors; all have brought about needed changes in barrel dimensions. As we can easily understand, the less choke, the less change over the years. Cylinder is still no choke and improved or quarter choke is given now as .005 or .006 instead of the former .010 inch.

♦ With different manufacturers using different specifications as a result of their testing and field trials, and the changes from ammunition improvement, **measuring the end of a barrel to determine choke is usually futile.** There is even a huge variation between bore diameters of the same make and model in the same gauge. Test firing is still required.

♦ Browning shotguns are utilizing what they call back-boring. The bore diameter ahead of the chamber is opened up about .745 inch in a 12 ga. instead of the usual .725 inch. Browning claims better patterns because of less pellet deformation and improved velocity with reduced recoil. From a standpoint of ballistic theory, this is reasonable and has been applied earlier with success.

There are numerous choke designs and many are very good if used for what they are intended.

♦ A Russian style choke has proven effective in producing a dense and even spread for skeet and other uses at 25 yards or less. It starts as a normal 12 ga. but out toward the muzzle it tapers from the standard .725" bore to about .780" shortly before closing it down to about .760". This is then rapidly tapered out to about .795" at the opening. Several choke styles open back up to the bore size or larger at the muzzle instead of just closing in for the choke and remaining there. All of these alternative methods are more costly to produce.

♦ Rare and unusual is the *Paradox* shotgun barrel, which has 2" to 3" of spiral rifling at the muzzle. The rest of the bore is smooth. It is reported to be good with both shot pellets and slugs. *Webster's Dictionary* defines a paradox as something that is contradictory, unbelievable or absurd but may actually be true. The name is appropriate.

PERCENT OF REDUCTION

This is the difference between the choke diameter and the bore diameter in inches (the constriction). It is easy to figure, but it is easier to refer to a chart. The three place decimal numbers are thousandths of an inch. In other words, .018 would be read as 18 thousandths.

	10 ga.	12 ga.	16 ga.	20 ga.	28 ga.	.410 bore
1%	.004	.004	.003	.003	.003	.002
2%	.008	.007	.007	.006	.006	.004
3%	.013	.011	.010	.009	.008	.005
4%	.016	.015	.013	.012	.011	.007
5%	.019	.019	.017	.015	.014	.010
6%	.024	.022	.020	.018	.017	.012
7%	.028	.026	.024	.022	.020	.015
8%	.032	.030	.027	.025	.022	.017
9%	.035	.034	.030	.028	.025	.019
10%	.040	.038	.034	.031	.028	.021

CHOKE LONGEVITY

It is generally believed that a choke can be shot out of a barrel. In other words, the abrasive action of the pellets will wear down the choke. Winchester tested this with a standard production Model 97 in 12 ga. Beginning in 1914 and for about the next 29 years, the weapon fired 1,247,000 shotgun shells. Winchester stated in 1943 that the gun, by actual test, was above the passing pattern percentages for a new gun of the same gauge. Not all barrels will be as good as this Winchester, but most chokes should last for many years.

It is also believed that chokes are worn down more by one type of shot than another. If this were true, fine pellets would cause more pressure and friction than larger pellets or buck shot. As Winchester showed us in the previous paragraph, a good barrel will take more wear than most shooters will ever give it.

% IN CIRCLE AT 40 YARDS

The following chart should be considered a rough guide only. There is too much variation between companies, countries and constantly changing improvements for this to be anything but an approximation. All charts are different, and to make it worse, they are frequently changed. Most charts say theirs is the standard. Of course, they cannot be standard if they are all different. **This is an average approximation.**

	Also called ⬇	constriction ⬇
full choke		70% to 75%......030"
three-quarter choke	improved-modified...	.60% to 65%.........020"
half choke	modified	50% to 60% ....011"
quarter choke	improved-cylinder..	45% to 50%..........007"
skeet no. 2		50% to 60%
cylinder-skeet no. 1		.30% to 35%..........000"

PATTERN TESTING

To paraphrase, all chokes and barrels are not created equal. There is considerable variation between manufactures and also between guns made in different countries. The actual choke amount and more important, the effect, can only be determined by test firing (patterning).

The amount of choke is based on the percentage of pellets fired into a 30" circle from 40 yards. The distance is important and it should be measured. As would be expected, the diameter of a pattern will get larger with increased distance. The area it has to cover increases with the square of the distance. In other words, the pattern thins out twice as fast as the diameter increases. (When comparing two different shot charges with a different amount of pellets, the distances of equal pattern density are the same as the square roots of their shot loads.) Normally, all gauges will be patterned at 40 yards except the .410 bore. Pattern it at 25 yards. The 40 yards can also be changed to suit a specific need. For example, skeet patterns may be tested at 25 yards, the standard skeet distance.

Don't draw the circle first. Fire the shotgun into a large sheet of paper or cardboard and then draw a 30" circle around the center of the thickest part of the pattern. This also should have been the aiming point.

For a large quantity of tests, some people use a large steel plate. Before each shot, brush on a coat of paint made from a medium consistency mixture of motor oil and white lead. The oil prevents the coating from drying hard and keeps it pliant. Use oil thick enough to prevent the mixture from running down before the shot is fired and the pellets counted. After each shot and circle is drawn, the plate can be repainted in seconds and ready for another test. Vegetable oil with about a 4 to 1 mixture of an oil based paint will also work well. (Heavy on the oil and light on the paint.)

Patterning a shotgun takes a lot of time to do correctly. The person who doesn't have the time or energy to perform the task properly will pay for the lapse at the trap-shoot or hunt or wherever the gun is used.

You have to know the number of pellets in the shell. Pellet count is important and can vary by a moderate amount even in the same box. Several in a box should be opened, counted and the average used. Divide that number into the number counted in the circle. This gives the percentage. Correctly, a number of shots should be fired. Use an average of them. Five would be a minimum for testing a new gun or shell. Ten shots is best and considered standard. This will give 95% reliability in spotting differences down to about 6% or 7%.

This is very time consuming but it is worth the effort.

Example: 218 holes in the circle from a total of 460 in the charge. 218 divided by 460 equal 47%.

Testing has proven the pattern of shot pellets does not travel out in a straight line but instead spreads further out as the range is increased. While this is known to happen, experts can't agree on why it occurs.

EXAGGERATED FOR CLARITY

VERTICAL PATTERN TESTING

Shot patterns from vertical firing have been shown to be the same as from horizontal firing. *The American Rifleman* reported back in June 1965 that test patterns had been taken on paper mounted 100 feet up and fixed to extend outward from a tower. The patterns were basically the same as from the same guns fired horizontally. With the use shotguns receive during bird hunting season, this is good to learn. From a ballistic standpoint, there is no reason to suspect the result would have been otherwise. While the direction of the pull of gravity is different, it is minor force (everything considered) that may change the point of impact but not the pattern.

TARGETING - IMPACT POINT

At the time patterning is being tested, the center of impact should be checked. Very few shotguns shoot where the owner believes they do and just throwing it to the shoulder and looking down the barrel is not a good indication. True, it will occasionally tell something, but not as often as expected.

Check for point of impact at the distance of intended use. Test it both with a careful aim and a fast point. For best results, it should be tested by the person that will use the gun. The pattern will be the same from one gunner to another. The impact point can vary with technique and body build.

There are many reasons besides shooter error for a shotgun to not hit where it is expected. A few are choke type and quality, reamer drift during manufacture, rib position and alignment. Furthermore, there are all the problems that go with fitting a stock to a shooter's size and style.

Targeting a shotgun is different in that an aiming point is put on the paper first. Shoot at the aiming point in the same way that you would if hunting. Do not aim carefully as you would with a rifle. Take several shots and check if the pattern's center is near the mark. Most hunters prefer the center of impact to be about 3" above the aiming point. Clay target shooters prefer about 9" high. Some trap-shooters will want even more height, but it is always a matter of personal preference. Any major errors to right or left as well as height will require either a change in technique or a trip to the local gunsmith, or both.

THE WANNSEE TARGET

The 16 section Wannsee target is a German invention from the 1920's. It consists of an outer circle of 30" diameter divided into 12 equal sections and an inner circle of 15" diameter divided into 4 equal sections. All 16 sections are an equal area of 43 sq. in. This is supposed to have a direct relationship to game hunted by shotgun. Two adjoining full sections equal a rabbit, two-thirds of a section equal a mallard or pheasant and one-third equal a partridge, etc. The target indicates the direct result a pattern would have on different types of game.

This chart is somewhat similar to a Russian design by A. Zernov which has more rings and sections.

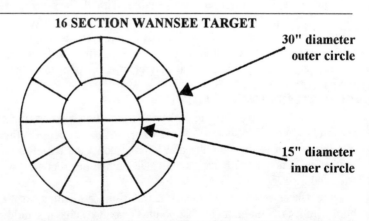

Strikes in each section can be counted and the result is the relationship between sections and the pattern as a whole. The usual pattern percentage can be determined from the percent of charge in the 30" circle. The *thickening* can be figured by the density of inner circle strikes compared to the density of strikes in the outer circle. There is three times the area in the inner 15" circle, so for comparison, the total strikes in the inner circle is multiplied by three. Divide the answer into the total strikes in the outer circle for a quantitative number.

This 16 section pattern is also useful in other comparisons as well.

To use a direct comparison in a 30" circle and thereby eliminate the multiplication, the inner circle should be 21.2 inches. This gives the inner and outer circles equal areas. The illusion created prevents them from appearing equal, but nevertheless, they are. While not as useful as a 15" circle, it is much easier.

A PRACTICAL TEST

Another interesting test, and one that is perhaps more practical, is shooting at something that resembles field shooting. After patterning, try shooting at targets of game drawn life size and placed at the range that possibly will be used. Don't aim carefully but raise the gun quickly and without careful aiming, fire. The idea is to reproduce field conditions. The result can tell a lot about aiming, range, gun fit, pattern, etc. The outcome can be a real surprise.

DOUBLE BARREL

Doubles have their own special problems. They are usually set to impact both barrels at the same point at 40 yards, but this is only with a perfect gun. And we all know how hard perfection is to achieve.

Double barrel guns, whether side by side or over-under, should be tested to see that each barrel will pattern as expected and shoot where pointed. Sometimes a change in ammo and loads can create a situation or solve a problem where the barrels don't shoot together. It is rare and the reason can remain unknown. The owner of a double who has a problem of this type should try different loads, components, etc. It may not help, but then again, it might.

Some shooters believe that an over-under or side-by-side double will change point of impact because of heat from rapid fire. The theory is based on the two barrels heating unevenly and one barrel pulling or distorting the other; i.e. a warp or bending. One barrel will heat and expand more than the other. The hot barrel will not be able to simply expand straight but will be pulled or bent some by the cooler unexpanded barrel. Some experienced shooters say that is baloney and it will not happen.

If we stop to think about it, the answer is obvious. From a purely technical standpoint it would be possible because of the attachment of the two barrels. Heat is an ingredient that cannot be ignored. But from a practical standpoint, rarely would enough shots be fired rapidly to create enough heat. More important, few people would be able to tell if a pellet group was pulled off by an inch or two at 40 yards. In this case, what would be a disaster for a rifle would be unnoticeable with a shotgun.

PELLET SIZE CHART

This chart is very complete and includes a lot of odd and unusual sizes. All buckshot sizes are listed in the Eastern system, now called the American system. The Western system, in use until 1950, is not given. Buckshot sizes are allowed + or - .015 inch tolerance.

			Pellets in 1 oz.	
Size	Dia. in inches	Dia. in mm	Lead	Steel
12	.05	1.27	2385	
11	.06	1.52	1380	
10	.07	1.77	868	
9	.08	2.03	585	
8 1/2	.085	2.15		
8	.09	2.28	410	
7 1/2	.095	2.41	350	
7	.10	2.54	299	
6	.11	2.79	225	316
5	.12	3.05	170	243
4	.13	3.30	135	191
3	.14	3.55	109	153
2	.15	3.81	87	125
1	.16	4.06	73	103
B	.17	4.32	59	
Air rifle	.175	4.44	55	
BB	.18	4.57	50	72
BBB	.19	4.83	42	61
T	.20	5.08	40	53
TT	.21	5.33	35	
TTT (or F)	.22	5.59	30	
TTTT (or FF)	.23	5.84	27	
4 Buck	.24	6.10	21	
3 Buck	.26	6.35	18 1/2	
2 Buck	.27	6.86	15	
1 Buck	.30	7.62	11	
0 Buck	.32	8.13	9	
00 Buck	.33	8.38	8 1/2	
000 Buck	.36	9.14		
00-0	.323			
2 Federal Buck	.285			
1 Special	.289			

PELLET SIZE

There are at least 24 known charts and systems of shot sizes in use by different countries. All but three assign the largest number to the smallest pellet size. Several countries use more than one system. France has eight, which is somewhat exceeding a reasonable amount. Worldwide, some methods are similar and others are as different as night and day. The Polish and Austrian systems are the same. The English and Belgian disagree on only the coarse shot. The German system is probably the simplest and most straightforward. They do not use conventional shot size numbers. Instead, the sizes are labeled by the pellet diameter in millimeters; direct and readily understood.

The importance of choosing the best pellet size for the job cannot be overstated. The heavier the shot, the better the penetration for a quicker kill but also less pellets in the shell and a weaker pattern. It is not good, in bird shooting for example, to use a shot size heavier than needed and have a thin pattern that the bird can fly through. Usually it is best to use the smallest size that will kill the game that is hunted. This will give the best pattern out to a suitable range. Beyond that range, it will be a chance kill.

PELLET SIZE RULE OF THUMB

A simple rule of thumb for finding a shot pellet's diameter (sizes 1 through 12) is to remember the number 17. The size number subtracted from the constant 17 will give the shot size diameter in hundredths of an inch. **Example:** 17 minus 6 equal 11 and the chart shows .110 as the diameter for No. 6 shot pellets. Another example: number 8 shot, take 8 from 17 for an answer of 9. The 9 is then read as .09 because the answer is the diameter in hundredths. Very little thought is required to reason out where the decimal point goes.

PELLET DEFORMATION

MISSHAPED PELLETS USUALLY *FLYOFF*

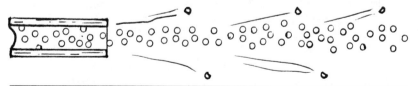

The spherical shape of pellets gives them poor ballistics even when undamaged. The pellets that come in contact with the bore are damaged by the heat, friction and abrasion. Pellets also bounce off of each other and push and hit each other. All the pellets in the load are competing for the same space. Lead pellets are softer than steel pellets so the harm is more.

Pellets that are misshaped usually fly off and reduce pattern density. The air resistance acts on them in an uneven manner causing them to gyrate and twirl from the string. These damaged pellets, even if remaining in the pattern, suffer from energy and velocity loss. Low velocity shells with lower pressures usually give a tighter pattern. This is one of the main reasons the old saying to that conclusion is true. Lower velocity can frequently cause less damage to the pellets and therefore more pellets will stay in the pattern. Besides velocity, the shot deformation can be reduced by the use of lead that is harder or *magnum grade*. The trip down the bore can be harmful to the shot and in turn, on the pattern. It is worth the extra cost for the harder shot.

The lower velocity shells also give a lower recoil and the reduced kick can be an advantage. Another detail about lower velocity shells is they lose their velocity and energy at a slower rate. In other words, while the shot may leave the barrel at 10% less than the higher-velocity load (example only), at 50 yards the difference may be down to as close as 2%. Air resistance is the primary cause. The slower shot pellets hold their speed better. If you wish, refer to Chapter 13 where velocity and aerodynamic drag are discussed. Drag increases almost as the square of the velocity. If we double the velocity, the drag increases 4 times. Triple the velocity and the drag increases 9 times. This holds for subsonic velocities. The drag at the higher speeds increases even more rapidly.

The use of as heavy a shot as suitable will also help to maintain muzzle velocity and energy.

Pellet deformation, and thus erratic and unpredictable flight, can be caused by hand-loading errors. The most common is not using proper wad seating pressure. This prevents the wad from springing and absorbing some of the shock. The bottom pellets can thus be deformed during the rapid acceleration. While this is not as fussy with todays materials as in years past, it still should not be ignored.

AIR FLOW

In Chapter 13 there is information on airflow, resistance and the pressure flow around a moving bullet. Shot pellets and other spherical projectiles pick up a *boundary layer* of air that is next to the object and moves with it. The air directly next to the object moves at the same velocity while air further out also moves but at less velocity. This is similar to bullets and other projectiles. Readers familiar with aircraft and aerodynamics will be familiar with the term boundary layer. The meaning is similar.

SPREADER LOADS

Modern chokes and ammo make long-range shots easier than in the past. The short range shot at ruffled grouse, for example, demands a wide spread with a pattern that is even and nicely spaced.

With the improvement and popularity of choke tubes and other methods to vary the constriction, the use of loads that would spread the pellets out in a full choke is not as popular as in the past. Once factory loaded and now only loaded by hand, spreaders will cause a full choke gun to perform as a modified or improved. It isn't that precise or simple, of course, but that is the basic idea and it does work.

There are two methods of loading spreaders. One is to cut 2 pieces of cardboard so they can be interlocked and placed in the shell at right angles to each other in the shape of an X. This divides the pellets into 4 separate compartments as viewed from the open end. The pellets are then poured in to the four areas.

The easier and faster method is to pour one third of the pellets into the case. Next place a cardboard wad or divider and another third of the pellets. Then add a 2nd. cardboard divider and the final third of the pellets. The result is a load split into three sections. An exact equal division is not required

Testing is as easy or as difficult - take your pick - as any patterning job and should be done if someone is considering their use. In a full choke gun, the spreader loads work amazingly well, at least in the author's experience. They are a help if using a full choke barrel at close range, but not much benefit for barrels that are open. The pattern may not be as good as an improved cylinder, for example, but it does serve a useful purpose.

Another method of spreading the pellets is to load or use the softest shot available at the highest velocity. This will spread the shot because of deformation. The success of this method proves that distortion and misshaping can and does affect the ballistics of the pellets.

LEAD vs. STEEL PELLETS - - - and BISMUTH ?

Lead shot has been prohibited by the U. S. Fish and Wildlife Service for waterfowl hunting for environmental reasons since the 1991-1992 season. The objective was to lower the loss of waterfowl from lead poisoning. Lead shot and steel shot have different weight and hardness characteristics and therein lies a problem. Steel has a density about 70% to 74% less than lead and the result is a capability to cripple instead of kill.

A few alternatives are being tested. Some, such as bismuth, are very promising. As this is written, Bismuth has been approved on a trial basis through the end of the 1994-1995 waterfowl season by the Fish and Wildlife Service. Note that Federal approval does not always give state approval.

Steel pellets and lead pellets are ballistically different and the difference shows up while the shot is still in the bore. While the soft heavy lead is the best, it is not compatible with the hard, and sometimes hot, shotgun bore.

Steel shot is harder than lead shot. It is not misshaped by the bore as much so it stays with the pattern better. A shot string is usually about one third shorter and the pattern is tighter compared to lead. The lack of bore damage protects it from energy and velocity loss. Steel, on the other hand, is much lighter than lead, and has a ballistic disadvantage because of the lighter weight. It loses its energy and velocity faster. At 40 yards, it requires a No. 2 steel pellet to produce the energy of a No. 4 in lead. Standard lead is, therefore, better than steel for shot. (Environmental issues aside.) It is made of low carbon steel that is treated with a rust preventative. It is annealed from 150 D.P.H. (diamond pyramidal hardness) to 75-90. By comparison, lead shot is about 25-35 D.P.H. This is hard enough that only screw-in choke systems approved for steel shot should be used.

Lead shot can be protected from the damage done to it by the bore with the use of a material that is put in the shell with the pellets to cushion and protect them. A buffering agent is the term applied to the cushioning material. Bone meal was used some years ago but it is now replaced by plastic powder. The buffered or protected pellets average around 45 f.p.s. faster. Buffering helps with steel shot as well as lead shot. This indicates benefits beyond the protection from deformation. Possibly in helping the fluid characteristic discussed earlier. **Caution note to hand-loaders:** Buffering adds weight and can increase pressure. Use only approved material and techniques and follow instructions with care.

Hard shot forms a different pattern than soft shot. A lot different. In a modified barrel, very hard shot will pattern as if the barrel was full choke and very soft shot will pattern as open or cylinder choke.

The addition of antimony hardens the shot pellets. A silver-white, brittle element, it's main use is in alloys to increase hardness. It is also used in pigments and medicines. Soft shot will have about .5% and hard shot about 3%. Some premium grades will go to 7%. (Note: the .5% has a decimal point in front and the others do not: an important difference.)

A rule says that for the same energy as a lead pellet, to go two sizes larger with a steel pellet. If the ballistic tables are checked, it will appear to be close in most sizes. In other words, use no. 4 steel pellets instead of a no. 6 lead pellet, etc. The reason is because a steel pellet will weigh about 30% less than lead when the diameters are the same. Therefore, it will take more space in the cartridge if the charge remains equal. Worded differently, there are more steel pellets in equal weights. The catch is that the lead will deliver more impact energy than the steel shot because of its heavier density. The projectiles are **not** equal in density by using two sizes larger. **In most cases, to get an equal performance to lead, including impact energy, the change would have to be bigger than two sizes.** The earlier mentioned University College London Ballistics Research Laboratory says that the deceleration rate of steel # 4 is exactly the same as hard lead # 7 1/2. As the reader has probably learned from earlier in this book, the deceleration rate will influence the flight time and the lead on a moving target.

The amount to shoot ahead of a moving target with steel shot will be slightly less at ranges below about 35 yards and longer at ranges over that amount. The reason is because steel shot usually (not always) leaves the muzzle faster. But, it can't hold it's velocity and it rapidly decelerates. Of course, the change in lead will only be a few inches and affect only the best shooter. To be honest, most of us are not good enough to notice a few inches difference in lead. (Not the metal this time, but meaning *ahead of*. Both are spelled the same.)

LEAD vs. STEEL COMPARISON CHART
Source: SAAMI Exterior Ballistics Tables Adopted 4/23/81.

Shot type	Wt.	Shot Size	No. Pellets	Muzzle Velocity (f.p.s.)	Retained energy per pellet (ft. lbs.)	
					40 yards	60 yards
Lead	1 1/4	6	281	1330	2.3	1.3
Steel	1 1/8	4	215	1365	2.5	1.4
Lead	1 1/4	4	169	1330	4.4	2.7
Steel	1 1/8	2	141	1365	4.4	2.6
Lead	1 1/2	4	202	1260	4.1	2.6
Steel	1 1/4	2	156	1275	4.1	2.4
Lead	1 1/2	2	130	1260	7.0	4.6
Steel	1 1/4	BB	90	1275	8.3	5.2

A lot of false information has circulated concerning the effective range of steel shot. A chart that is being distributed by both the private organization that created it and by the U.S. Fish and Wildlife Service promotes use well beyond proper range.

Bird shooting should be approached with the same concern for crippled and wounded game as used by most big game hunters. Most hunters are responsible and will follow sensible guidelines. Unfortunately, long-range use with steel shot will not give effective results. The ballistic performance is just not available.

Two extensive tests have been made that prove the ineffectiveness of steel shot on birds at range. One was conducted by Winchester and surprisingly, the other was by the U.S. Fish and Wildlife Service.

BISMUTH

It was just mentioned that bismuth is now approved for use in the U.S. on a trial basis. By the time this is printed, it may be approved permanently.

Bismuth has a ballistic superiority over steel shot and is nontoxic to humans and waterfowl. The most talked about type has 3% tin as an alloying agent to reduce fragility. It is 91% as dense as lead while steel varies from 70% to 74%. This is a major improvement. Another is called TBT because it is a combination of 36% tungsten, 44.5% bismuth and 16.5% tin alloy. This mixture has almost the density of lead.

Pure bismuth is too brittle to use without an alloy. This leads to shot that is not just bruised but broken. The poor patterns that follow are expected. The addition of tin softens the shot. It should not damage bores, but it is too new for a definite statement at this time.

As an ore, bismuth is an elemental metal and can be found in deposits of lead, silver, gold, tin and copper. Interestingly, it is an ingredient in Pepto-Bismol.

PELLET MAXIMUM RANGE

Shotguns are similar to other firearms in the angle between the bore and the horizon for longest range. Not 45 degrees as some people may believe, but 20 to 30 degrees. 45 degrees would only be correct in a vacuum; a no atmosphere condition with no air resistance. A problem in safety is evident because shotguns, due to the nature of bird hunting, are frequently fired at the angles that give the longest range. (Chapter 19 discusses maximum range and the laws apply to shotgun pellets as well as rifle bullets.)

The basic rule for shotgun range goes back to the early 20th. century, and is still valid today. A Frenchman, Journee, discovered that the maximum range of shot pellets in yards can be found by multiplying the diameter of lead shot in inches by 2,200. **Example:** No. 9 shot is .08 * 2200 = 176 yards.

There have been several other methods or rules created, and all give answers that are close to Journee's rule. These different methods and charts, including the U.S. Army, agree with Journee in some areas and disagree in others. Unfortunately, the areas of agreement and disagreement are different among them. In other words, they don't agree on what to agree on.

Until something better comes along, Journee's rule can still be used. Remember, this is for lead shot.

An interesting extra note; velocity was not included because it has very little effect on the outcome. Again, air resistance is the culprit. Also, choke has such a little effect that it is not included.

EXAMPLE: 00 Buckshot   .34 * 2200 = 748 yards
             # 2 shot   .15 * 2200 = 330 yards
             # 6 shot   .13 * 2200 = 286 yards
             # 9 shot   .08 * 2200 = 176 yards
             # 12 shot   .05 * 2200 = 110 yards

A muzzle velocity of 1,230 f.p.s. is a good average for shot pellets. (A check of the latest Winchester Shot-shell Information showed the lowest at 1,040 f.p.s. It was all by itself with no other below 1,125 f.p.s. The fastest was 1,422 f.p.s. and it was the only one - out of 66 - over 1,375 f.p.s.) The spherical shape has such poor aerodynamic qualities that even if the muzzle velocity is double that amount, it will be reduced to 1,200 f.p.s. in about 30 yards. Therefore, if we doubled the velocity we could not increase the range by more than 30 yards.

There is more on shotgun range and Journeé's formula on page 236.

One problem that should be mentioned about pellet range is *clumping*. It has been known to increase pellet range and endanger people. Clumping, sometimes called *shot balling*, is where some of the pellets are fused together by the escaping hot gas. The wads of today's shells seal very good and this is now a rare occurrence. Many years ago, it was more common but it still happens occasionally.

BORE & CHOKE FOR RANGE

If the bore is **reduced** by a small amount, by two or three thousands of an inch. (.002" - .003"), the barrel will require slightly less choke to produce the same pattern percentage.

For longer than normal range, the best luck will be a very smooth polished bore that is .002" or .003" **oversized**. Remington Arms reportedly used this technique in special long range Model 11's.

Normally, less choke is needed if the shot size is increased and more choke if the shot size is reduced. As stated so frequently the reader is probably tired of reading it, test your gun. It may not be in the *normal* group.

BUCKSHOT

Buckshot, which is really nothing more than larger size pellets, kills by a multiple strike pattern the same way that bird-shot kills, unless the shooter gets very lucky and one pellet hits a vital spot. As with shells with smaller pellets, buckshot loads must have an adequate number of pellets to form a pattern density that will do the job. The large buckshot, such as 00 and 000, may not have enough pellets for coverage. They are popular with shooters who don't realize how few pellets they have in a shell.

That is not to say that they should not be used. Of course they should. Just be aware of the density involved. A chart is included on this subject.

Smaller bird-shot pellets are dumped in the empty shell at random, but buckshot has to be placed in the shell in layers. The size of the bore (gauge) and the size of the buckshot determine the arrangement.

There has been a lot of improvement in buckshot ammunition in the past years, but it is still entirely for short range use. Only someone who cares not whether he cripples and maims, will shoot deer with buckshot at a 100 yards.

Smaller buckshot have a larger total pellet weight because they fit more efficiently in the space available in the shell. This higher weight will sacrifice a little in muzzle velocity, but it is usually not enough to be important.

BUCKSHOT LOADINGS

Ga.	Length	Velocity f.p.s.	Buck no.	Layers	Pellets / layer	Total pellets	Total weight in oz.
10	3 1/2 "	1100	4	6	9	54	2.54
12	2 3/4 "	1325	4	4	7 *	27	1.27
12	M. 2 3/4 "	1250	4	5	7 *	34	1.60
12	2 3/4 "	1250	1	4	4	16	1.46
12	M. 2 3/4 "	1075	1	5	4	20	1.83
12	2 3/4 "	1275	0	4	3	12	1.32
12	2 3/4 "	1325	00	3	3	9	1.11
12	M. 2 3/4 "	1290	00	4	3	12	1.48
12	2 3/4 "	1315	000	4	2	8	1.28
12	M. 3 "	1210	4	6	7	41	1.93
12	M. 3 "	1040	1	6	4	24	2.19
12	M. 3 "	1210	00	5	3	15	1.84
12	M. 3 "	1290	000	5	2	10	1.60
16	2 3/4 "	1225	1	4	3	12	1.10
20	2 3/4 "	1200	3	5	4	20	1.07
20	M. 3 "	1200	2	6	3	18	1.19

Notes; M = Magnum and * denotes one less pellet in top layer which explains the total number of pellets being one short of the expected total. The number of pellets may be less than listed, but will never be more because of the space involved. Velocity is taken 3 feet from the muzzle of a full choke barrel with a tolerance of + or - 50 f.p.s.

BUCKSHOT LOADING

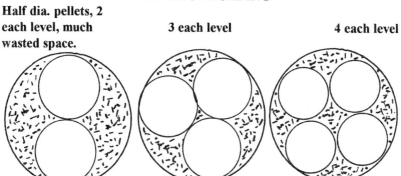

Half dia. pellets, 2 each level, much wasted space.   **3 each level**   **4 each level**

Buckshot will normally give better pattern percentages than birdshot. Although the problems of distortion from setback forces that affect smaller pellets also affect buckshot. Hardened buckshot is an improvement and buffering agents can help in some cases. (Setback forces comprise the rapid acceleration following ignition. It is normally figured while the projectile is still inside the barrel.)

Not only should pellet loads be patterned, but buckshot loads as well. It is the only way that the hunter can determine what is the best for his gun in terms of load, range, choke, etc. Normal patterns are tested at 40 yards, but a clean deer kill with buckshot at 40 yards is not likely; with slugs, yes; buckshot, no. Try testing at a more reasonable 25 yards and also at 40 yards. Remember that the standard 30" circle is much larger than the clean kill area on a deer. A 20" circle is closer to the average deer's vital area. Some state laws restrict deer hunting to buckshot. Clean kills are still wanted but harder to obtain. If a hunter is planning to hunt deer with buckshot he should always pattern his shotgun. Patterning may be more important in this situation than in most others.

Contrary to what some hunters believe, a one pellet hit will rarely kill a deer. Penetration and shock are required and one buckshot pellet will seldom do the job. Three is marginal and five is more probable. While the larger pellets, such as 000 have a high striking energy they also have a small amount of pellets and are less likely to put any in a vital area.

Penetration is also a concern. The larger pellets may give a marked improvement in depth over the smaller pellets, but once again, fewer hits.

A good starting point? Perhaps 00 buckshot with a modified choke at 25 yards. Yours probably will be different and that is why it should be tested.

BUCKSHOT BALLISTICS Compiled from various sources.

Gauge	Size	Muzzle velocity f.p.s.	Velocity 50 yards f.p.s.	Mid range trajectory in inches
12-2 3/4" HV&M	00	1325	965	0.9
12-3" M	00	1250	940	1.0
12-3" M	4	1225	870	1.1
12	0	1300	945	0.9
12	1	1250	920	1.0
12	4	1325	900	1.0
16	1	1225	910	1.0
20	3	1200	860	1.1

TRACERS FOR SHOTGUNS

The nature of some shotgun shooting is instinctive. In other words, just shoulder the gun and shoot. The rifle and handgun sight alignment is a luxury not available or beneficial to most shotgunners. A tracer in a shotshell can be a big help in practice and training.

Most tracers have the nasty habit of starting fires unless used only over water. One product on the market claims theirs burns for such a short time that it will be cool by the time it reaches the ground.

SLUGS

The word slug has several meanings (see glossary), but here we are concerned only with the soft lead projectile used in shotgun shells.

Many local and state laws prohibit hunting deer or other large game with anything other than slugs. The reason is their limited range makes them safer for use in congested areas. Very few hunters would choose slugs for hunting if it was not required by law. All except the smallest and weakest rifles would be better, and so would many of the higher powered handguns. By better, we are including energy, accuracy, stopping ability, and effectiveness at range.

Conventional slugs are not suitable for game beyond about 75 yards. Sabot slugs, with proper use, to perhaps 100 yards. Even with a 12 ga., longer ranges will usually only wound the game. With a 16 ga. or smaller, the effective range will be reduced.

A slugs maximum range is normally about 810 to 820 yards. Note the short effective killing range with a lot of yardage yet to go. As covered in Chapter 19, the distance a bullet or slug is capable of traveling has little to do with its **useful** range. The slugs maximum range is kept low by a low ballistic coefficient.

Even when elevated upward to about 30 degrees, a rifled slug will have a range of about 817 yards. A shotgun zeroed for 75 yards and fired horizontal will probably hit the ground at no more than 200 yards. The ricochet, while still having a deadly potential, will not go far. The soft lead slug deforms considerably while still in the bore and is further distorted by hitting anything. It also has a low striking velocity at range.

Slugs are undersized for the shotgun bore. Standard bore size for a 12 ga. is .729" and most slugs for 12 ga. are smaller than .700" and some even less than .685". The expanding gas pressure in the bore is sealed by the rear skirt. They must pass through the choke without too much restriction. The soft lead and the thin walled base help when needed. A full choke can still be used and the slug will squeeze through, if necessary.

The following chart was compiled by the Sporting Arms and Ammunition Manufacturer's Institute.

MAXIMUM RANGE OF SLUGS

Gauge	Slug weight in oz.	Muzzle Velocity in f.p.s.	Maximum. range in feet	Maximum range in yards
12	1	1600	2450	817
16	7/8	1600	2450	817
20	5/8	1600	2450	817
.410	1/5	1830	2530	843

The 12 ga. slug with a muzzle velocity of about 1,600 f.p.s. drops to about 950 f.p.s. in the first 100 yards. That isn't to say it is safe to shoot slugs with homes or people in the near distance. Clean and efficient kills on game are one thing; accidents against people and property are another.

Rifled slugs have a poor ballistic coefficient and this reduces their long range and velocity holding abilities. While the coefficients will vary, they will average about .057. This does not compare well to a handgun or rifle projectile. A 12 ga. rifled slug loses 27% of its muzzle velocity in the first 50 yards and 41% in 100 yards. The 16 ga. rifled slug loses 28% of its muzzle velocity in 50 yards and 34% in 100 yards. The 20 ga. rifled slug loses 27% in 50 yards and 41% at 100 yards. The .410 loses 28% of its velocity in the first 50 yards and 44% at 100 yards. (This is based on an analysis of current Winchester shotgun slug ballistics.)

Even with a slugs limited ability to make clean kills at beyond about 75 yards, that is still the range that most experts believe is best to zero a shotgun. Based on an average of most ammunition, a 12 ga. slug from a gun with a 1.2" line of sight above the bore center line would be 1" high at 50 yards and 3" low at 100 yards. On out to 125 yards it would drop to 8.1" low. This is not a bad trajectory, all things considered, but the velocity and energy have dropped low enough that a clean kill would be unlikely. At 125 yards, energy is only about 46% what it was at 25 yards (not at the muzzle) and about 68% of velocity. That is based on a standard 12 ga. rifled slug.

The spiral ribs on Foster style slugs help a little in causing the slug to rotate or spin, but not much. The gyroscopic effect is very small. The grooves are compressed by the bore but they still impart a slight rotation which helps improve the groups. The thin skirt at the back creates a hollow base with a heavy front end. This keeps the slug in position and prevents tumbling in a similar way as an arrow with the weight at the head. (Of course, the feathers add stability, but that is another matter.)

Tests show that slugs without spiral grooves still hit nose first without tumbling. The grooves do have some effect because the slugs with the grooves have proven better accuracy.

Tests by several groups, including *The American Rifleman* (NRA) and various manufacturers, all agree that for slugs, open-chocked barrels are significantly better for accuracy. Modified barrels are not as good, but better than full choked barrels which are last in choice. To put it plainly, the choke squeezes the soft lead out of shape too much.

Winchester-Western conducted tests on slugs and found that a rifled slug rotated at one turn in 24 feet of travel which is about 3,600 r.p.m. While that is a slow spin, it still contributes to accuracy. The slug, as stated, is a nose heavy projectile, sometimes with fins and a slow rotation. The fins have a helical angle to help rotation. The effect is similar to a rocket or mortar shell which is fin-stabilized.

SABOTED SLUGS

Sabot is a french word which is pronounced *say-bo* or *sab-o*.

The saboted slug has a plastic carrier that fits around the slug and then falls away in one or two pieces after the slug leaves the barrel. Actually, from about 10 to 40 yards down range. (Note: Sabots can damage chronograph equipment. Use caution.) The slugs diameter is smaller than conventional and due to the plastic protection, it is not deformed by the bore. The plastic shields both the bore and the slug which can be of a harder material because it does not need to be crushed or squeezed by the barrel or choke. Muzzle velocities range from 1,200 f.p.s to 1,550 f.p.s., depending on which company made it and how accurate they are in their claims.

Down range, the sabot slug holds velocity and energy better than a conventional slug. They may, in some loads, leave the muzzle slower but will usually be about equal with the standard type at 100 yards.

An older style slug that would be about .70 caliber will be about .50 caliber in a sabot. It will expand to about .90 caliber. Sabots perform well in the few shotguns with rifled barrels and a 1 in 35 twist will give rotation and some gyroscopic stability. This is helped by an hour glass shape with a hollow base. When used in smooth-bore guns, a fair amount of yaw is possible because spin is needed for gyroscopic stability as with other projectiles. An examination of the plastic sabot fired in a rifled barrel shows that it does pick up the rifling rather than stripping.

Lead slugs can leave lead residue in the bore. The plastic sabot prevents this.

No projectile, or any other body for that matter, can accelerate without being acted upon by a force. When a bullet leaves the muzzle, if it is accelerating it may continue to do so for a short distance; **a very short distance**. At which point deceleration begins. A saboted shotgun slug, for example, cannot and will not accelerate after it leaves the muzzle and the sabot drops away. A few people wrongly believe that it will.

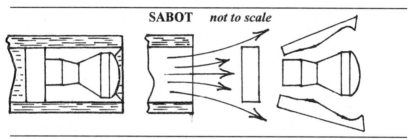

SABOT *not to scale*

RECOIL

Recoil in a shotgun follows the same rules as other firearms. The weight of the powder has an effect as does the weight of the wad. A wad that requires less powder to reach standard velocity will reduce recoil as will a lighter wad. Chapter 6 has a lot of information on recoil and it would be redundant to repeat it here.

SHELL & CHAMBER LENGTH

The length of a shotgun shell is considered the length of the shell after firing. That may be a surprise to people that are new to shotguns. In other words, the length after the closed end has been blown open and unfolded. The length of an unfired shell is shorter than the length it is called. The older rolled crimp and the modern folded crimp both are noticeably longer after firing. This open length determines the length of the chamber and the length that will be purchased. If bought factory loaded by different companies, they may be of slightly different lengths.

Most modern shotguns made in the U.S. are built for either 2 3/4" or 3" long shells. The .410 is made in 2 1/2" or 3". Old guns and some foreign weapons may not be standard.

Factory testing has shown that use of a 3" 12 ga. shell in a gun made for a 2 3/4" 12 ga. will cause about a 500 p.s.i. to 1,000 p.s.i. pressure increase. About the same increase will be found with a 20 ga. by use of a 3" in a 2 3/4".

The extra 1/4" in a .410 (2 1/2" or 3") means that to fire a 3" shell in a 2 1/2" chamber, the shell has to be forced in with some effort. On firing, the pressure increase can be dangerous at 45% to 55% above normal.

Never use a shell in a gun that is not made for it. Even if it works safely for many shots, it may blow up on the next one.

WIND DEFLECTION

Shot pellets are deflected by a cross wind the same as bullets. The larger and heavier pellets are deflected the least and the smaller and lighter pellets the most. The same as bullets, the path is a curve leading away from the wind direction and getting further off course the longer the range. The basic principle of shotguns, that is a pattern of pellets instead of a single bullet, make wind deflection less of a problem when compared to rifles or handguns. The shorter distances involved, also reduce the problem as compared to the other firearms. Still, under conditions of wind and long range, it must be considered. For example, No. 7 or 8 shot can be expected to be deflected about 7" or 8" in a 10 m.p.h. wind at normal range. A 20 m.p.h. wind would increase the drift to 14" to 16". That is enough to be a problem. The deflection would be more at a higher wind speed, but a wind over 20 m.p.h. is enough to alert almost any sportsman.

The same data and mathematical formulas referred to earlier also apply for shotguns.

VELOCITY NOTES

The velocity of pellet patterns or slugs do not offer the variety available with handgun or rifle ammunition. For example, checking the latest Winchester factory chart, it shows centerfire pistol and revolver ammunition varying from a low muzzle velocity of 680 f.p.s. for a .32 Smith & Wesson to 1,790 f.p.s. for .30 Carbine. (In this example, the .45 Winchester Magnum at 1,400 f.p.s. would be a truer representation.) Rifle centerfire ammunition varies from a low muzzle velocity of 1,160 f.p.s. for .38-40 Winchester to a high of 3,750 f.p.s. for the .22-250 Remington. And, there is a nice even spread between the two extremes.

Shotgun loads extend from 1,075 f.p.s. to 1,425 f.p.s. for a variation of only 350 f.p.s. Slugs run from 1,200 f.p.s. to 1,830 f.p.s. The fast slug is, as expected, a .410 although the 12 ga. 3" Magnum at 1,760 f.p.s. is still very impressive.

An interesting fact about velocity for shot pellets is that increasing it by 100 f.p.s will only extend the effective range by about 1 yard. That is correct, one yard. With factory ammunition already at the limit of safe all-around peak pressure, greater range is not readily obtainable.

Increasing the velocity is difficult, but if it can be accomplished, and it will tighten the pattern. How wide the pattern is at any range is a direct result of the time of flight, if all other factors stay equal.

VELOCITY & BARREL LENGTH

Barrel length affects velocity in shotguns for pellet use the same as in rifles and handguns. Long barrels in the vicinity of 30" or more have an advantage in pointing and recoil reduction but a disadvantage in weight and ease.

Controlled tests of velocity reported in July, 1965 in *The American Rifleman,* a publication of The National Rifle Association, showed a 28 f.p.s gain from a 30" barrel to a 36" barrel. In testing all popular gauges by cutting off 2" and then measuring the velocity, removing another 2" and measure velocity, etc. a very small decrease was noted. The decrease was about 7 f.p.s. per inch in 12 ga., 16 ga. and 20 ga. The .410 was a little more at 12 f.p.s. per inch of barrel removed. The drop was about the same at 28" as it was at 20". (linear)

12 ga. 2 3/4 " slugs fired in a 30" barrel test at 1,600 f.p.s. and 2,488 ft. lbs. of muzzle energy. In a shorter barrel of 20" length, the figures are about 1,500 f.p.s. and 2,070 ft. lbs. In this case, the velocity loss is exactly 10 f.p.s. per inch.

Both are small amounts. Long barrels are good for the swing and follow-through and the muzzle blast is less. Short barrels are faster, easier to point and lighter. It all boils down to handling qualities and personal preference. The longer barrel helps very little from a ballistic point of view. The difference in barrel length is not as important as was believed at one time. Readers who have followed the subject will notice that this is similar to results with rifles and handguns. Barrel lengths make a difference, but not as much as most people expect.

VELOCITY READINGS BY CHRONOGRAPH

Home chronographs to determine projectile velocity are now both cheap and accurate. For shot-shell velocities it is still very difficult for any but a large company or government agency. Rather than a single bullet, the shotgun pellets are numerous. They can move together in a cloud which can be all bunched up together or strung out in a line or spread out in about any configuration possible. While many of these patterns are not desired, they must be recognized and checked. Special high speed photography with velocity measurement can be used.

Some chronographs can be used to take a reading on the center of the mass which can be considered the velocity of the entire shot charge. (Although, recent testing by The University College London Ballistics Research Laboratory has shown lower velocities toward the rear of a shot string.) Best results are found by placing the first screen 1 1/2 feet from the muzzle and the second at 4 1/2 feet from the muzzle. This obtains a 3 foot reading.

Protection may be required for the frames as the pattern may be larger than expected. Some amateur tests are successful with shot patterns, many are not. With bullets, the use of proper equipment and proper procedures will guarantee good results. Unfortunately, this is not correct with a charge of pellets.

A shot string can have a few pellets that, for reasons unknown, may be ahead of the pattern. They can cause an error, usually on the high side. As you have read and heard many times, follow the chronograph manufacture's directions.

TEMPERATURE EFFECTS

High air temperature gives a closer pattern. The air is thinner and offers less air resistance. High altitude has about the same effect. The velocity is also higher with more energy and penetration. Not by much, but enough an observant person can tell if inspections are made.

As would be expected, cold temperature has the opposite effect. Temperature influences on shotgun shells are well known and basic. Shotgun shells lose some velocity in the cold. The temperature for proper ignition is harder to reach and the chamber pressure is lower. Winchester has conducted tests that show chamber pressures at 0 degrees F. are lower by 700 p.s.i. to 1,600 p.s.i. as compared to 70 degrees F. Velocities are also lower by 50 to 100 f.p.s. These are just average rounded-off figures and each shell will be different as will each temperature, but the picture is clear. Lower temperature means lower pressure and lower velocity. Besides ignition, resistance by denser air and other factors are involved.

U.S. Army testing demonstrated a velocity loss of from 5% to 8% at -40 F. For readers who are not inclined to mathematics, 8% of 1300 f.p.s. would be 104 f.p.s. which would leave a velocity of 1,196 f.p.s. This is a small loss at a very low temperature. In the normal winter temperatures that most people would hunt or shoot trap or skeet, the velocity drop would be less and could be disregarded.

Generally, patterns are slightly tighter with closer pellet distribution in the colder weather. This effect has been both casually noticed and properly tested. The reason is not as clear. Many experimenters believe that less pressure reduces the setback (pellet deformation). This is discussed elsewhere in this book, and is usually taken as the theory behind cold weather patterns being better. More testing needs to be done at cold temperature with shot of different hardness and deforming characteristics.

SWAMPED BARRELS

A swamped barrel is a barrel where the outside diameter is larger at the muzzle end than for the remainder. The amount will vary from .008" to .020" on the diameter and 3" to 9" on length. Used mostly on top quality shotguns, such as Holland & Holland Ltd., its purpose is to enhance the appearance. There would be little ballistic benefit, except for the negligible extra weight at the muzzle.

FORCING CONE

The forcing cone is the area forward of the chamber that is reduced down to the bore diameter. It is tapered somewhat like a shallow funnel and helps the transition of the shot charge into the barrel. The idea is to not deform the pellets while they make the transition into the bore. The proper forcing cone length is important.

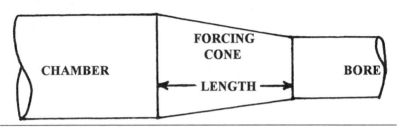

If it is too short it fails to do the intended job. The pellets will be mashed and jammed together. Pellet damage, poor patterns and *flyers* will be the certain result. Also slightly higher chamber pressure. Many people believe there is increased recoil from a short forcing cone. They are correct in their physics but in error with their terminology. This is unimportant as long as both people in a conversation have a meeting of the minds. In this book, we must be as accurate as possible. As explained in detail in Chapter 6, there is a difference between recoil and the felt kick. The forcing cone length will not affect actual recoil, but the felt kick may be harder with a short and abrupt forcing cone.

If it is too long, it permits the hot gas to leak into the shot and disrupt the pattern. The velocity will be lower and so will the energy.

Notice the forcing cone is not designated in terms of degree of taper or an angle, but as a length. The longer the length, the shallower the angle and the smaller number of degrees involved. A half an inch (1/2") is generally considered the shortest. Two and a half inches (2 1/2") would be a long forcing cone.

Manufacturers vary in their opinion of the best length, but all are usually short to moderate. In cheap guns, the surface is sometimes very rough and a little polishing may improve the performance. Some experts believe that many guns - actually the patterns - can be improved by lengthening the forcing cone. This probably is true, but considering that it is an action that once done, is almost impossible to reverse, caution is urged; consult an expert. (This book does not want to be in a position of recommending any firm or product, so the following is included for academic purposes only. Walker Arms in Selma, Alabama is nationally known for work in this area. Brownell's, Inc., of Montezuma, Iowa, a gunsmith supply firm, sells the reamers along with instructions. Both companies are well known and have an excellent reputation.)

A similar area in a rifle is frequently called a forcing cone. Note that it is only similar, not the same. It is more correctly called a throat or leade. The term forcing cone is used incorrectly for any firearm other than a shotgun.

SLUG & PELLET TRAJECTORIES TOGETHER ?

Although there is an opinion to the contrary, most shotguns will shoot to about the same point with both pellet charges and slugs. All guns are not alike, so testing will tell if one meets this criteria. If a study of both pellet and slug ballistic tables is undertaken, it can be seen why most guns will shoot both slugs and pellets together. They both have similar ballistics at the short to moderate ranges where a shotgun is used.

MATHEMATICS

For the readers who are interested in the formulas associated with shotgun ballistics, the following is presented. Basically, it is similar to other areas of this book with some necessary changes. While all the rules are still the same, a few subjects are looked at differently.

Pellets are like bullets in their need for energy. Light small pellets may give a higher hit count in a given area (pattern) but not have enough penetration. Most shells give about the same velocity so the pellet size is the deciding factor.

Very heavy drag on shot pellets cut their range in comparison to bullets. There is no *design to cut drag* as in rifle and handgun bullets and even, to some extent, slugs. Technically, the ballistic coefficient is based on only the diameter because of the all-over round shape. The sectional area of shot is proportional to the diameter squared (2^2) and the weight to the diameter cubed (2^3). Simplified, that means if one shot size has twice the diameter of another, its frontal area is 4 times as big and its weight will be 8 times heavier. In part, this is the reason that shot is strongly affected by a head or tail wind. Range can be increased or cut by as much as 15% in a direct 10 m.p.h. wind and 30% in a 20 m.p.h. wind. These figures may appear unusual, but they can happen and show the caution that should be used, especially if shooting with a tail wind toward a populated area.

This book is on ballistics, not safety, but this is a good place to remind the reader that the two go together. A shooter with a good knowledge of ballistics can be safer as well as more accurate.

♦ Ballistics coefficients are explained in detail in Chapter 14 and they can be found easily for shot pellets with a simplified formula.

$$C = w / i\, d^2$$

Where: C = ballistic coefficient
w = pellet's weight in pounds
d = pellet's diameter in inches
i = pellet's form factor

The shape of undamaged pellets is always spherical, so their sectional density, governed by the weight and diameter, is the important ingredient.

◆ For pellet weight, as required in the ballistic coefficient formula, the charts tell how many pellets to the ounce. Take this times 16 for number per lb. Divide this number into 1. **Example:** No 4 shot has 135 pellets to the ounce times 16 = 2,160. 1 divided by 2,160 = .0004629 lbs. for each pellet.

The information and math can be very helpful to anyone interested in ballistics. It also is both easy and practical.

◆ The Ingalls-Sciacci method for retardation of a projectile was universally used by ballisticians up to WW I. It is still seen in some text books, but has generally been discarded. It was found that any such mean value method was inadequate.

◆ Lead (not the metal) is the extra amount a gun is pointed ahead of a moving target. The objective is for the shot pattern (or bullet) to arrive at a given point at the same time as the target which could be a bird, rabbit, clay target, etc. It is not necessary to go into details of *sustained lead* and *swing through,* etc. as that is technique, not ballistics.

This formula will determine the lead for a shot pattern where the target is crossing directly right to left or left to right at an angle of 90 degrees to the gun. Wind correction and trajectory are not considered. (Of course, it is not practical to figure this in the field, it is given, as is the other math, for academic purposes.)

$$L = TV * PT$$

Where: L = lead in inches
TV = Target velocity in inches per sec.
PT = Pellets flight time for distance in fraction of a second.

Notes: To find PT, it is necessary to know or calculate the projectiles median velocity at the range involved. Lead in inches can be changed to feet by dividing by 12. **Miles per hour * 1.467 gives feet per sec. * 12 for inches per sec.**

The chart of gauges from 1868, still in use today, became practical after the discovery of methods to accurately measure sizes. Before this time, it was common to measure a gun bore by the weight of the ball that would fit the opening. This was frequently used for both shotguns and smooth bore rifles and pistols.

FULL LIST OF GAUGES in inches

A	2.000	8	.835	28	.550	48	.459
B	1.938	9	.803	29	.543	49	.456
C	1.875	10	.775	30	.537	50	.453
D	1.813	11	.751	31	.531	51.05	.450
E	1.750	12	.725	32	.526	54.61	.440
F	1.688	13	.710	33	.520	58.50	.430
I	1.669	14	.693	34	.575	62.78	.420
H	1.625	15	.677	35	.510	67.49	.410
J	1.563	16	.665	36	.506	72.68	.400
K	1.500	17	.649	37	.501	78.41	.390
L	1.438	18	.637	38	.497	84.77	.380
M	1.375	19	.626	39	.492	91.83	.370
2	1.325	20	.615	40	.488	99.70	.360
0	1.313	21	.605	41	.484	108.49	.350
P	1.250	22	.596	42	.480	118.35	.340
3	1.157	23	.587	43	.476	129.43	.330
4	1.052	24	.579	44	.473	141.95	.320
5	.976	25	.571	45	.469	156.14	.310
6	.919	26	.563	46	.466	172.28	.300
7	.873	27	.556	47	.463		

The above chart is from the 1868 British Gun Barrel Proof Act. It is still in use in the U. S. Note: The U. S. considers these numbers as the minimum with a tolerance of plus .020". **Example:** the 20 ga. can be from .615" to .635". Other than the shift in method at # 50, any variation between this list and others is caused by a lack of agreement on what the diameter should be. It can be carried on down to .220 which is 436.85

At the time this gauge system came into use, lead was believed to have a density of .41058 lbs. per cubic inch. Today, the figure is considered to be .4097 lbs. per square inch. As you can easily imagine, the change is so small it can be overlooked. The original formula is gauge = $4.6516 / D^3$ and it now is gauge = $4.6616 / D^3$ where D = bore diameter in both formulas. (Author's note: I am happy to once again thank Mr. William C. Davis of Tioga Engineering Co. and *The American Rifleman*.)

The bore measurement is made 9" down the barrel from the breech end. This follows a British plan established long ago. Some types of shotguns can't be checked in this location without great difficulty. A measurement from the muzzle end can be used if it is in far enough to be away from any choke that may be present.

GERMAN SYSTEM

Old German shotguns are marked in the obsolete Teschner method.

Teschner	Gauge	Teschner	Gauge
0	10	5	18
1	12	6	20
3	14	7	24
4	16	8	28

RIFLED & SABOT SLUG TRAJECTORY TABLE
An excerpt from the Winchester Ammunition Product Guide, 1994

Gauge	Muzzle velocity f.p.s.	Slug weight ozs.	Slug type	Trajectory height in inches range in yards				
				25	50	75	100	125
12	1550	1	sabot	0.7	1.7	1.6	0.0	-3.2
12	1300	1	sabot	1.4	2.8	2.4	0.0	-4.7
12	1450	1	sabot	0.9	2.2	1.9	0.0	-3.8
12	1200	1	sabot	1.8	3.2	2.7	0.0	-5.1
20	1400	5/8	sabot	1.2	2.5	2.2	0.0	-4.4
12	1760	1	rifled	0.2	0.0	-1.5	-4.6	-9.7
12	1600	1	rifled	0.4	0.0	-1.9	-5.9	-12.1
20	1600	1	rifled	0.3	0.0	-2.0	-5.9	-12.1
16	1600	4/5	rifled	0.4	0.0	-2.0	-5.9	-12.1
410	1830	1/5	rifled	0.3	0.0	-1.9	-5.8	-12.0

NUMBER OF PELLETS IN A STANDARD CHARGE

Shot size	Lead 3/4 oz	Lead 1 oz	Lead 1 1/4 oz	Steel 1 oz.
9	439	585	731	
8	308	410	512	
7 1/2	263	350	438	
6	169	225	281	316
5	128	170	213	243
4	101	135	169	191
2		88		125
1				103

CHOKES-ITHACA PRIOR TO 1941

No. 0 - cylinder
No. 1 - improved cylinder
No. 2 - modified choke
No. 3 - improved modified choke
No. 4 - full choke

CHOKES-CHARLES DALY

+   full
+ -   modified
++   improved cylinder
++-   improved modified
ss   skeet

CHAPTER 22

HANDGUNS

This section covers only the ballistic points that are different from other firearms. Most technical items are the same. And why not? But, there are still a few ballistic problems that are unique to handguns. The shorter barrel removes some problems and creates new ones. And while the ranges are less, they are longer than they were in past years.

Handguns are now in common use for hunting at ranges up to 100 yards and beyond. Ballistic tables can help in sight-in and preparing for that distance. Although not as valuable to hand-gunners as rifleman, the tables can be a benefit even at the shorter ranges. A handgun should be test fired to zero at a suitable range and then tested at 25, 50, 75, and 100 yards. This will show the owner what to expect at these ranges and tell him if a change in zero is required. Most guns and calibers used for hunting will strike close at all ranges if the sights are properly adjusted. It may not be fun to fire a hard-kicking handgun 5 or 10 times at each of 4 or 5 separate ranges, but if the weapon is difficult and hard to control, perhaps the shooter should consider using a different or less powerful firearm.

LOADS AND TRAJECTORIES

Many intelligent people believe false information about hand gun trajectories. Incorrect data and misunderstandings are common. Therefore, an explanation is in order.

All else being equal, handguns shoot lower on the target with light loads and higher with heavy loads, no matter the velocities. At first thought, that may sound backwards and that may be the reason for the confusion. Nevertheless, it is a correct statement. The cause is involved with the barrel jump and gun movement as discussed in interior ballistics. Also, this is figured at normal handgun ranges that are not long enough for trajectory drop to be involved.

There are two main reasons for this. The barrel jumps up from recoil as the gun is fired. This jump begins the instant the bullet starts to move out of the case. A decent amount of the muzzle jump is while the bullet is still in the barrel and moving toward the muzzle. The sights are set (see sketch on page 281) so the barrel is pointed below the intended impact point at the time of firing. The muzzle jumps and the bullet leaves the barrel as the barrel comes up past the proper point.

If two bullets are fired at the same velocities, but one is much heavier than the other, the recoil will be heavier for the heavier bullet. With the same velocities, they will reach the muzzle at the same time, but with a heavier kick and jump, the heavier bullet departs with the barrel at a higher angle. The impact is, therefore, higher.

If we add powder behind the lighter bullet so the jump from recoil is the same as the heavier bullet, will it now hit at the same point? No, because the lighter bullet now has increased velocity as well as increased recoil. It will now move down the barrel faster and exit the muzzle while it is at a lower angle. The lighter bullet will again impact the target lower. The two are equalizing in that the low-velocity bullet and the high-velocity bullet, if the weight is the same, will impact at about the same place. But, a change in bullet weight will change the impact point regardless of the velocity.

Therefore, the two reasons are the heavier recoil of the heavier bullet which raises the barrel higher and the slower acceleration which will keep it in the barrel longer. The two combine to make it leave at a higher elevation. With some gun and ammo combinations, the difference between heavy and light bullet weights can be large. As much as 10" at 25 yards.

Well known ballistics expert and author, General. J. S. Hatcher, conducted some early experiments which confirmed this viewpoint.

It is noteworthy that this effect is different in rifles because handguns have more jump and the longer rifle barrels go through much bending and flexing. Refer, if you wish, to the internal ballistics section.

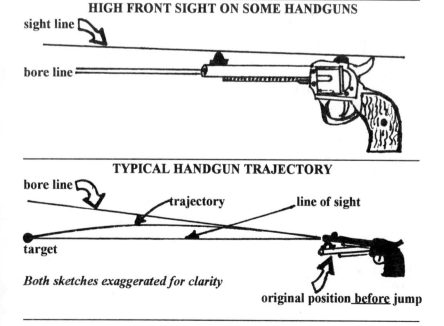

Both sketches exaggerated for clarity

HAND GUN SIGHTS

Some handguns, such as the Colt .45 caliber revolver, have the front sight higher than the rear sight so the gun is pointed below the strike point. The reason is recoil and muzzle jump. The recoil brings the barrel up so that by the time the bullet is leaving the muzzle, the barrel is on the proper line. Generally, the heavier the bullet, the more this effect. Trial and error testing is the only way to establish how the sights should be set.

SIGHT CHOICE, FIXED OR ADJUSTABLE

While adjustable sights are necessary for target shooting and hunting, for self defense they can be a hindrance. There are 3 points to consider. **(1)** Most defensive shooting is done at short range where very exact sight adjustment is not necessary. **(2)** Also, as any police officer with combat experience will relate, in the heat of a fire-fight, few people take time to properly align the sights. **(3)** Perhaps most important, any sight that can be adjusted can accidentally become unadjusted. (In reference to no. 2, we should all try to remember, even under stress, that the only shots which do any good are the ones that hit. Shot placement is not just important, it is vital.)

FIXED-SIGHT ADJUSTMENT

Sight adjustment on handguns can be easy if adjustable sights are installed. If permanent sights are used, several methods are available. Filing away a small amount from one side of the front sight or adding metal to it is one option. Widening the rear sight notch on one side or the other is also frequently done. Whatever method is used, two rules must be followed. **First**, be extremely careful and think out your action before starting. Adjustable sights can be simply readjusted. Metal removed is difficult to replace. **Second**, the short barrel on most handguns means that a small change in the sights will cause a big change in the point of impact. Take off very little and then test it. Following both rules can prevent an unpleasant surprise.

Right or left correction may be solved by moving the front sight the direction the impact is off. In other words, if the bullet strikes too far to the right, the front sight should be moved right. Another way of saying; move the front sight in the opposite way the strike point should move.

Elevation adjustment on fixed sights can be accomplished by carefully filing on the front sight to raise the impact point. If the impact point should be lowered, the front sight will have to be changed. It is a lot of work, but metal can be added to it. Also the rear sight can be modified. None of these methods are easy, and on some types of sights, it will be very difficult. It still will be worth the trouble.

BALLISTIC DATA
FOR MODEL 1911 PISTOL CAL. .45 ACP 230 GR. BULLET.
DATA FROM U. S. ORDINANCE DEPT.

(Inst. Velocity at 25 ft. = 800 f. p. s. - Elevation 0^0)

Range in yards	Time of Flight in seconds	Drop in inches	Left Drift in inches
5	.018	0.1	0.0
10	.037	0.3	0.1
15	.056	0.6	0.1
20	.075	1.1	0.2
25	.094	1.7	0.2
30	.113	2.4	0.3
35	.132	3.3	0.3
40	.151	4.4	0.4
45	.170	5.6	0.5
50	.190	6.9	0.6
55	.209	8.3	0.7
60	.229	9.9	0.8
65	.248	12.0	0.9
70	.268	14.0	1.0
75	.288	16.0	1.1
80	.308	18.0	1.3
85	.328	20.0	1.5
90	.348	23.0	1.7
95	.368	25.0	1.8
100	.388	28.0	2.0

ADJUSTABLE SIGHT ADJUSTMENT

An adjustable rear sight is moved to the left to move the hits on the target to the left. As expected, a right move of the sight moves the hits to the right. If the rear sight is raised, the target hits will be raised. The front sight movements change the hits in the opposite direction as the sight is moved.

TRAJECTORY

The bullet's trajectory will be the same as a rifle and cross the line of sight after flying up to meet it.

The sights are adjusted to get hits at a set range and the bullet will cross the line of sight from below because it leaves the muzzle from below. At very close range, the aim should be a little high. All guns are different, but a good average for the crossing point would be about 15 feet down range. Very accurate short range shooting, such as a snake's head at 6 feet, will require a slight high aim. Try it and see how your favorite handgun shoots close-up. It won't be much higher, but that diamond-back rattler may need a darned good hit with the first shot.

The sights and the sight line are higher than the bore and this creates an angle where the bullet's flight path and the sight line intersect. This angle is more acute at the longer ranges. At rifles ranges, it is not considered. At short range, the barrel is aimed higher and creates a larger angle between the sight line and the projectile's path. A handgun zeroed at 40 yards will shoot low at 10 yards. If zeroed at 10 yards, it will shoot high at 40 yards. (If all else is equal.) **For extreme accuracy at short range, a handgun has to be zeroed for the exact range involved or an adjustment made based on testing.**

BARREL LENGTH & VELOCITY

There have been literally hundreds of tests to determine velocity differences with various barrel lengths. They all show there is a definite loss in velocity with shorter barrels, but it is a small amount. Testing of a .38 Special with a 158 grain bullet showed a velocity drop that increased as the barrel got shorter. The drop from 8" to 7" caused a typical 10 f.p.s. decline.

One test was conducted by the National Rifle Association and reported in *The American Rifleman*. They used a Smith & Wesson .357 Magnum and showed an average drop of 14 f.p.s. per inch of barrel removed. The drop was slightly less at the longer lengths (7" or 8") and more at the shorter lengths (around 2"). This is typical of most testing.

With some guns and calibers, the drop is higher than average. For example, the following velocities were measured at 20' from the muzzle of a revolver in .44 Remington Magnum. The same gun was used with an inch removed each time for the same 240 grain ammunition. Ten rounds were fired at each length with the average listed.

Barrel length	Velocity (f.p.s.)
8"	1515
7"	1465
6"	1420
5"	1380
4"	1310
3"	1235
2"	1105
1"	987

While testing typically gives answers in velocity readings, that directly influences energy (or power, if you prefer that word).

While a change in barrel length alters velocity, it is more noticeable with handguns than with the rifle's higher velocity. Handgun velocity is greatly affected by even small changes in barrel length. No two will be alike, but compared to rifles, there will generally be a two or three times greater change for an equal variation in length.

The longer barrels not only have a longer sight radius and are therefore easier to aim accurately, but also obtain higher velocity. (Up to a point.) The added weight of the longer barrel also helps in balance and muzzle rise. The velocity increase is small and may not be worth the increased weight and size. As with many things, a compromise is called for.

With some cartridges and fast-burning powder, a barrel as short as 6" is the longest for maximum velocity. For normal use, the maximum barrel length is about 9" and the only advantage to extending it longer is the increased sight radius and the public image.

Barrel length in a semi-automatic pistol includes the chamber while in a revolver, the cylinder which holds the cartridge is behind or extra from the barrel. This must be considered in any calculations. The gap, however small, between the cylinder and the barrel and the larger than bore diameter of the cylinder will reduce the effect somewhat and must be considered. Also, even in the best factories, no two guns will be identical, and after some use, the difference will be enhanced. Testing has been done by many companies, organizations and hobbyists. Velocity testing on handguns of the same barrel length can vary by a large amount. 200 f.p.s. is not unusual although the normal will be between 20 f.p.s. and 135 f.p.s.

RIFLING

There is a long discussion on this subject in Chapter 12. Here, it is only necessary to say that handguns frequently have rifling with a faster twist than rifles of the same caliber. The reason is the lower velocity usually found in handgun ballistics. To give a bullet proper gyroscopic stability, it must rotate a required number of revolutions per second. With the handguns lower velocity, it travels at less feet per second than a rifle bullet of the same caliber, so it must rotate faster. (It must be noted that this is an old theory which is not always followed today.)

Most U.S. made handguns have a right hand twist. Colt guns use a left hand twist. The number of grooves varies, usually between 4 and 8. The depth of the grooves will be between .0035" to .005". (3 1/2 to 5 thousandths of an inch)

Most .22 caliber handguns use 14 to 16 inches per turn. One exception is 12. .38 caliber revolvers are 16 to 18 3/4. .45 ACP semi-autos are 14 to 16. Most .44 calibers have a 16 or 20 twist. As you can see, the experience and thoughts of the different manufacturers arrive at different results.

LONG RANGE SHOOTING

Not all handguns are suitable for long distance shooting, but some can do a surprisingly good job. Even short barrel handguns can be excellent with both accuracy and repeatability. The correct twist is important to give the bullet gyroscopic stability.

A few expert handgun marksman, such as Ed McGivern and others, have recorded excellent shooting out as far as 600 yards; even keeping all hits in a man size silhouette at that range. While it is true that few people can do this, it can be done with the proper gun and ammo combination and a lot of practice and skill.

REVOLVERS

In a revolver, the gap between the barrel and the cylinder produces a loss of pressure and velocity which can reduce the advantage of longer barrels. Of course the gap can vary a great amount from one revolver to the next.

The gap can even cause injury. Frequently the hot gas contains tiny particles of powder and sometimes it may include slivers of bullet material. The gas and light weight bullet particles accelerate and escape at a higher velocity than the heavier bullet but they cannot maintain speed for any distance.

A revolver's gap can be verified by wrapping tape around the area and observing any cuts or damage. Of course it may be different with each chamber and can be unexpectedly worse on occasion.

The gap mentioned above is a common fault with some movies and TV shows. Although not as common now as in the past, viewers frequently saw a revolver with a silencer on the muzzle. Revolvers cannot be effectively silenced. The gas escaping from the gap not only influences velocity, it also makes noise.

Revolver accuracy is influenced by each chamber. It is not unusual for a revolver to have better accuracy with some chambers than with others. In a perfect world, they would all line up with the barrel properly; be straight and of the proper size and with no burrs or errors. It is said that all men are created equal. That may or may not be true, but all chambers are definitely not equal.

The holes in revolver cylinders are normally a little larger than the groove diameter of the barrel. A cast bullet will instantly expand to this size. (Not quit, of course, but it appears so.) As the bullet jumps the gap from the mouth of the cylinder into the barrel, it is *mushroomed* or expanded a small amount by the hot expanding gas. This larger area at the bullets base lasts for just a tiny fraction of a second before it is swaged into the rifling of the barrel. Hand-loaders will get the best results with bullets the size of the smallest cylinder hole. Yes, the smallest. They frequently will vary in size by a small amount, and the cheaper the gun, the wider the tolerances in manufacturing.

The bolt notches on a revolver are the odd shaped openings on the outside of the cylinder. The locking piece (bolt) engages in this to hold the cylinder in place so it will not rotate until required. On six shot revolvers, this notch is placed over the chambers which is the thinnest and weakest location. Five shot revolvers have the notch between the chambers where there is more steel and therefore more strength. In many examples, this will result in a firearm that can withstand higher internal pressure. Caution is advised when hand-loading to any amount higher than normal.

SEMI-AUTOMATIC PISTOLS

Semi-automatic pistols do not have the gas loss of revolvers so the maximum barrel length for highest velocity is figured like a rifle and based on bullet size, bullet weight, powder weight, powder characteristics and several other factors. Of course, there is still a practical length that overshadows the desire for velocity.

Technically, the semi-auto should be more accurate than the revolver because the barrel and chamber are made in one piece. There is no gap and more important, no potential alignment problem between the barrel and cylinder. The top quality revolvers by S & W and Colt are aligned perfectly when new, and are therefore, hard to beat. Semi-autos are always made a little loose to guarantee that they will always function. The revolver problems are overcome by precision manufacturing and the semi-auto advantage is lost through the need to add some inherent looseness. They tend to cancel each other out. So far as accuracy is concerned, most will be about equal.

COMPARISON OF REVOLVER VS. SEMI-AUTOMATIC

A revolver can be used with either hand in both single action and double action. They will handle any load, both light and heavy, up to their maximum peak pressure. Semi-automatics are made to fire bullets with weights and pressures within a small range. This is controlled by the recoil spring. For the most part, revolvers can operate in extremes of temperature and dust, dirt and mud that would jam many semi-autos. The revolver is safer for used by inexperienced people. It has a disadvantage in the requirement that the chambers must align properly with the barrel compared to the semi-auto where they are together in one piece.

The semi-auto will usually carry a large amount of cartridges and can be reloaded faster with a clip. (Although an expert with a speed-loader can load a revolver faster than most people would expect.) A bad cartridge can jam a semi-auto and two hands are needed to clear it. Some semi-autos will not fire if the opponent is close enough, (and cool enough) to press the forward end back out of battery.

The semi-auto will fire faster. The revolver, even when triggered fast by an expert, cannot fire as fast as a semi-auto. The wait for an automatics slide to cycle is a myth. Have you ever seen a semi-auto malfunction and fire full auto? A short, quick spurt and the magazine is empty. A revolver shooter has to wait for the trigger to return forward. To some readers, the word *wait* may seem incorrect, but to an experienced expert, it is sensible.

The safety on a semi-auto can work both for and against its owner. If it is forgotten in an emergency, then the weapon may not fire when needed. If the gun is taken away and turned against him or her, the opponent can forget the safety or not know how to release it. This may save the owners life. Also, the owner can forget the safety after firing when it should be applied once again. The answer is obvious. If you own a semi-auto, practice with the safety, both removal and applying, so that it becomes a trained response. In other words, second nature; habit.

PLUS "P" AMMUNITION

Plus "P" (+ P) loads are cartridges loaded to higher pressure for higher velocities. The normal .38 Special has a maximum chamber pressure of 18,900 C.U.P. (Copper Units of Pressure). + P loads are made up to 21,900 C.U.P. Many shooters don't believe the increased noise and recoil is worth the small increase in velocity. The biggest improvement is in the short barreled revolvers (2") where the standard charge doesn't have the time to build velocity. The extra push amounts to about 145 f.p.s. in a .38 with a 2" barrel. A six inch barrel will deliver more velocity without + P and will be helped by about one third as much.

Some shooters believe + P ammunition in a .38 makes their gun like a .357 Magnum. The maximum + P pressure is, as stated, 21,900 C.U.P. The .357 Magnum is 46,900 C.U.P. This gives a velocity increase of about 350 f.p.s. for the same bullet weight.

+ P ammo is better than the regular .38 Special ammo. If the gun can safely handle the pressure, fine, but don't expect too much.

.22 CALIBER

A .22 caliber handgun with a short barrel will not obtain much benefit from the use of cartridges that produce a high velocity such as the CCI Stinger. High velocity cartridges of this type are wonderful in a firearm that can use their potential but they require a much longer barrel to reach their peak pressure and velocity. The little .22 short cartridge will sometimes have a higher velocity in a barrel as short as 2". (See the discussion of barrel length and expansion ratio in this book.)

BARREL FLEX

Barrel flex and vibration are not limited to rifles. Even handguns are affected and that is further proof of the phenomena. Testing has been conducted by individuals, private laboratories and the U. S. Military. This is usually done with a machine rest holding the gun to ensure precision and prevent variations caused by the shooter.

First, we need to realize that among semi-automatic .22 caliber target pistols, there are different barrel types in common use. Both the non-circular and the very heavy bull barrels have enough strength to prevent most flexing. A small diameter round barrel is more springy. Also, a long barrel, if it is flexible, will flex more. All of these statements sound logical and sensible and the tests show them to be true.

Guns of equal barrel length shoot tighter and more accurate groups if the barrels are stiffer. Weights attached to the front of flexible barrels helped to increase accuracy as a direct result of reducing flex. The weights can improve a flexible barrel to about the level of a stiff barrel if it is done properly. A few counterweighting methods do not increase stiffness and therefore they do not increase accuracy. Most help by improved rigidity and mechanical dampening of the vibration and flex.

Longer barrels increase accuracy in handguns with stiff barrels, as we would expect. However, in thin flexible barrels, there is more spring and it wipes out or cancels the benefit. Remember, the testing is with a machine rest so the longer sight radius is not involved and offers no benefit.

In summary, it is correct to state that increased stiffness, due to either shape or size or a combination of the two, will improve accuracy. Properly installed counterweights will improve stiffness and therefore, accuracy. A longer barrel will only improve accuracy if it has enough strength to overcome the increased tendency to flex.

BULLET WEIGHT

Frequently, target shooters that hand-load rifle cartridges check each bullet's weight. It is known that even a grain variation can affect group size. The actual weight is not as important as consistency. All bullets in a batch must be equal for uniform performance. The problem is not confined to people that cast their own bullets. Few factory bullets are made to the exacting standards required for competitive long range use.

The problem is not as much with velocity as with accuracy. At ranges as short as 25 yards, a variation of even 1 grain can make a slight change in accuracy in controlled testing. A 2 grain variation is, as expected, even more noticeable. Several controlled laboratory tests have been conducted on this subject, but it is an easy test for sportsman to do themselves. As with all testing, all other factors must be carefully controlled so they will be the same from one shot to another.

CHAPTER 23

AIR GUNS to MUZZLE-LOADERS to CANNONS

AIR GUNS

Although considered toys by many people, air guns can be deadly at ranges up to about 50 yards, although the actual range will be much longer. For example, the Daisy Pistol model 790 fires a .177 pellet with the aid of a CO_2 cylinder. The factory manual claims a high velocity of 425 f.p.s. and a maximum distance of 232 yards. The velocity has been chronographed as high as about 500 f.p.s. This is excellent for an air gun with a 8.3" barrel length.

Pellets are not made to the close tolerance applied to bullets. The bore fit disparity and weight difference accounts for much of the velocity variation. A loose fit permits too much pressure to leak past the pellet and permits low velocity. A tight fit requires too much pressure to overcome friction. This is basically the same as a bullet in a rifle or pistol.

Pellets have their weight forward in the head because of the thin skirt at the rear. This is the opposite of bullets which have the center of gravity behind the mid-point. Therefore, a bullet will tend to tumble if everything is not correct while a pellet will remain stable. The skirted shotgun slug has a situation similar to the pellet.

Air guns are, in most examples, not as accurate as other firearms but they can still do a fine job if the owner takes the time to practice.

CO_2 DETAILS

The CO_2 cylinders are of liquefied gas, not just compressed air. The pressure is under the control of the temperature and it is quit high. At 55 degrees F. it is about 700 p.s.i. It goes up to 1,125 p.s.i. at 88 degrees F. Above 88 degrees it is a gas instead of a liquid and not as efficient. While the gun will still operate, it is best to keep the cylinders out of the sun.

Compressed air will descend in pressure and give less velocity with each shot while the CO_2 cylinders remain about the same.

Air that is compressed is affected very little by temperature and varies directly with the temperature above absolute zero (-459.6 degrees F). A pressure of 1,130 p.s.i. at 90 degrees F. (549.6 absolute) would change to only 1,200 p.s.i. at 125 degrees F. (584.6 absolute). Assuming, of course, that the volume of the gas chamber remained the same.

CO_2 cylinders intended for use in air guns include a lubricant to help gun life and performance. They should not be used for any purpose involving food or drink, unless someone likes oily seltzer water.

Some airguns, such as the Beeman R-1 spring-piston air rifle, brought out in the mid 1980's, have a muzzle velocity of 940 f.p.s. with a pellet of .177 caliber. This velocity will exceed the .25 ACP which, depending on manufacturer, will be from 760 to 815 f.p.s. The trajectory will be a curve, as expected, but with a zero at about 30 yards, it will not be over 2" high or low to 50 yards. Beyond 50 yards, not only does the trajectory drop rapidly, so does the energy and velocity. This is not intended to be a comparison with the .25 auto on anything except velocity. The little .25 pushes 45 and 50 grain bullets compared to the air guns 8 grains.

Other spring guns produce muzzle velocities in the area of 750 to 800 f.p.s. The standard air gun operating with CO_2 or pump-up pressure will deliver muzzle velocities around 500 to 600 f.p.s.

A Sheridan rifle, using 8 pumps as recommended by the manufacturer produces a velocity of approximately 580 f.p.s. It is interesting to note that the velocity drops only about 3 or 4 f.p.s. between the muzzle and 15 feet.

Gas guns are low powered in comparison to conventional firearms, but there is still a danger involved. They can cause serious injury or blindness. All the same safety rules apply that are used for other weapons. Point the gun away from people, keep safety engaged, etc.

Target shooting with air pellet guns is normally done at 15 feet for smooth-bore guns and 25 feet for rifled-bore guns. (Note that throughout most of this book, ranges have been in yards. This is in feet.)

Technically, the term BB is incorrect as BB shot for shotguns is .180 diameter and the pellet for air guns is .175. Close, but not the same.

The Diabolo is a skirted pellet for use in rifled barrels of either .22 or .177 diameter. The base is open and hollow. The smaller front rides in the bore while the base or skirt engages the rifling.

The Sheridan air rifle uses a .200" (5 mm) pellet with a conical point and a long bearing surface similar to a regular bullet. A small band of larger diameter at the base is for engagement of the rifling.

MUZZLE-LOADING NOTES

All the rules of ballistics also apply to antique and reproduction muzzle-loaders. Here are a few notes that are not mentioned elsewhere.

Even with their crude methods, some early weapons were made with great skill. In the hands of an equally skilled shooter, our ancestors could reach out to 500 and even 1,000 yards with accuracy that many of us wish we could achieve today.

◆ Should powder be added to increase velocity? This is a subject every shooter can answer best by testing with his own gun. Keeping safety in mind, there frequently is a velocity that is difficult to exceed even after increasing the powder charge. At times, the ball is deformed by the heavier pressure. Increasing the charge beyond normal also will increase recoil and may not be as accurate. It also can damage the gun and even the shooter. It is best to use a chronograph to see if the attempt is worth it.

◆ The extra part on the round ball shape of the projectile left by the opening in the mold (the sprue) is seldom removed completely or properly. The result is radial weight that is not evenly distributed. The balance is off center from its axis. This can cause erratic trajectories.

◆ Round balls in flight appear in shadow photos to be flattened on the forward end by air pressure. The ball is not flattened but appears that way because of refraction. The light rays are bent as they pass obliquely through the bow wave of compressed air.

◆ Even modern round balls made with great care and fitting the bore properly, have most of the same old problems. While the ball may not *bound around* as it rushes down the bore, the contact with the barrel will be only tangential. In other words, it will touch only in the thin area where its curve and the straight bore meet. Rifling may impart a spin, but usually the rotation is small and ineffective.

◆ In percussion revolvers, a round ball can cause the next chamber to also fire. This is called a flash over. There is so little contact with the cylinder or bore that the hot gas and flame can ignite the next chamber.

◆ Many of the early smooth bore muskets used round balls that fit the bore very loose. Gas escaped around the projectile and reduced the velocity. The direction of departure was haphazard and could be a large angle away from the bore line. It depended on where the ball last hit the inside of the barrel. Also, many of the balls were home made and not spherical. They had unpredictable trajectories and a hit on anything over 100 yards away was more luck than skill.

◆ Not all calibers will be the same, but an average twist for early rifled barrels for patched round balls was one turn in 66". When conical bullets became popular, a faster twist rate of about 48" was required. Conical bullets, if the caliber is the same, are naturally longer and heavier.

MILITARY CANNON & ARTILLERY

It is not expected that many readers of this book will need this information, but it is valuable to serious students of ballistics. Here we are dealing with targets that are not in sight of the gunner due to either a great distance or because it is over a hill or tree line. In other words, artillery ballistics. Placing the projectile on the target is called *laying*. If the target can be seen, it is called *laying direct*. If the target cannot be seen, it is called *laying indirect*.

A learned gentleman with a Doctor of Science degree who taught this subject at the United States Naval Academy at Annapolis, has claimed in writing that this is a serous science but that using a rifle up to 1,000 yards needs "little instruction", requires "little practice", and "requires no recourse to theoretical principles." Any reader who has glanced or read through this book knows that is a false and inaccurate statement.

PROJECTILES

Early cannons were not limited to round balls and similar projectiles. For use against sailing ships and their masts and rigging, cannons fired 2 balls connected together by a chain. It was discovered very early that this must be fired as one load by one cannon. When attempted by two separate cannon with the chain laid between them, it was impossible to fire both simultaneously. The outcome was more disastrous to the gunners than to the enemy.

A lot of smaller balls or shot was fired bundled together as *grape shot*. This was held together by a cloth bag. If loaded in a metal canister, it was called *canister shot* and in a long round case it was called *case shot*. These projectiles are called *langrage* or *langridge*, if you prefer.

MILS

Artillery and cannon fire use a different system than the familiar minute of angle (MOA) which was explained in detail in Chapter 18. The mil is used for high angles of fire and in other military applications. It is based in the metric system, which explains the lack of interest in the U.S. for all except artillery.

Briefly, it is based on the idea of a circle having 6,283.18 equal arcs. This is rounded off for convenience to 6,400.0 and a right angle is said to have 1,600 mils. This number was chosen because, without changing its value by much, it is easily divisible by various other numbers. There are charts that list angles by degrees and minutes and give the corresponding mils. Example: 30 degrees is 533.33 mils. 5 minutes is 1.48 mils. 0.4 minutes (4 tenths of a minute) is 1.19 mils.

In technical wording; the mil is the angle whose tangent is 1/1,000 or the angle subtended by 1 unit at a range of 1,000 units. (Feet, meters, yards or whatever.)

Or, linear mils (straight down range) is based on one linear mil being equal to one thousandth of the range. At 100 yards, this would be 0.1 yard or 3.6 inches, 1 foot at 1,000 feet and 1 yard at 1,000 yards.

1 mil = 0.000982 radian

90 degree right angle = 1600 mils (milliradions = 1570.8)

The fundamental formulas which relate mils and radian are

$$1600 \text{ mils} = \pi / 2$$

or

$$1 \text{ mil} = \pi / 3200 = 3.1416 / 3200$$

so that

$$1 \text{ mil} = 0.000982 \text{ radian}$$

and

$$1 \text{ mil} = 0.982 \text{ milliradian}$$

With regard to the milliradian, this gives an error of less than 2% which is not considered significant.

It is generally assumed that a mil is a milliradian so 1 radian = 1,000 mils and the number of mils in an angle is 1,000 times the number of radians. This gives us 3 basic formulas that can be useful.

$$\text{mils} = 1{,}000 \text{ arc} / \text{radius}$$
$$\text{arc} = \text{mils} * \text{radius} / 1{,}000$$
$$\text{radius} = 1{,}000 \text{ arc} / \text{mils}$$

For example, if an object is know to be 5,000 feet away and extends an angle of 5 mils, how wide is it?

$$\text{arc} = \text{mils} * \text{radius} / 1{,}000 = 5 * 5{,}000 / 1{,}000 = 25 \text{ feet wide}$$

In military manuals, rifle elevations are frequently given in mils. It is therefore important to know that an artillery mil is 3' 22.5" of arc. That is also 3.375 minutes. To convert mils to minutes, simply multiply the mils by 3.375. **(There is more on MILS including a chart on pages 224 & 225.)**

TRAJECTORY

A cannon fired at a 45 degree angle in a theoretical vacuum would follow a parabola curve to reach its maximum range (or *random*). In a normal air environment, the air resistance causes the projectile to not rise as high or have as wide a range as the parabola. It follows an unsymmetrical line called a ballistic curve. The elevation for greatest range will be less than 40 degrees. (For small arms, it will be closer to 28 to 32 degrees. See Chapter 19 for more.) For short range, it was and is common to fire heavy rifled mortars up to 16" at angles up to 85 degrees. It is interesting to notice that they still land point first. At angles between 85 and 90 degrees, they may fall back to earth tail first.

With military gunnery, the variables that affect dispersion and probability are similar to the variables with small arms fire. Any 2 of the first 3 will determine the parabola if based on a theoretical vacuum. **All** are used in a real world atmosphere. **(1)** Quadrant elevation **(2)** muzzle velocity **(3)** range **(4)** characteristics of the atmosphere **(a)** humidity **(b)** wind **(c)** pressure **(d)** temperature **(5)** projectile weight **(6)** projectile shape **(7)** rotation of earth **(8)** curvature of earth **(9)** trunnion tilt.

There are 5 chance variations that can affect the outcome. **(1)** Quadrant elevation; minor departures from the precise value. **(2)** atmospheric characteristics; **(3)** projectile; minor differences in shape, weight, distance rammed and location of center of mass. **(4)** powder charge; minor variations in temperature, grain size, weight, burning rate and moisture content. **(5)** erosion; effect of chemical action from the hot expanding gases on the bore lining and the chamber.

Range tables are used to pick out the angle of elevation and corrections to firing angle.

GRAVITY INFLUENCE AT LONG RANGE

While it is seldom utilized in small arms shooting, even at the longer ranges of up to 1,000 yards, the curvature of the earth and its gravitational attraction may be considered in long range artillery shooting. Over an extended trajectory, the curvature of the earth not only is affecting the height of site, since a horizontal plane through a point at sea level cannot pass through a distant point at sea level due to the intervening land or sea, but also is producing a convergence in the lines of gravitational force. In the rectangular coordinate system of computing trajectories, the gravitational force is regarded as directed parallel to the vertical axis. As the trajectory is extended, the direction in which gravity acts is approximately toward the earth's center. The long range makes it more noticeable that this will not be parallel to the initial axis. A horizontal component of gravity is introduced, horizontal that is, with respect to a rectangular system which at the initial point has vertical and horizontal axes.

BALLISTIC TABLES

Ballistic tables for use in cannons and artillery do not give a certain projectile, gun or powder charge as standard. The caliber of the gun, weight and figure of the projectile, angle of elevation, standard velocity, and other firing-table constants never appear in ballistic table work. The only factors of the problem considered are the actual velocity, angle of departure, ballistic coefficient C, and the meteorological and more generally, geophysical forces which act along the path of the projectile.

The tables show no difference between the gun and the target, even if they differ in altitude by a large amount. Thus the tables are adapted equally to firing on the level and to antiaircraft fire, drop bombs, mountain firing, etc. The familiar theory of rigidity of the trajectory, as covered in Chapter 17, becomes wholly obsolete for these tables. For firing on a level, no supplementary correction table of the earth's curvature is needed because it was used in constructing the original table. (Remember, the long ranges that are involved in this type of ballistics. That is the reason for the earth's curvature to be involved.)

CHAPTER 24

TERMINAL BALLISTICS, GENERAL

Terminal ballistics is the term usually applied to how a projectile is affected when it strikes an object and how the object is affected during and after the strike. In other words, shocking power, penetration, expansion, wound channel, caliber, etc. As both the glossary and Chapter 12 explain, terminal is not the best word for this situation. Nevertheless, it is universally used and through constant use, it has became acceptable if not proper.

Target shooters may believe this section does not apply to them. Many do not hunt game or carry a gun for self-defense. But if you have ever been awakened in the middle of a dark night by a strange noise; a noise which was repeated after you became fully alert. A sound that seemed to not only be in your house but moving closer, some parts of this section may not only be interesting, they may save your life.

This is an excellent place for a safety warning. In a situation as just described, don't fire until you can be certain you are not shooting a family member who is wandering around because he or she can't sleep. Also remember the key word is self-defense. It is illegal to shoot someone to protect property, only lives.

Terminal ballistics can be divided into two basic groups. Hunting game with shotguns, rifles and handguns; and self-defense against other humans. These both may include the same three weapon choices. Of course, there is an overlap of the essentials because the end goal is the same; to incapacitate or kill with as few shots as possible. There are differences, of course, but the basic idea is the same except for one point. With game, the purpose is to kill and with humans it is to incapacitate. Military actions, wars and an occasional SWAT team project are unfortunate exceptions.

Both the police and citizens attempting self-defense can have major problems with stopping a person whose body is pumped-up with drugs. Hunters do not have to worry about rabbits or bears on cocaine.

There are differences in the psychological reaction to being shot between lesser animals and humans. Where animals will respond from an innate capability possessed at birth, the human response will be more toward learned behavior. But fear and pain and the desire for self-preservation can be strong in both and equally unpredictable.

Another difference; the average human has a strong basic desire not to kill another human. Even the sportsman that thinks nothing of shooting an elk will hesitate to defend himself against a violent criminal who may not have the same moral feelings. There is an *attitude* that is important to proper self-defense. It is an important subject, but not in the realm of ballistics so we should move on.

WORD CHOICES

Many sections of this book have words that need to be explained so the author and reader have the same thought when the term is used. Stopping power, killing power, disabling energy, knock-down ability; all are just words which each reader will, based on his or her own experience and knowledge, interpret in a different way. Let us not get hung up on a word choice.

Briefly, though, *stopping power* and *killing power* are "the ability to incapacitate an opponent immediately." They generally mean close to the same thing. Stop means, come to an end; the finish.

Disabling is not as severe or effective. It means to cripple or be unable to do normal activity. A disabled bear or a disabled man can still do a lot of damage and will be able to kill an enemy.

Another frequently used term is *wounding power*. It is not the same as killing power because some very bad wounds will not stop either a wild animal or a human. This book will not go into this area because it is unnecessary. **Stopping power and killing power both have their place in hunting and self-defense. Physical damage that does neither should be avoided.**

Stopping power and killing power are unprecise comparisons of how effective a bullet stops or cleanly kills a dangerous charging animal. For police, military and self-defence, the same stopping and killing power is used with regard to humans. Velocity and energy are factors, as well as bullet weight, range, type of bullet, caliber, and even to some extent, the personal preference and confidence of the shooter in the choice of gun and caliber. Supposedly, the moose in the hunters sights or the crazed killer will not be stopped quicker by the confidence of the shooter. Perhaps. But most people will perform better if they have confidence in their firearm, cartridge and training.

CHOICES OF CALIBER & WEAPON

Probably no area of firearms and ballistics is more hotly disputed than the choice of calibers for best killing power. Whether for game or self-defense, if 20 experts are interviewed, they will all have a different opinion. In this atmosphere of different views and changing ideas, only general statements will be made; comments that may guide the reader and help him or her to form an opinion.

We have people who believe that the large caliber bullet is best even if the velocity is low. Then we have the group that believe high velocity is the king even if the bullet is very small and light. Both of these two groups, and others in between, have test reports to back up and enforce their theories. For self-defense, a lot of words are written by medical doctors and others who base their information on human anatomy and the results of emergency room work and autopsies. This is not always as accurate as it is impressive sounding. The problem is in trying to determine who, if anyone, is correct.

Not long ago, a national gun magazine ran an article on this subject. It was written by a person who stated he was an ex-police officer with many years experience. He used half of his allotted space to insult every expert who had a different opinion than his. The word expert was in quotation marks and they were all called various vulgar names and told how stupid they were. Then he claimed that *he* was an expert and the balance of the article stated that his choice of self-defense caliber was the only one worthy of any consideration.

It is good for the public that he has retired from the police force. We do not need irresponsible hot-heads like him carrying a badge and a loaded gun.

There are a lot of good firearms and good calibers in the world today. Some are better than others. That does not mean that only one or two will do the job. When you make a decision on a caliber, whether by technology or emotion, it does not mean that every person with a different opinion should be treated with scorn and ridicule.

Both the law-enforcement officer and the home-owner have an interest in stopping a violent person intent on causing physical harm. The hunter's goal is less crippled animals to go away and die a painful death when a good clean kill would put meat in the freezer. Not to mention the sudden charge of a mother bear against a dear hunter who has stumbled on her cubs. All of the experts agree on one thing. Both the bear and the man can be incapacitated, but if the gun and cartridge aren't up to the job, your adversary can kill you even if he is hit in a vital area. But if someone does not agree with your choice of firearm, try to convince him with facts similar to what are used in this book, not crude insults.

MEDICAL VIEWPOINTS

Hydrostatic shock looks dramatic as a water filled can explodes or a gelatin block expands. In live tissue it may do the same. A body is affected by a bullet in a fashion similar to water. The bullet, as explained in Chapter 13, creates a bow wave in the air. It will also create a wave in water as a ship creates a wave, by fluid displacement. The major difference is that in the air, which is considered compressible for this purpose, the waves quickly dissipate. Fluid, as in water or a body, is considered incompressible and the waves are propagated without fading. The faster the bullet, the stronger the waves propagated through the body. The exit wound, then, is not a mere puncture but an explosive release of the wave before the bullet has reached the surface. This was shown by high-speed photography as far back as WW 2.

A well known theory states that a high velocity impact creates a hydraulic action in the body which disrupts the blood vessels and other body parts that are yielding. This action affects areas of the body not directly impacted by the bullet. Nerve and bone damage is also increased by the higher velocity and abrupt shock. This effect is strongest at velocities at and over about 3,000 f.p.s. The higher the velocity the more explosive effect. This is most pronounced in areas of the body that are primarily water like in kinematic viscosity. (Explained in detail beginning on page 301.) As an example, the brain is one of many areas that are primarily water like. There is a violent expansion in the microsecond that the high velocity bullet moves through. The force is outward in all directions. This is sound reasoning and no doubt has an influence on the outcome. It remains to be discovered or proven exactly how much.

Some experts disagree on the end result. They believe that the body returns to its original position and the larger cavity created by the explosive impact does little toward stopping power. In all probability, the end result would depend on many variables and be unpredictable.

Shock, as it applies to a bullet wound in the medical meaning, is the physiological state of rapid respiration and pulse, low blood pressure and potential loss of consciousness. It is generally caused by severe blood loss, not the force of impact.

The body can recuperate and recover astonishingly fast. The depth and width of the wound is more important. The bullet that makes the biggest wound will have the best stopping power, not the heaviest shocking power. (This disregards location. Obviously a .22 short in the brain will be more effective than a .45 ACP in a fatty area.)

In a broad general sense, the stopping effect and also killing efficiency of a wound, whether in a human or wild game, is a combination of the extent of the wound and its location. Hemorrhage is a significant element so any damage to large vessels in the chest and the base of the lungs is effectual. The stopping achievement may not be instant, but movement will only last until the blood pressure drops to a point where circulation through the brain is drastically lowered. Arterial pressure and vessel size drop with distance from the heart. Bleeding will cause collapse of the person or animal depending on how quickly the blood is lost.

Shots striking the brain, upper spinal cord and major blood vessels are quickly fatal. Sometimes with less bullet weight or energy because of the interruption of essential elements such as respiration and circulation.

A spinal cord hit may cause instant paralysis because the area below the point of injury will be cut off from the brain and all muscle control will be lost. It will not be fatal unless it is in the neck area which will cause paralysis of respiration. Paralysis may be described as a loss of voluntary muscular action.

Chest cavity wounds can be instantly fatal if the heart is stopped. Lung wounds to the central part can be quickly fatal due to hemorrhage. Hits in the outer areas of the lungs may not be fatal and may not even stop the game or assailant. In this event, death may come later from gradual blood loss.

Of primary importance is the magnitude and location of the wound. Obviously a small magnitude wound in one location may cause disability while a large magnitude wound in another may do little more than aggravate things.

PENETRATION

While expansion is beneficial, penetration is essential. The projectile's size is of no importance if it does not reach a vital area.

Bullet weight helps penetration because of its involvement in momentum. In bullets of the same velocity, the heavier bullet will penetrate deeper if all other things are equal.

Penetration is a matter of momentum versus resistance. Sectional density is when bullets are of equal weight but different caliber; the smaller caliber has more sectional density. (Refer to Chapter 14 for sectional density.) The reason sectional density is involved in penetration is a matter of momentum versus resistance. The larger sectional density will be with the small caliber. It takes less effort to stick a pin in our body than a big nail. The smaller caliber will go deeper because of the smaller area to resist against. The bullet with a larger sectional density will penetrate deeper than a lighter bullet, even if the lighter bullet is somewhat faster. Of course, deeper can also mean "passed clear through". If a person is on the far side, they can be in danger from a bullet that exits. Although in that case, the shot should never have been fired.

The bullet must be properly chosen to match the velocity. A bullet that expands and performs well at low or medium velocity may blow apart at a high velocity impact. At the other extreme, many bullet designs require a high velocity for expansion. The bullet must have the shape, core, jacket thickness, jacket stiffness, and in a hollow point, the proper angle and depth. All this and more must match the velocity to penetrate and expand.

There is some debate on whether it is best for a bullet to remain in the body after deep penetration or exit on the opposite side. Generally, if the bullet is properly placed, the wound is what counts. The damage to organs and tissue and bone are the most important. The exit wound supplies little except in increased blood loss. Exit wounds bleed externally a lot more than entrance wounds, which in some cases will bleed little to none. While this can leave a stronger blood trail for tracking a wounded animal, the ideal situation would be an instant kill with no tracking involved. **It is best to avoid the need to track an animal rather than making it easier.**

From a ballistics point, over-penetration is not always caused by a bullet that is too fast. If the velocity is too low for its design, there may be no expansion. This will maintain energy and increase depth. Most bullets designed for expansion will not perform well over the full velocity range. In some cases, the velocity range for good expansion is narrow.

PENETRATION OF SPHERES

For shot pellets, spherical projectiles like musket balls and round nosed bullets, penetration equals sectional density times a constant times the difference between muzzle velocity and striking velocity. It is discussed in Chapter 14, but briefly sectional density is dividing the square of the diameter in inches into the spheres weight in pounds. For pellets, add .033 to the diameter first. This is necessary because of the "boundary layer" of air discussed in another chapter. For projectiles of this shape, the constant is .233. Other shapes will require a different number.

This equation is good for velocities up to about 1,000 f.p.s.

BULLET EXPANSION

One type of expanding bullet will have the lead tip of a soft point crush back and split the jacket. As the penetration continues, the jacket peels back further as the lead expands outward. The bullet's expansion is partially controlled by the thickness of the jacket.

Fast expanding bullets do not penetrate deeply. An average expanding bullet will go deeper and a slow expanding bullet will go deepest, if everything else is equal.

KINEMATIC VISCOSITY & BASIC MATERIAL TYPES

Targets are classified in 3 basic groups for the study of depth penetration: primarily cohesive, primarily viscous and primarily like water. Of course it is not that simple because in the real world, whether the target is an elk or a felon barricaded behind an automobile door, all 3 groups may be represented in different amounts and grades and blended together in infinite combinations.

Kinematic viscosity is the term used for this subject. The resistance of a substance to flow if a liquid or yield to stress if a solid. It is primarily in the abstract without a reference to force or mass.

PRIMARILY COHESIVE

If the impacted material is primarily cohesive, it is united by tough adhesion that resists breaking apart. Examples are hardwood, masonry and to some extent, bone.

The opposition to penetration is constant and the depth of penetration can be obtained by dividing the projectile energy by the square of its diameter. (The depth is proportional to the square of the energy.) In other words, if the velocity is doubled, the energy goes up 4 times and the penetration goes up 4 times as well. This law can be traced back to Ben Robins in 1742. He is mentioned in other parts of this book for his contributions to science and ballistics.

PRIMARILY VISCOUS

If the impacted material is primarily viscous it has a molecular attraction that resists a tendency to flow. The material is sticky and thick or syrupy as is honey or heavy syrup. Animal and human tissue closely fit this group.

The opposition to penetration is in direct proportion to the projectile's velocity. In simple terms, the slower the bullet or shot, the less resistance. The faster the projectile, the more resistance.

PRIMARILY WATER LIKE

If the impacted material is primarily like water, the opposition to penetration is proportional to the velocity squared. A projectile at low velocity will penetrate water easily. Resistance is slight and less than in a viscous material like syrup. As the velocity increases, so does the resistance only at a faster rate than in a viscous substance. The resistance is not in direct proportion as in syrup but proportional to the square of the velocity. In this case, doubling the velocity will **not** double the penetration. Some body areas, including the brain, are in this group.

ENERGY & MOMENTUM

This may be a good place to remind the reader that there is a major difference between momentum and kinetic energy.

The lethal effect, or killing power, of a bullet is closely associated with kinetic energy. The knock-down or shove of a bullet at impact is related to momentum. It is easy to confuse and mix-up the two, but they are like vinegar and oil in a dressing. They are associated together and may even help each other, but they are still entirely different things.

KNOCKDOWN & ENERGY

An animal or opponent can go down instantly from a strike on a weight bearing bone, a brain or heart shot or a hit in the upper spinal cord. The right caliber in the right place can do enough instant damage and cause enough induced shock to put down a charging bear or a crazed human. Sometimes hunters believe it looks like the game was thrown over by the bullet. Unless it was a squirrel hit by a .30-'06 - 220 gr., probably not.

We need to discuss momentum, but first let's consider energy. As given in the charts, energy is frequently misunderstood. There are too many other important items. Remington ballistics show the .458 Winchester

Magnum as having 4,712 foot pounds of energy at the muzzle and 3,549 at 100 yards. This is in a 500 grain bullet and is tremendous energy by anyones standards. Do not fall into the trap of believing that, because a bullet has 4,000 lbs of energy, it will deliver a blow of 4,000 lbs. and will knock a 1,000 pound elk clear off its feet. It won't. That is not to say this is not a powerful cartridge, but the 4,000 lbs. is sometimes called *paper energy*. It is a mathematical number that in physics means loosely, *the capability to do work*.

As we discussed in an early chapter, 4,000 ft. lbs. of energy means that the bullet has the energy to raise a 4,000 lb. weight 1 foot or a 1 pound weight 4,000 feet. That energy is used or consumed on impact but it will not move the object hit as expected. If a dead deer were hung from a tree and shot with a bullet with 4,000 ft. lbs. of energy, it would move it very little.

The energy is used in creating heat, deforming both the bullet and the object hit, creating noise, breaking bone, penetration, etc. Only a small amount is used in knocking or moving the object struck. Usually any sudden movement of a living target is caused by an urge to flee or an instinctive muscle or nerve reaction to the abrupt shock and pain. Both are caused by the victims own physical reaction instead of the push or punch of the bullet.

All too often, the bullet with the high energy is not suitable because of bullet type, weight or velocity. Use it as a comparison of just one facet of a multifaceted issue.

For example, the impact point is one of the most important items in killing and stopping force. A small bullet at high velocity may have a high muzzle energy but when hitting an animal quartering away it may only wound the animal. A heavier bullet at a slower velocity and less energy may have caused a kill. A good heart shot may have been effective with either caliber, but the hunter can't always get the perfect shot.

Ballistic tables give energy at the muzzle as well as at different ranges. The muzzle energy is important for some math problems and in the recoil that is felt by the shooter. But what really counts is the energy that can be delivered to impact. This has a direct effect on killing power and a perfect hit does more harm than good if the animal is crippled.

For a few examples, let's start with the old favorite, a .30-30 Winchester Super-X with a 150 grain bullet. The muzzle energy is given at 1,902 ft. lbs. but it drops to 1,356 after just 100 yards. Not many people would try a 300 yard shot with a .30-30, but if they did, the energy would be down to 651 ft. lbs.

Another example is .30-'06 Super-X in 165 grain. The muzzle energy is 2,873 ft. lbs. with 2,426 ft. lbs. at 100 yards. All the way out to 300 yards, it still has a respectable 1,696 ft. lbs. Of course, the point is not what these two cartridges have in down range energy, it is what the cartridge you plan on using will have.

A lot of energy is used in bullet expansion. If the bullet had little energy to start with, the energy used for expansion won't leave enough for suitable penetration. Also, if the expansion is on impact, while the expanded bullet would do great damage pushed into depth, it will take increased energy to push it in. In many cases, that energy is not available. Too much initial expansion is not good unless accompanied by high energy. Both penetration and expansion are required. Neither is satisfactory by itself. Usually when one goes up the other goes down.

The military, NATO in particular, regards the killing power of bullets to be best expressed in kinetic energy. They consider 108 ft. lbs. to be the **minimum** to put a man down. Some NATO cartridges still have this energy at a long range. The 7.62 NATO at 2,600 meters (2,844 yards) and 7.62 x 39 at 1,500 meters (1,641 yards).

Although this is not a ballistic item, it is interesting that the military consider a wounded enemy soldier to be more beneficial than a dead enemy soldier. This is because other soldiers are put out of action to care for the wounded man while the dead require little attention.

KNOCKDOWN & MOMENTUM

An excellent way to show the blow or knock-down effect is not as strong as it seems is by General Julian S. Hatcher and quoted from the Aug. 1958 issue of *The American Rifleman*. He wrote that if a man were holding up a 1/4 " thick steel plate, 14" square, he would have no problem if it was shot from close range with a .45 ACP. As the General said, it wouldn't be safe because of bullet shatter, vibration, etc. but the plate would only move back about 3/4 of an inch at 2 feet per second. This is based on a 230 grain bullet at 800 f.p.s. and a plate weight of 13.6 lbs.

The mass times velocity (momentum) would follow the principle discussed in Chapter 13 about the ballistic pendulum. The bullet would impart its momentum to the plate at a kinetic energy of .84 ft. lbs. Gen. Hatcher figured that to stop the plate in 3/4" would require about 5 times the pressure required to stop the recoil of the pistol in the same 3/4".

This author conducted an unscientific but interesting test that further proves this point. A wooden box about 1' x 1' x 2' was attached to 4 legs made of 2" x 4" boards 3' long. The box was filled with dirt and the complete unit weighed 203 lbs. The object was to see what effect, if any, the impact of a bullet would have. It may have been comparable to an upside down ballistic pendulum only without the accuracy and measuring scale.

The first shot was by a young man with a .22 Magnum who said he bet even his small cartridge would knock it over. It made no detectable movement even though fired from as close as about 10 yards. (Remember, this was for fun. We didn't measure the range.)

A .30-30 - 170 grain bullet and a .30-'06 - 220 grain bullet both made a noticeable movement of perhaps 1/4". No more. Both the .30-30 and the .30-'06 penetrated and expanded well and if impacted in a vital area of a deer, would probably have been a quick kill. Knocked over? No way.

Knockdown, the actual push or shove, is a matter of mass times velocity (momentum), not kinetic energy. The momentum figure will be a much smaller number than if we incorrectly used kinetic energy.

$$(k.e. = 1/2\ MV^2)$$

Momentum remains the same before and after impact. In technical terms, the velocity before and after impact are inversely as the masses.

$$(MB/MT) * VB = VT$$

Where: MB = mass of bullet in grains

MT = mass of target in grains (lbs. * 7,000 = grains)

VB = velocity of bullet at range in f.p.s.

VT = velocity of target after strike in f.p.s.

Example: A 300 H & H Magnum, 180 grain bullet with 2,412 f.p.s. velocity remaining at 200 yards. (Winchester Super-X) Weight of game animal 550 pounds.

$$7,000 * 550\ lbs. = 3,850,000$$
$$(180 / 3,850,000) * 2,412 = 0.113\ f.p.s.$$

As anyone can see, 0.113 f.p.s. is almost no movement from the impact. The location of the hit will determine the final outcome.

VELOCITY WEIGHT & MOMENTUM

High velocity helps stopping power and so does bullet weight. One can be substituted for the other or combined for higher effectiveness. In a substitution, we must remember the physics of momentum. Light bullets at high velocity may have insufficient momentum to penetrate through animal hide, bone and muscle. Mathematics will show us that if we use a 150 grain bullet at 3,000 f.p.s., a 50 grain bullet would have to achieve 9,000 f.p.s. to have the same momentum.

Momentum gives more importance to a projectile's weight.

Momentum is the reason a bullet that disintegrates on impact has such poor effect. If it were split in half, each half would have the same velocity (disregarding minor differences) but only half the weight and half the momentum.

CROSS SECTIONAL AREA

A bullet's cross sectional area is more valuable than its diameter in figuring stopping power. This is found by squaring its radius and multiplying the answer by π. This can give a truer picture of the bullets frontal area than just its diameter.

BULLET SPIN

People frequently wonder if bullet spin has any effect on killing power. The simple answer is yes, but the amount is very small and the explanation is not simple.

Many hunters have noticed the wound path created by the bullet and one or more smaller paths created by a bullet particle which has broken off and spun out by centrifugal force. The spin causes additional damage from the fragmenting similar to a piece braking away from a fast spinning flywheel.

Imagine the particles being broken off and moving away from the original path at a right angle and also driven forward at the bullets velocity. The speed the fragment will have as it flies off will be the same as its circular path surface speed. (All reduced, as it were, by the tissue and bone, etc. Deceleration will be heavy because of reduced momentum.)

The spin adds to the killing power in the extra projectiles that damage tissue and bone and the *cone effect*. Very little is added in additional power. The energy at impact will be greater, but by a tiny amount.

THE MATHEMATICS OF SPIN

For the reader who is interested in mathematics, it can be proven easily. The chapter on rifling has 8 equations that deal with rotational speed and related subjects. Several may be of benefit here, but in the interest of saving space, please refer to Chapter 12. The equations for surface speed, resultant speed and spin energy will be useful.

One route not shown in the rifling chapter that will give the same answer is:

$$S = R\,\pi * 2 * T / 12$$

Where: S = surface speed in ft. / sec. circular
T = bullet turns / sec. (revolutions)
R = radius of bullet

EXAMPLE: To show the small increase, we can work out a short series of equations. Given circumference of bullet, (.942") .0785 feet. Velocity: 2,900 f.p.s. from a 10" twist barrel. Using the equations in the rifling chapter, we find 208,800 r.p.m. From the same chapter we find $S = C * R$. This gives us .0785 * 3,480 = 273 f.p.s. (The 3,480 is revolutions per sec.) The spin is integral with the forward velocity which, in this case, is 2,900 f.p.s. Again from the rifling chapter we find:

$$RS = \sqrt{v^2 + s^2}$$

Where: V = muzzle velocity in f.p.s.
S = surface speed in ft. / sec.
RS = resultant speed

This worked out to 2,912.8 minus 2,900.0 for 12.8 f.p.s.

For the math. fans among our readers, another formula may prove interesting.

$$E = 1/2 \ I A^2$$

where: E = spin energy in ft. pounds
A = angular velocity
I = moment of Inertia about the long axis

This also will give a similar very low number.

Even if you skipped all the mathematics, you can see that 12.8 ft./sec. and a similar energy increase in foot pounds is very tiny. Most all cartridges will give low amounts. To compute the entire bullet energy use:

$$E = 1/2 \ M V^2 + 1/2 \ M K^2 W^2$$

Where: E = bullet energy
K = the radius of gyration of the bullet about its longitudinal axis in feet.
V = velocity in feet per second.
W = rotational velocity in radians per second

The rotational velocity adds such a tiny amount to the final result that it is normally omitted. This leaves:

$$E = 1/2 \ M V^2 \quad \text{or} \quad E = W V^2 / 450{,}400$$

Where: E = energy in foot pounds
V = velocity in f.p.s.
W = bullet weight in gr.

YAW & PENETRATION LOSS

Penetration test conducted by the U.S. Army have shown less penetration at very short ranges than at moderate ranges. The lack of penetration at short range is blamed on initial yaw which prevents the bullet from striking straight on. When the bullet stabilizes, the depth of penetration increases. Then, as velocity, energy and momentum decrease with range, the penetration depth gradually drops again.

KEYHOLING & YAW AT IMPACT

Some people believe that a keyholing bullet has great stopping power. If we really think this out to a logical conclusion, we can see this is wrong. Perhaps in a few rare cases it could work out that way, but not normally. A tremendous amount of work has gone into designing bullets for proper expansion at depth. A bullet that has yawed or tumbled so that it strikes on its side may appear to perform well because of the larger surface area in contact, but the penetration will be short and expansion almost zero. All other things being equal, the effect will be less than desired.

While many people would complain about the cruelty to animals (perhaps while eating a hamburger), The Aberdeen Proving Ground tested bullet effect on live animals that were sedated. Horses, pigs and goats were shot and killed to study the reaction of the impact and the wound. Several people that were well known, or would be later, were involved in one way or another. This included Gen. Julian Hatcher and Col. Frank Chamberline. This was a military test, so only military ammo was involved. That is disappointing because with the protesting from animal-rights people, no civilian company would dare to test their sporting ammo in this fashion.

Later, Col. Chamberline expressed concerns that no two bullet wounds were alike and perhaps not as much was learned as expected. Nevertheless, the experiments did show that a bullet that is spinning without proper gyroscopic stability or with a slight yaw will turn sideways on impact. (If necessary, the reader should refer to Chapter 8 and gyroscopic stability. Remember, the center of gravity of most bullets is behind the center of its longitudinal axis. In other words, the center of gravity is toward the base. This will give it a strong turning effect if a force is applied behind the tip.)

In the Aberdeen tests, the bullet would yaw after as little as 2" of penetration and present its larger side area forward. The wound would be much worse and, the same as an expanding bullet, penetration less. While this test showed how catastrophic a yawing bullet can be, it is still not as good as controlled expansion with proper energy. Although, we can only write of generalities because as tests of this type show, no two examples are the same. What is great in one case may fail miserably in the next.

TESTING FOR EXPANSION, PENETRATION, ETC.

It is both fun and interesting to do bullet expansion and penetration testing. It is an excellent excuse to spend a day enjoying the gun hobby. Of course, if it is a business or a job, then it can suffer the fate of all employment; dullness and tedium.

Many hunters, perhaps even most, never test a bullet or load except by killing game. Even if the gun is sighted in to zero at a set range, the bullet expansion and penetration are not tested. Frequently not even considered. Shooting animals in the field is never a satisfactory way to tell performance because no two shots are alike. One may hit bone and another soft tissue. One straight on and another at an angle. This does not mean that nothing can be learned hunting. Quit the contrary. But a controlled test with watermelons (too expensive, but fun), ordinance gelatin, wet newspaper stacks; all can give a lot of information if done properly.

Some people don't believe that any information can be learned by shooting into gelatin blocks or water soaked paper. People with that attitude are right; they won't learn anything. The rest of us can usually learn a lot. Bullet expansion is not as simple as it may appear. Testing, even if done in an amateur way, will still demonstrate it better than 10 books on the subject, including this one. The idea is to compare different calibers, bullet types and loads with a lot of shots. It is true that not much would be gained by firing 5 rounds from the same box and quitting.

A large number of shots must be fired and without preconceived views and opinions to distort the thinking. In other words, a lot of shooting and then believe the results.

TESTING MATERIALS (target mediums)

Ductseal can be used but it is expensive. It also is hard to examine the bullets path and it can become full of bullet fragments. (**As the reader can imagine, several of the following methods have these same problems.**)

Clay is a useful test material but the bullets path is difficult to examine. Some people have tried filling the cavity with plaster, but this is simply more trouble than it is worth.

Wooden boards give an interesting comparison from one bullet to another. For farmers and ranchers with extra wood laying around, it is cheap and worth the effort. But there is no resemblance to an animal or the human body and bullet recovery can be difficult. It is frequently done and fun and interesting, but there is very little scientific benefit.

Gelatin is used extensively in testing because it is believed to respond similar to living tissue only without the bones, muscles, etc. While some people disagree on this point, it remains a reasonable test that is practical. It is not easy or cheap for amateurs because the large blocks require a refrigerator or freezer and also an investment of time. Ballistic gelatin, sometimes called ordinance gelatin, is similar to a thick gelatin desert. The ingredients are made by a division of Knox Gelatin Co.

Newspapers and phone books make a very good test if tied together about six inches thick and soaked in water for at least a night. It will take a number of big-city books or papers even though their size will expand with the water to almost double their original size. For that reason, don't tie them too tightly with weak string. They should be used soon after removal from the water, if possible. This is an excellent home method. The bullets path can be examined by opening the bundle. Depending on the bundles width, from two to six bullets can be fired before the stack is opened. Long range is no obstacle, as long as the accuracy is available to hit the stack in from the edge. A strike near the edge may be forced out by the higher pressure inward in the stack.

Plain water works to some degree but as many people have found, one steel barrel is frequently not deep enough. Two will have to be welded together as water tight as possible. A piece of pipe, capped or plugged at the bottom, is fine if the diameter and length are big enough. The container can be tilted at an angle to the ground. In any case, to avoid a soaking, the shot should be fired from a ladder or elevated position or the container needs a cover with a hole just big enough for the gun-barrel to enter. A screen basket or similar method will be needed to recover the bullet. A big disadvantage is that penetration cannot be measured.

Water filled plastic bags is another method of testing bullets that is both inventive and novel. It is nothing more than a long narrow wooden box filled with plastic bags that are filled with water and then sealed by squeezing and sliding the top closed. The box can be very simple. Just something to hold the bags in place. An open top is suitable. The length should be at least 3 feet long with an opening in one end to fire the bullet. A few smaller openings in the bottom will be a help in draining the water from the broken bags. The width and height can be anything suitable to fit the size of the bags. Penetration length cannot be measured exactly, but it can still be figured closely by the number of bags broken.

Note on water testing: Both kinds of water testing will expand bullets that will not expand in flesh. Also, it cannot show the bullets path, sometimes called a temporary cavity or wound channel, which can give details that are not available from the recovered bullet.

Barrier materials can and should be used for tests on human ballistics, but not for game hunting tests. Deer don't normally hide behind car doors and they do not wear clothing. This is one of the major differences between self-defense and game shooting. (Barrier materials will be covered in Chapter 26.)

Regardless of what method is used, after each shot, dig out the bullet and measure the penetration, if possible. The cavity can be checked for diameter at various depths and the maximum length determined. The shape can be important as a check for expansion depth, tumbling and keyholing. Note this information and bullet type, load, range and any other variable that applies. Keep them all separate with their notes. Little plastic bags are excellent. Later, they can be cleaned, inspected for expansion diameter and weighed to determine lose. Also measure the length of the remaining body. Core and jacket separation can also be observed. Early separation will reduce effectiveness, but late separation usually has little adverse effect.

NOTES

◆ Velocity and placement are vital, but the bullet's performance on impact makes the difference between success or failure. Many bullets of a similar weight and aerodynamic shape will have a similar trajectory and then be different as night and day in the way they act from impact onward.

◆ Bullet construction is a science. Some are solid lead; others solid brass; some have hollow points; lead cores with copper jackets; partitions where the jacket extends across and divides the core into two pieces, etc. All are intended to penetrate deep, expand big and retain most of their weight. (That is, not shed little pieces along the wound channel.)

◆ If all things such as weight, velocity, etc. are equal, the bullet that expands the most will penetrate the least and conversely. A compromise is usually required.

◆ Kinetic energy is a good point to consider in choosing a cartridge, but not the dominant point. The ammunition with less energy may suite the situation better. A bullet that goes clean through has plenty of energy but little may be imparted to the body if it passes through only nonvital tissue and fat. No amount of energy will make up for poor placement.

◆ While bullet weight and sectional density can be overdone, most experts believe that for killing power, it is best to error on the heavy side. Weight. That is the receipt for breaking bone. High velocity is not usually thought of as being as important as weight for bone smashing. **It is important to note that this is controversial and not everyone agrees with this statement.**

◆ Velocity is what causes a wound channel to expand beyond the bullet's diameter if we do not have expansion. In other words, a very slow moving bullet can punch through the body and leave an exit wound no bigger than the entrance wound. Higher velocity, all else being equal, creates a larger wound. This is easily and frequently proven.

◆ **Rule:** If two bullets have the same velocity and weight and neither exits the animal (or person), they will both impart the same kinetic energy. Nevertheless, the actual killing or stopping power may be different. If the diameter is the same, the wound volume will be about the same. The bullet that expands the fastest will penetrate less and will release energy quicker.

HISTORY NOTE

It is interesting that in the early 1600's and even as late as 1720, several European gun makers made barrels with strange shaped bores. For example, square and triangular bores, along with their square and triangle shaped projectiles, were made in Germany by Augustinius Kotter. Their inventors believed the sharp corners were more lethal than the round balls in use at the time. Perhaps in a way, that would have been true, except that it was a nightmare to properly make the projectiles to fit the bore; especially with the technology of their time. Also, the projectiles could not be made to spin and would have poor ballistic qualities.

CHAPTER 25

TERMINAL BALLISTICS for HUNTING GAME

This chapter will deal with terminal ballistics for the hunter. Of course, everything that was discussed in the last chapter applies here. The facts on kinetic energy, momentum, bullet spin, medical information; they all fit here as well.

The goal is to produce hunting trips that are successful. One way to help in that respect, at least from a ballistic point of view, is to use the proper caliber and bullet for the animal hunted. Also the best barrel and twist, properly sighted in for the correct range and the trajectory involved. Much of that is covered elsewhere in this book. In this chapter we will discuss the terminal ballistic factors that can help to make a hunting trip a success story.

Some hunters base a lot of their knowledge on one memorable example. If they merely wounded a small deer with a .30-'06, then from that moment on, the cartridge is no good. The reason for failure may have had nothing to do with their choice of cartridge. If on the next trip, they get lucky with a .30-30, from then on, it is the best.

There are frequent exceptions, but the eastern hunter normally shoots over short distances and through brush more than the western hunter. The western hunter needs a flatter shooting bullet for longer ranges and more open spaces. The proper gun and cartridge for one hunter may be a fast light bullet for small deer. For a different hunter or a different trip, the best may be a heavier bullet with controlled expansion for large animals. A .22 Hornet is not good for hunting moose and neither is a .375 H & H Magnum a good rabbit gun. One problem is that most people don't own enough guns to always have one available that is perfect for the job.

LONG RANGE PROBLEMS

Long range shooting is not limited to target work. While hunters should never fire at ranges that are not practical, there are times when only long range shots will do the job. An example of stupid long range shooting would be a deer hunter with a .30-30 lever action at 300 yards. On the other hand, elk, mountain sheep and goats and a few other species are routinely taken at that range. This requires well placed shots from guns with sufficient caliber, bullet and trajectory. Not always, but usually a bolt action rifle with minute-of-angle accuracy and heavy bullets with high velocity.

Most experts, such as Elmer Keith, agree that for long range hunting, bullet weight, density and a velocity of 2,700 or 2,800 f.p.s. are more important than a higher velocity. This is an old theory that is still good today. One secret is expansion. The bullet used for 300 yards has to expand at 300 yards. That sounds obvious, but a lot of bullets that expand well at the shorter ranges don't have enough energy or the proper design to expand at the long ranges. Remember that velocity and energy deteriorate as the range increases and both are usually required for proper expansion and penetration.

If a bullet does not expand well, the larger calibers have an advantage because the hole is larger. Of course this is basic, but some hunters forget it and try to make clean kills with unsuitable cartridges.

For a few examples, the minimum for lighter long range game such as antelope, goat and sheep is a good expanding 150 grain bullet at 2,700 f.p.s. Once again, that is the minimum. A little more weight and velocity would be an improvement. The .270 Win., .30-'06 and 7 mm, each with a 150 or 160 grain bullet is a minimum.

For larger game at long range, such as elk, grizzly bears and moose, nothing is too big if it can be carried for a long time and can handle the range. A start would be a minimum caliber of about .35 and a bullet weight of at least 250 grains.

The bullet's weight, velocity and expansion properties should be in proportion to the range, size and strength of the game. The big animals with heavy hides need to be hit with a bullet with strong penetration. For this, hunters prefer a heavy bullet moving slower. (As we discussed in Chapter 24, increasing the velocity may or may not increase the resistance to penetration. It depends on which of the basic material groups of kinematic viscosity is contacted.) The smaller animals need less penetration but more and quicker tip expansion, as with a hollow point. Larger game need the tip expansion after deeper penetration.

Some readers may have made clean kills with calibers smaller or slower. That is fine for hunters with exceptional skill, but most of us know our limitations and should use a caliber of suitable type, size and velocity.

At the longer ranges, an error of as small as 10% in estimation of the range can almost guarantee a miss. At the shorter ranges, say 200 yards, an error of 20% with an old .30-30 may be a problem but it would not be with a .30-'06. Gravity pulls the bullet down the same amount per second of flight. Time of flight is important as is range and velocity. The longer range shots are normally made with a magnum or high velocity firearm. Even so, the velocity drops and so does the bullet.

Dispersion increases in proportion to range. A group that is 2 minutes of angle or 2" at 100 yards will become 6" at 300 yards and 12" at 600 yards. (Actually, slightly more.)

Velocity and energy loss at long range are major considerations in hunting. The charts should be consulted for the ranges involved. Some western hunting, for example, may require figures out to 500 yards. With even the best cartridge, scope and wind conditions, 500 yards is about the maximum range a kill can be assured. Formulas and computer programs are available for the readers and sportsmen who like mathematics

Cartridges that are suitable for game at 200 yards may only cripple an animal at long range. All game animals deserve to be killed suddenly rather than wounded. As we have covered in other parts of this book, and will again in the next chapter, penetration and expansion are both a result of velocity and energy and as the velocity slows down-range, so will the ability to penetrate. Bullet weight can be a help in this regard. The heavier bullet may arrive a bit slower but may still be more effective. If two bullets have both slowed to the same down-range velocity, the heavier will have more penetration and be capable of doing more damage, all else being equal.

A hunter that takes all shots he can get, no matter what position the game is in relation to himself, needs a bullet with deep penetration. Deep for the game he is hunting. Penetration is always important, but large animals such as big bears demand it.

High velocity gives more expansion but with less penetration. It is important for a flat trajectory and long-range hits, but if the hunting is not done in long-range country, perhaps a flat trajectory is not as important. Much game is wounded because a hunter had a flat shooting rifle that he thought would hit the game at a long distance. Hitting the game and killing it are two different things. Many flat shooting cartridges lose too much energy at the longer ranges.

Some readers may notice that, so far, most of this chapter mentions details that have been discussed in detail earlier in this book.

MATHEMATICS OF STRIKING ENERGY AT RANGE

◆ Two items are needed; bullet weight and the remaining velocity at the range involved. **Take half of the square of the velocity multiplied by the mass and divide it by the acceleration of gravity.**

In other words, find from the charts or work out by math the velocity at range and square it. Then divide that by 2. Take the bullets weight in grains times 7,000 to change it to pounds. Divide again by 32.17 to convert weight to mass. Multiply this times the half of the square of the velocity number you found earlier. The answer is the energy.

OR

The 7,000 times the 32.17 times 2 equals 450,380. So, it can be simplified by squaring the velocity and multiplying it by the bullet weight in grains and then divide by 450, 380 which is usually rounded of to 450,400. (Refer to both Mathematics and Newton's Laws, Chapters 2 & 3.)

Striking energy is not changed a great deal, from a practical point of view, by either an increase or decrease of up to 250 ft. lbs. of energy.

SKILL FACTORS

A hunter's skill is an important factor. A skilled woodsman will get closer to the game by stalking and his shots will be at a shorter range. The hunter that hunts just a few weekends a year will need a heavier cartridge. This is not intended as an insult to the casual sportsman. Not at all. But, hunting is like most other endeavors; experience increases skill and the higher the skill, the more can be done with inferior equipment.

A major cause of crippled animals is poor shooting. Another is poor bullet selection. Still another problem is a poor selection of gun and caliber. (Unless we are rich enough to own a large collection of modern guns, we are all guilty of this, at times.)

NOTES ON EXPANSION

If two projectiles are made completely different in shape and materials, but they are both moving at the same velocity and have the same mass, they will both impact with the same kinetic energy. They will both have the same potential for doing damage, but the damage will be different according to the bullets diameter, construction, expansion capabilities, etc.

Bullet expansion is primarily a product of the bullet's construction and its velocity. The velocity does not have to be extremely high, just as high as is required by the bullet in question.

It is important to remember that expansion can be controlled by the design, at least to some extent. Some bullets will expand rapidly at impact and others will expand at depth. Some combine the two and expand as they move through tissue.

As we mentioned earlier in this book, some hunters prefer a bullet that exits the far side with a large hole from both expansion and hydrostatic shock. This usually causes major blood lose and leaves a blood trail for tracking. Other hunters prefer a bullet to expand fully and travel to almost the other side without exiting. This transfers all of the bullets energy to the game but the wound channel will become smaller as the bullet decelerates inside the body.

It is important to remember things that have been mentioned earlier in this book. One is that expansion is a result of velocity, but the velocity has to fit the bullet's design. Some are designed to expand at lower velocities, others at high velocities. Some will come apart in flight if pushed faster than intended. Some will expand at impact and others at depth. There is probably one that is just right for the job intended, but it may take some research to find out which one it is.

While Americans generally prefer expanding bullets, most African hunters, whether European or native, seem to prefer the full-metal-jacket.

They believe the expanding bullets do not penetrate deep enough to hit vital organs on a high enough percentage of shots. The hard nosed bullets keep their jackets and deliver more punch at the depth required. Separation is common with many expanding bullets and it leads to wounded game that have to be tracked down. Expansion is good only if the bullet has enough energy to penetrate and remain intact. It should be noted that the African hunters we are discussing hunt animals that are similar in many respects to some North American game. Very large African game, such as elephants, are a different story, but the basic requirements of placement and penetration remain the same.

◆ The bullet's jacket provides strength and the mass is provided by the core. At impact, the two should remain together.

◆ Small game require bullets that expand rapidly, almost at impact. Big game requires a bullet that penetrates deeper or has enough energy to punch through after expanding early.

Expansion should never be judged on the basis of just one or two good shots that produced a kill with a bullet that was recovered properly enlarged. Too many other factors are involved. The angle the shot was fired from, whether the bullet hit bone and muscle or just some tissue, the range that was involved, the type of animal and size, and on and on. There are too many variables. Testing, as outlined in the last chapter, is the only way to be certain. Of course, reading books and magazines about tests performed by other people can also be a help, it just isn't as much fun.

RECOIL

Muzzle energy will affect the recoil on a hunting firearm as we discussed in Chapter 6. A heavier gun will reduce the felt recoil or kick, but is more tiring to carry all day. The lighter weapon will recoil more, but is only a problem during the time involved for a few shots. Most hunters will pick the cartridge that is best for the game hunted and a light gun; choosing the few harder kicks to the all day fatigue that goes with carrying a heavy weapon.

CHAPTER CLOSING THOUGHT

Back in 1951, the famous hunter and gun writer, Elmer Keith, was quoted in the *American Rifleman* as saying, "I have long maintained that if a man cannot handle the recoil of an adequate rifle, then he should not torture game by using a small one."

CHAPTER 26

TERMINAL BALLISTICS for SELF-DEFENSE

This section will discuss both the special needs of police officers in the line of duty and private citizens who need self defense. It is a continuation of the material given in Chapter 24 about terminal ballistics general information. Energy, velocity, and everything that was covered in that chapter certainly apply here.

The same problems are involved in self-defense against humans as in killing game: bullet expansion, penetration and placement. It takes energy to push an expanded bullet to the depth that will be required. If the bullet does not expand it will go deeper. Perhaps too far and exit on the opposite side and become a down-range hazard. As expansion goes up, penetration goes down. The secret is to have enough energy to do both or a bullet with enough frontal area to do the job without expansion.

The experts don't agree on which is the most important, velocity, weight or expansion. Perhaps the best is a good balance of all with the realization that for each item we gain, we lose a little of something else. Since the trend to semi-automatics with large capacity magazines, police training has put more emphasis on the volume of shots and less on accuracy. Very few people believe this is right. Whether for game or self-defense, bullet placement is still the single most important item.

In the discussion of best caliber, it is understood that most police officers do not have a choice. They carry the weapon their department has issued them. For a back-up gun, a hide-out so to speak, many officers can carry whatever they wish. Private citizens needing self-defense usually have a choice.

The trick to self defense is not necessarily having the best gun, but just having one and knowing how and when to use it. Many people believe the key is the use of the *right* gun. It is true that if attacked by a drug crazed mad-man, a Colt .45 semi-auto is far superior to a .22 revolver. Still, most people would prefer, if given a choice, being shot at with the .45 by a person who couldn't hit a barn door at 10 paces than by an experienced gun-man with a .22. Let us hope we are never in either situation.

It is best to work out these problems now. There is a huge difference between a person reflecting on the subject while he sits in an easy chair in the comfort of his home and the man who is walking alone in the dark of night in a dangerous place, his nervous hand resting on the butt of his handgun.

The best advice is to get as powerful a gun as you can handle and conceal, if concealment is necessary and legal. Also, it should be simple enough to understand and use safely under stress. Automatics are great inventions, but they are not for everyone. Most people will never need a gun in their entire life. If they do, it will be a life or death situation with their death if they don't have one. Or perhaps worse, have one they don't know how to use.

LEGAL NOTE

It is assumed that all police officers know this, but sadly, most citizens do not. We are speaking of when it is legal to shoot and when you shouldn't. Has the man stolen you grandfather's heirloom gold pocket watch? Has he turned his back and is running away? Are you mad as hell and want to stop him? You had better not shoot him or you will end up in jail charged with a serious felony. State laws are different from one to the other, but in almost all cases you have to believe that either your life or someone else's life is in jeopardy. You cannot kill to save property, only to save human life.

This is not covered under the umbrella of ballistics, so we should move on to scientific areas. There is an excellent book on this subject titled *In the Gravest Extreme* by Massad Ayoob. It should be read by everyone that is involved with firearms.

DRUG-CRAZED CRIMINALS

Before we delve deeper into the ballistics of self-defense, we need to discuss a problem that has always existed to a small extent, but in the drug culture of today, it is all too common. That is the problem of trying to stop or kill a person who is high on drugs.

Many of us reading this book have minds that think in scientific terms. In other words, we believe that in nature there are certain actions that are followed by certain reactions. We believe in proven facts and theories that sound logical and sensible. The sections on gyroscopic action and Newton's laws are easy because they make sense. This next little section defies comprehension, yet it is true.

If Chapter 24 titled *Terminal Ballistics, General* has not been read, do so. Especially the area on *Medical Viewpoints*.

We will begin with an incident that took place in Indiana. The scene opens in a restaurant that is busy with the dinner crowd. A robber with a pistol approaches the cashier and demands all the money in the cash register. Saying the crook is high on drugs is like saying the Pope is religious. Apparently he was in *another world*. What he would do with his handgun or how many people he would kill was unknown and unpredictable. What happened next was also unpredictable.

The restaurant owner was armed with a firearm that, from a standpoint of terminal ballistics, would be considered excellent. He had a short barreled .357 magnum revolver loaded with 125 grain jacketed hollow point (JHP) bullets. (A chart printed a few pages later will show an RII of 44.4 and a ranking of no. 4. While some people will disagree with this ranking and some later test results come out slightly different, it is still an excellent self defense load.)

A normal person would not be able to withstand many hits from such a weapon, but drugs change a person until the reaction is abnormal.

When the crook was ordered to drop his weapon he continued to threaten the life of everyone around him. He then absorbed 18 hits, mostly in the chest, before he dropped his gun and went down. And no, that is not a typo or a mistake. It was reported that the restaurant owner reloaded his six shot revolver twice. The criminal was so high with drugs that his body did not respond to anything, either internal or external.

More details of the above story would not prove helpful. But there is no reason to believe it is anything other than the truth. Especially in light of the following story. It is also very shocking and a real eye-opener. Actual names, dates and the location will be given so that if the reader finds it also hard to believe, as many will, it can be checked for truthfulness.

This occurrence took place in Baton Rouge, Louisiana on August 1, 1977. The facts are well known and well documented. A video-tape made by Officer Steve Chaney, a survivor of the incident, is used in police training classes. It also has been written up many times in various publications. The account can be told as a long and interesting story, but we will keep to the points that have a relationship to terminal ballistics.

Field Training Officer Steve Chaney was accompanied by Rookie Officer Linda Lawrence when they responded to a burglary call at an apartment in the Broadmoor Plantation. They were confronted in a bedroom by John James Mullery, a 6 foot, 200 pound, 42 year old Caucasian with a long criminal record that included kidnapping and assault. He was also, at the time, high from cocaine and PCP. At the beginning of a brief struggle, Mullery made a grab for the Officer's 4" barreled S & W model 64 which was loaded with 125 grain .38 Special + P ammunition. While this is not the most powerful cartridge around, it is a good defensive round and generally performs well.

A fight for the Officer's gun ensued, and Officer Chaney realized that Mullery was going to win. He told Linda Lawrence to shoot Mullery. She was armed with an identical weapon which she promptly used. The bullet hit Mullery near the wrist and did tremendous damage as the hollow point expanded. <u>After</u> he was hit, Mullery fired 2 shots into the wall.

Officer Chaney managed to again get control of his gun and fired 2 shots directly into Mullery's chest with no noticeable effect.

Mullery then turned his fury on Linda Lawrence and grabbed her service revolver and shot her once in the middle of her chest. She collapsed immediately. At this point, the struggle between the two men continued for the two weapons. The fight was both long and hard. At one point, Officer Chaney fired a shot into the left side of Mullery's abdomen from zero range. There was no effect and the fight continued. (At this point, if you have been counting, Mullery had been hit 4 times with 3 in vital areas.)

The fight for life went on. Chaney had fired a number of rounds into the floor during the struggle to prevent Mullery from firing them into him. Now he was holding an empty gun. As Mullery stabbed him and beat him in the back, Officer Chaney used a Dade speedloader to reload 6 fresh cartridges into the S & W. He then shot Mullery again, this time in the upper middle part of his abdomen. After a few more seconds, he shot him again in the top of his head. Mullery finally went down.

Over? It should have been over a long time earlier, but as Officer Chaney backed away and sought support for his tired and wounded body, Mullery revived and stood up. At this point he had been shot many times and lost so much blood that he looked like a bad scene in the scariest of horror films. And yet he was not only alive, but ready to continue the fight to kill Officer Chaney.

Four more .38 + P hollow-points hit Mullery. Two in the chest with no noticeable effect. Another in the abdomen also had no effect. The last cartridge in the officer's gun put a bullet into Mullery's pelvis and broke his hip. This removed a main part of his skeletal support and he went down again. He was still not dead, just down. Mullery tried to crawl but slowly died as the last of his blood drained from his body.

Rookie Officer Linda Lawrence had died almost instantly from a bullet to the heart. She was the first female officer in Louisiana to be killed in the line of duty.

The drugs in Mullery's body kept him going in a narcotic induced rage. This enabled him to continue with wounds that would have eventually killed him even if medical attention had been available. Officer Chaney fired nine shots at him and each scored a hit. Officer Lawrence scored a hit with her one shot. A review in proper order shows strike no. 1 in the wrist; no. 2 and no. 3 in the chest; no. 4 in the left side of the abdomen at zero range; no. 5 in the upper middle part of the abdomen; no. 6 in the top of the head; no. 7 and no. 8 in the chest; no. 9 in the abdomen, and no. 10 in the pelvis and hip.

Many experts write and talk of the need for a gun and caliber that has enough initial cross sectional area, energy, velocity, expansion and penetration to stop a violent person who is high from narcotics. This true story should convince any reader that a drugged up person can require a lot to put down and much more to keep down.

And the lessons that can be learned from Officer Lawrance's and Officer Chaney's nightmare go far beyond ballistics.

HATCHER'S FORMULA - 1935

Much effort has been spent trying to determine a simple way to ascertain stopping power. One of the more famous is General Julian S. Hatcher's Relative Stopping Power formula. Published in 1935 in Hatcher's *Textbook of Pistols and Revolvers*, it was an improvement on his earlier formula published in 1927 in his *Pistols and Revolvers and Their Use*. While Hatcher's RSP formula has been replaced by modern methods, it deserves discussion because it does have some validity. It's main fault is omission because not all important points are considered.

Essentially, it is momentum times shape times area. The momentum can be obtained by multiplying mass of the bullet times the velocity. This is a little involved and a detailed explanation is elsewhere in the book. Hatcher also suggested, if the energy is known, the momentum can be obtained by dividing the energy in foot lbs. by the velocity. Energy is $1/2\ MV^2$ so this would yield only $1/2$ of the actual momentum. This error does not matter because the figures are all arrived at by the same formula and are a comparison. For example, the relationship of 20 to 40 is the same as 40 to 80.

To briefly explain the $1/2$ error, let us use a .270 Win. Mag. 150 grain by Weatherby. The muzzle velocity of 3,245 times the muzzle energy of 3,507 yields 1.080. The 150 grain bullet divided by 7,000 gives .0214 lbs. divided by 32.17 as the gravity constant gives .0006652. Multiply this times the velocity of 3,245 and receive 2.1585. This is about double the 1.080 received by the other method.

Shape was listed by Hatcher as 5 different multipliers.

BULLET SHAPE	MULTIPLIER	EXAMPLE
Jacketed w/ rounded nose	.90	.45 ACP
Jacketed w/ flat point and Lead w/ rounded nose	1.00	.38 S&W Special
Lead w/ blunt round point or small flat point	1.05	.45 Colt
Lead w/ large flat point	1.10	.32-20
Lead w/ square point and wadcutters	1.25	.38 Special Wadcutter

To bring the result to whole numbers instead of a three place decimal, the shape multiplier is itself multiplied by 1,000. Thus the 1.05 becomes 1,050.

Area is the cross sectional area in square inches. In other words, the hole size the bullet makes. It can be obtained by πr^2 (3.1416 times half the bullet's diameter squared.) This is where the larger calibers, all else equal, pay off big.

CALIBER	BULLET DIAMETER IN INCHES	CROSS SECTION AREA IN INCHES
.22	.223	.039
.25	.250	.049
.30	.308	.075
.32	.314	.077
9mm Lugar	.354	.098
.38	.358	.101
.41	.406	.129
.45 ACP	.450	.159

As you can see, if the diameter is doubled, the cross sectional area does not double, it increases about 4 times. The shock effect is increased thus.

Hatcher's formula is called Relative Stopping Power and *relative* is a key word. It is relative from one number to another meaning comparative. It is not definite or positive. An RSP number of 50 will not stop with 50% of the shots but it will have about twice the effect as one that has a 25 RSP.

Hatcher's formula omits the result of expansion but this can be estimated with a fair degree of accuracy. For example, a bullet with good expansion may rate a shape multiplier of 1.25 even though the actual shape would rate a 1.00.

We could express Hatcher's formula by putting it in different terms than at the start of the section, yet the result would be the same. The product of mass times velocity times the coefficient of bullet shape and material times impact area.

RELATIVE INCAPACITATION INDEX - 1983

There is also the U.S. Army Ballistic Research Laboratory's Relative Incapacitation Index (simply called RII). The number arrived at is not based on a formula but on complicated gelatin tests and knowledge of how wounds affect the human anatomy. The report is very detailed and complete. The NIJ Report 101-83 dated November 1983 is not the full report yet it is 119 letter size pages. (An earlier report on police handgun ammunition by the U.S. Dept. of Justice was thorough in 31 pages.)

The report was sponsored by the National Institute of Justice to provide the law enforcement community with criteria for deciding what ammunition is most suitable.

A computer simulation plotted the path of bullets through the body and arrived at a vulnerability index at 6 meters for the average shooter.

Bullet geometry, construction, mass, and velocity were tested for penetration, wound formation and relative stopping power.

Bullets were fired from Mann test barrels rather than regular guns so velocities and chamber pressures higher than standard could be evaluated.

A 20% gelatin solution in blocks was used to simulate tissue. The cavity was filmed by a high-speed motion picture camera that was capable of 10,000 frames per second on 16 mm roll film. The data was recorded and stored by computer.

The relative incapacitation index was based on the formula:

$$RII = \sum_{X=1}^{X\ MAX} P_i R^2 \quad \text{(vulnerability index)}$$

The value for R^2 for 1 cm. increments of penetration are totaled and multiplied by π.

The tests on bullet behavior demonstrated that it made a temporary cavity that was several times larger than the actual bullet diameter. This extra large cavity (wound channel) was of very short duration and then the cavity would close to a size closer to the bullet diameter. The kinetic energy of the bullet controlled the size of the temporary channel. The RII is primarily determined by the size and shape of the maximum temporary cavity produced in the block of ordnance gelatin.

The chart shows a comparison of Hatcher's RSP and the RII. Hatcher's figures give more advantage to the bigger diameter calibers than the RII. The shape multiplier is also probably not as accurate or complete as it could be. In all fairness to the late Gen. Hatcher, the RII index was developed with the aid of advanced technologies that did not exist when Hatcher created his RSP. Even today, Hatcher's RSP is still useful for many quick comparisons.

From the U. S. Justice Dept. RII summary report: "With the exception of the full-metal-jacketed bullet, the onset of deformation occurs at a given velocity for each bullet construction type (a thru f): i.e., a hollow-point bullet will begin deforming, at a velocity above 705 f.p.s and a lead round nose at a velocity above 1115 f.p.s. Unless the bullet's muzzle velocity exceeds this threshold value, bullet deformation is highly unlikely. Note that these threshold velocities were obtained by flash x-ray photography."

The following table compares 20 handgun cartridges with both the RII by the U. S. Government and Gen. Hatcher's RSP. Of course, it does not list all of them. The RII tests covered almost 150 different center fire cartridges.

RELATIVE RANKING OF RII VS. RSP

CALIBER	BRAND	BULLET WEIGHT (GRS.)	BULLET TYPE	STRIKING VELOCITY (F.P.S.)	STRIKING ENERGY (FT. LBS.)	RII	RSP	RELATIVE RANKING RII	RSP
.44 Mag	Speer	200	JHP	1277	724	54.9	82	1	2
.41 Mag	R-P	210	JSP	1260	740	51.9	78	2	3
.44 Mag	W-W	240	LSW	1330	943	50.0	129	3	1
.357 Mag	Speer	125	JHP	1301	470	44.4	56	4	11
9mm Lugar	R-P	115	JHP	1192	363	38.0	30	5	16
.38 Spl.+P	R-P	95	JHP	1187	297	28.0	25	6	18
.38 Spl	R-P	125	JHP	1108	341	25.5	31	7	15
.357 Mag	Fed.	158	JSP	1255	553	21.1	44	8	8
.45 ACP	R-P	185	JHP	895	329	21.1	59	9	7
.38 Spl.	W-W	158	LHP	915	294	18.4	32	10	13
.357 Mag.	W-W	158	LRN	1230	531	16.6	43	11	9
.45 ACP	R-P	185	JWC	821	277	14.7	67	12	5
.38 Spl.	W-W	158	LSW	924	300	14.3	41	13	10
.38 Spl.	R-P	148	LWC	741	180	12.4	30	14	17
.38 Spl.	W-W	158	LRN	919	296	8.0	32	15	14
.45 ACP	R-P	230	JRN	839	359	6.7	62	16	6
.45 Colt	W-W	255	LFP	821	382	6.6	78	17	4
.38 Spl.	R-P	200	LBN	730	237	4.5	34	18	12
.380 ACP	W-W	95	JRN	948	190	4.0	18	19	19
.22 LR	W-W	37	LHP	872	62	2.3	3	20	20

Abbreviations: JHP (jacketed, hollow-point); **JSP** (jacketed, soft-point); **LSW** (lead, semi-wadcutter); **LHP** (lead, hollow-point); **LRN** (lead, round-nose); **JWC** (jacketed, wadcutter); **LWC** (lead, wadcutter); **JRN** (jacketed, round-nose); **LFP** (lead, flat-point); **LBN** (lead, blunt-nose).

This comparison table was originally published in the August 1982 issue of the NRA's *The American Rifleman* and is the work or William C. Davis, Jr. of Tioga Engineering Company, Inc. It is published with their kind permission which is much appreciated.

CARTRIDGE RANKING BY TYPE

Over a wide range of velocities, the ranking in order of decreasing RII is:

    a.   lead hollow point (LHP)
    b.   Jacketed hollow point (JHP)
    c.   Semi-wadcutter (SWC)
    d.   wadcutter (WC)
    e.   jacketed soft point (JSP)
    f.   lead round nose (LRN)
    g.   full metal jacketed (FMJ)

FBI TESTS - 1989

Now we should move on to more recent testing that was conducted by the FBI in 1989. This was conducted on only the larger police calibers. No effort was made to test .22 and .25 calibers, for example. But the testing broke new ground and ascertained much new information. While the FBI results do not sink either the RSP or the RII, they do expose some flaws.

Regrettably, the FBI tests were the result of a deadly fire-fight between FBI agents and some criminals in Miami, Florida on April 11, 1986. This incident drove home what was already known but frequently disregarded. Essentially, that a man can continue to return deadly fire after receiving a hit or even multiple hits that will eventually lead to his death. The key word is eventually. **Incapacitation must be immediate** and the only way to do that is to stop the central nervous system, the heart or by extensive blood loss. We are discussing a high-stress situation where the criminal is, in effect, fighting for his life. The assailant must be rendered unable to use a weapon and pain alone will not do the job. Neither will a wound that will cause unconsciousness or death in a few minutes. In that few minutes he can kill the law officer or an innocent citizen. Hunting game can occasionally create a comparable situation. A charging mother bear who believes her cubs are in danger is a good example.

(Note; see Medical Viewpoints in Chapter 24 and Drug Crazed Criminals earlier in this chapter.)

The earlier RII test placed much value on the enlarged temporary wound cavity. **The newer belief is that human tissue is so elastic that after its return to the small cavity, no permanent damage will remain. Only permanent damage that is adequate will cause incapacitation.**

The testing also showed that kinetic muzzle energy isn't a dependable method of comparing ammo. This will require a lot of people to rethink their choices or argue the point in question.

The FBI tests show the larger diameter bullet opens a larger permanent wound cavity for maximum blood loss. This is old thinking to many gun experts who have preferred the heavier .45 ACP type guns to the lighter and faster 9 mm.

BARRIER MATERIAL

One of the most interesting elements of the FBI tests was the placement of barrier material between the gun and the gelatin blocks. The obstacles included heavy clothing, light clothing, auto sheet metal (2 pieces of 20 gauge steel 3" apart.), home type wallboard (2 pieces of 1/2" gypsum 3 1/2 " apart as in a normal home.), one piece of 3/4" plywood, automobile glass (from 10' at 45 degree and 15 degree angles and also 20 yards straight on).

The goal was to penetrate into the gelatin blocks a minimum of 12". The vital areas of the human body are deep and protected by muscle and bone.

Expansion of the hollow point was stopped by plugging the hole with matter in both plywood and wall board. Clothing caused a reduced expansion, but not as much as the more solid material. Auto glass had a major effect on the bullets. The amount that reached the gelatin was misshapen and reduced in size and weight as well as velocity. Most people, including police officers, will have little need to try to kill someone through automotive glass. Even so, the test is essential for a full and complete understanding of the bullets performance.

40 rounds were fired with each cartridge type. 5 in each of 8 different tests. For a 100% success rate, the bullets had to reach 12 " deep in the gelatin blocks. Only the 10 mm Auto Norm JHP 170 grain and the .357 magnum Federal Hydra Shock 158 grain reached 100%.

To briefly summarize the results, the .45 ACP did very well in most loads, as did the 10 mm auto. The .375 magnum also did well. Only one of the six .38 Special loads reached 90% and for the most part, that caliber didn't do well. The 9 mm performed about the same with none of the 5 tested doing better than 82.5%. Statistically, the 9mm as a group averaged 5.4% better than the .38 Special but that is still poor by comparison. A lone .380 ACP was tested at a 20% success rate.

TESTING BY OTHERS

The Justice Dept. study prefers high velocity bullets with high expansion. Another theory, first promoted by J. Hatcher and later by another well known firearms expert, Jeff Cooper, promotes the big bore weapons with not much concern given to velocity. Neither theory is all wrong or all right. There are many gray areas and both sides have ample studies to prove their point.

The 1989 FBI tests leaned more toward the heavy bullet theory.

◆ A report from 1975, was prepared for the U. S. Department of Justice by the U. S. Army Ballistic Research Lab. at Aberdeen Proving Ground, the Law Enforcement Standards Lab at National Bureau of Standards, and the Law Enforcement Assistance Administration. That is certainly an impressive list.

This report is quoted at several points in this text. Here, it is sufficient to mention that this paper does not recommend the police use any handgun with a RII of less than 10. They also said that an upper limit of between 20 to 25 "represents the upper limit required for reasonable reliability. This statement should not be construed to indicate, on an absolute basis, that an RII either higher than 25 or lower than 10 is unsuitable, undesirable, or unnecessary. It has been shown, many times over, that a hit in a vital spot by any bullet, whatever its RII, can cause death or incapacitation and should not be underestimated. RII deals with probabilities rather than absolutes, as is true with all biological measures of this type." End quote.

In reference to the statements in the above quote, few people today would agree with placing an upper limit at 20 to 25. The more recent FBI studies do not follow that line of reasoning.

◆ Evan Marshall, who was once a sergeant with the Detroit, Michigan Police Department and later a professor at a Michigan college, has researched and tabulated the results of actual shootings. He has calculated the percentage of *one shot* stops from different handgun cartridges. An example is his rating of the 9 mm ball at about 60% and the .45 ACP with 230 grain bullet at about 63%. Many people would agree with the 9 mm conclusion and disagree strongly with the low figure for the .45.

HOLLOW POINTS

The following is a quote from the U. S. Dept. of Justice RII summary report.

"Numerous arguments have been raised concerning the use of hollow point, semi-wadcutter and other variations of the traditional round-nose lead bullet. Opponents of change have quoted reports that the other bullets cause more severe wounds, are not permitted in international warfare, are dangerous to bystanders, etc. Some medical examiners have stated that they find no differences in the nature of the wounds caused by different bullet types, but these judgments are not widely known or accepted."

"The argument that these bullets are in violation of the Hague Convention has been countered by the assertion that only full-jacketed bullets may be used under these rules, and that the traditional police bullet is also not in accord with the Convention. The fact that full-jacketed bullets have a greater ricochet potential than lead bullets is rarely considered."

"The myths should be dispelled. State and local governments should be able to make their decisions on the basis of solid fact. Bullet selections should be made with due regard to their effectiveness against the criminal, as defined by their incapacitation potential (not their lethality), and with maximum safety to the general public." End quote.

As the FBI testing has shown, a bullet fired through a barrier seldom performs on the opposite side as it would if no barrier was involved. Steel, wallboard, plywood, cloth and glass are all problems. While a hollow point may smash through with ease and in some materials with little velocity drop, the hollow point will be plugged and expansion prevented. It is amazing how a thin piece of cloth, for example, can completely plug a bullet so it will not perform as expected. Even a shirt can change penetration and expansion; heavy winter clothing almost always will. Remember, we are not discussing the clothing stopping the bullet as with soft body armor (a bullet-proof vest), but changing it so it will not have proper stopping power for self defense.

We must remember that a bullet is treated to an enormous amount of force and stress as it crashes its way through most barriers.

The full metal jacketed bullet, as used in the .45 ACP, among others, has a dangerous potential for ricochet and, in some calibers, over-penetration. The 1975 Dept. of Justice report says, " ....full-jacketed bullets have a greater ricochet potential than lead bullets ...." Also, "A hazard to innocent bystanders can occur if the officer misses his target or if the bullet over-penetrates the target and exits with sufficient velocity to inflict a wound. With regard to the latter, over-penetration can occur if the bullet velocity is too low (absence of deformation) or it is too high." And also, "The hazard due to ricochet decreases as the frangibility of the bullet increases." The hollow point is decidedly more frangible than other types including the round-nosed lead bullet without a jacket. This last is paraphrased from the same study.

For defensive purposes, a hollow point is generally best and it is not a cruel and inhuman bullet as some anti-gun people will say. It will do better at stopping the intended person and is less likely to ricochet or penetrate walls. Imagine trying to shoot a drug-crazed rapist in an apartment and missing the victim and the bullet ricochets from an ash tray (or whatever) and kills the girl being victimized. Or just as likely with some bullet types, penetrate 2 or 3 walls and a refrigerator and still have enough energy to kill an innocent person several apartments down the hall.

Shotguns are excellent for home use. Rifles are likely to kill someone 5 blocks away after going through a few walls. (If that's all you have and someone is breaking in, use it with care, as we should with all weapons.) With handguns, it is hard to go wrong by following the FBI's choice in ammo selection. A defensive bullet requires some thought. Not all ammo is suitable for all uses.

One manufacturing firm has developed a bullet with a patented design built for maximum expansion and penetration at normal velocities. Based on tests and criteria from the FBI, this bullet design is a large improvement over the previous hollow-point and similar concepts.

This new design is a metal-jacketed bullet that has the jacket extended into the hollow point. The jacket is thicker at the nose. Three slits cut across the hollow-point form six sections. These slits reach down into the hollow end. It expands quickly and has good expansion even after going through most barriers. The six sections fold back during expansion so they look like petals on a flower.

The expansion does two things whether in a dangerous criminal or in game. It enlarges the wound channel which does more damage and it prevents the bullet - or it at least reduces the chance - of it exiting the other side. In regard to wound damage, remember the bullet will be spinning very rapidly and the pieces of bullet protruding out from expansion will be spinning as well. Although the rotation will stop very quickly, the damaged will be increased a tiny bit. (See Chapter 24.)

An interesting note on the patented design bullet just mentioned; the factory has discovered that each caliber has a velocity that is best for proper performance. A variation of as little as 70 f.p.s. above or below that optimal velocity, and efficiency will drop. Also, all of the best speeds are below 1,000 f.p.s. In many cases, a pistol bullet used primarily at short range does not require a high velocity.

Many hollow-point bullets, including the most modern jacketed types, won't expand properly **below** 900 f.p.s. Some won't expand well at 1,000 f.p.s. or **more**. Some factory ammunition has a muzzle velocity below this amount. Others leave the muzzle at a suitable velocity and drop below it in 50 yards. **Proper expansion requires proper velocity**, but this will differ with design and caliber. This expansion velocity is generally thought to be between 900 to 1,000 f.p.s. While this is true in many examples, there are always exceptions.

Accuracy is more important than caliber and stopping power figures. A .22 is more effective than a .44 Magnum if the shooter can put all his shots where they are needed, but can't hit the famous barn door with the hard kicking and loud sounding big bore. **It must be kept in mind that self defense is not like big game hunting. Proper defense requires at least 2 shots fired in rapid double action.** Even more are normal now that the *double tap* is no longer taught.

This text has frequently used the word compromise, and it is appropriate once again. There isn't a thing wrong with a good .38 Special with Winchester 158 grain lead hollow point ammunition. This will expand to about .50 to .70 caliber and is manageable by a 95 lb. woman. (No discrimination intended.)

HOLLOW POINT LEGALITY

The discussion of hollow point bullets brings us to a subject that needs clarification. Some people try to convince us that the Geneva or Hague Conventions make it illegal for shooters to use hollow point bullets. **That is wrong!!!!**

The Geneva Convention in Geneva, Switzerland, was held in 1864 and 1868 to discuss sick and wounded soldiers. Later, in 1906, the agreement was revised. The Red Cross brought into existence these meeting which had nothing to do with ammunition. Also having nothing to do with ammunition was the Geneva Conference of 1927 which dealt with naval armament and the 1932 World Disarmament Conference, held in Geneva, which considered limitations of military personnel, expenditures, etc.

Meetings in the Hague, Netherlands in 1899 to discuss peace and universal armament among nations dealt with poison gas, expanding bullets, submarine mines, aerial bombardment, etc. Another series of meetings were held in 1907. While there was some significance, they were generally considered a failure. What practical value is there in saying that war is ok? It is fine to go out and kill one another but it must be done in a humane and kind manner. **The United States did not sign the agreement.**

Nevertheless, this all dealt with military and war and had no bearing what-so-ever on the choice of bullet in a self defense weapon. Some bullets are banned by the U.S. because their purpose is to stop or kill a person and they do it too efficiently with less overpenetration and ricochet danger. Does this make sense?

BULLET VELOCITY

There are several reasons why many experts believe the most important property of a moving handgun bullet is its velocity.

(1) The size of the wound channel depends partly on the striking kinetic energy which is based on one half of the mass times the velocity squared. $e = (M/2) V^2$

(2) There is a threshold velocity, below which a bullet will not deform and deformation greatly affects the size and shape of the wound channel.

However, it is important to understand that the striking kinetic energy cannot be used as the sole criterion for ranking handgun bullets.

BULLET MASS

The mass of the bullet affects the size and shape of the wound channel. A lighter bullet will slow down more rapidly in the target medium and a heavier bullet will penetrate further. Once again, it is the location of the wound channel with respect to the vital organs that produces varying degrees of incapacitation.

NOTE: The next few paragraphs covering bullet velocity, mass, shape, construction and accuracy were paraphrased from the U. S. Department of Justice Evaluation of Police Handgun Ammunition Summary Report.

BULLET SHAPE

The effect of the bullet shape (bluntness of the nose) is important only in that it establishes the initial value of the hydrodynamic drag coefficient, C_D (O). This coefficient enters the RII formula for the Maximum Temporary Cavity (wound channel) and it is also a part of the formula for the threshold deformation velocity. At velocities too low for expansion to occur, C_D is a constant and the effect is that blunter bullets yield higher values of RII. The wadcutter has the largest value.

At velocities sufficient to cause expansion of the bullet, the drag coefficient changes as the bullet deforms. Bullets with smaller initial values of drag coefficients can deform in such a way as to out-perform those with a higher coefficient.

BULLET CONSTRUCTION

Deformation of handgun bullets depends strongly on both velocity and construction. Construction involves principally whether the bullet is jacketed or not, the length, thickness and hardness of the jacket material, the presence of hollow noses, cavities or hollow bases and the hardness of the lead. Construction also directly affects fragmentation of the bullet in both hard and soft targets.

SHOOTER ACCURACY

It is important to stress that no matter how perfect a bullet is from the point of view of ballistics, if it either misses the target or strikes it in a non-vital area, incapacitation will not happen. If the subject is high on drugs, this may be more difficult than anticipated.

MINIMUM TO BREAK THE SKIN

The human body's outer covering, the skin, must be penetrated before a bullet sized projectile can reach the fatty tissue and internal organs. The U. S. Army has determined that a velocity of 170 f.p.s. is required by a blunt or spherical projectile. Thin summer clothing will increase the figure to about 200 f.p.s. Winter clothing requires even more and the cloth can plug a hollow point, as was just explained, making it ineffective.

HANDGUN & RIFLE SHOT PELLET LOADS

Another self-defense load is a cartridge made for handguns and rifles that fires numerous bird-shot pellets instead of a single bullet. There are several brands available; some using just the shot charge for the projectile and others including a hollow point slug along with the pellets. The same legal problems and benefits to bystanders are involved.

From a ballistic point of view, most brands have ample velocity and energy to be effective at short range. (They are terrific protection in snake country.) They have a high cost and create feeding problems in some semi-automatics. Never change ammo without testing for proper gun operation, no matter how great the ballistics seem to be.

DEFENSE WITH SHOTGUNS

There is little dispute over the defensive capabilities of a shotgun but there is frequent debate concerning the best type of weapon or load. A study of ballistics may place some light on the choice of shells. The gun itself is a more personal decision.

Perhaps because of Hollywood movies and high adventure novels, most people believe that 00 buckshot is the best anti-personnel (self defense) round. It also is the standard of the U.S. Army Military Police. It does sound impressive. The pellets are .33" in diameter and in a standard 2 3/4" load, 9 are fired at a muzzle velocity of about 1,325 f.p.s. That will drop to about 965 f.p.s. at 50 yards. Many people believe 9 is not enough pellets to expect more than one or two to hit the opponent. Preventing him from returning fire is not as likely as it would appear. Remember that the one or two hits have to be in a vital area with good penetration to be effective. 00 buck at close range over penetrates and at range is spread out too much.

The 12 ga. 00 buckshot in a 2 3/4" Magnum shell has 12 pellets and a lower initial velocity of 1,290 f.p.s. The 12 ga. 3" Magnum carries 15 pellets and a still lower muzzle velocity of 1,210 f.p.s.

In a standard 2 3/4" gun, possibly the best is No. 1 buckshot. At a diameter of .30" and a muzzle velocity of 1,325 f.p.s., it is no slouch and with 16 pellets, the pattern is not as open. The 50 yard velocity drop, due to less pellet weight, will be a slight bit lower at about 900 f.p.s., but that is not so important. (The ballistics are poorer because 00 buckshot weights 53.8 grains and no. 1 buckshot weights 40.0 grains.

The modern trend is to use No. 1 buckshot in a gun that uses 2 3/4" shells. If 3" Magnum can be used, then 00 buckshot with 15 pellets per shell is practical; but No. 1 buckshot at 24 per shell is even better.

Open chokes are common for police and self-defense shotguns, but many people who have carefully studied the use of shotguns against humans prefer the full choke. It cuts down on danger to innocent people in the vicinity, which is very important, and will slightly extend the effective range. **There are many regions of terminal ballistics that are disputed and the choice of choke and shells for police and self-defense are debated strongly.** Still, a 12 ga. full choke with either No. 1 buckshot in 2 3/4" or in 3" Magnum, either 00 or No. 1 buckshot, would be an excellent choice.

PSYCHOLOGICAL FACTORS

There are several unusual psychological phenomena associated with a gun fight. The brain begins processing images and thoughts at a rate much faster than normal and creates the illusion of slow motion. Movie directors frequently try to create this sensation during scenes of violence.

Repeated training for emergency situations pays off in big ways. Police officers, pilots, military personnel; anyone can benefit from training that is done often enough to become *habit*. Under the stress and excitement of a fire-fight, things such as aiming, safety removal, hold and stance, reloading; all will be done as training has taught if it has became a habit.

Concentration on the act of survival will be so strong that all other sensory perceptions will be excluded. Most gun fight participants either don't hear their own gunfire or remember a .45 as sounding "quiet" or "like a mild pop sound."

CONCEALED WEAPONS

It is important, with a concealed weapon, to have enough firepower to stop an attacker. That is basic. Still, there is a valuable place for the compact .22 and .25 ACP. For much of the year, people in the southern states dress so that concealing a weapon is difficult. Even the smallest gun can become a problem to keep hidden. At times, the same is true for the rest of the country.

The small calibers do not have the stopping power of the bigger guns unless the bullet hits at least one of a few vital areas. That is true. But the experts who object to someone carrying what some people call a "mouse gun" would not want to be shot at with one. And if the shooter is a cool expert, so much the worse. The small caliber gun is not as effective as a larger one, but better than nothing in a defensive situation. Much better. In some instances, no shots need be fired. Just the presence of the weapon is enough to defuse a tense situation. From the stand point of simply staying alive, it is better to have a small gun than no gun at all.

Now that the good points about a small and easily concealable handgun have been mentioned, here are the bad points:

Hits in vital areas may not be as serious as would be expected. Heart shots from small caliber weapons may not be fatal and frequently are survivable; even from heavy caliber weapons in the .38 and above class. (Heavy as opposed to .22, .25 ACP, .32, etc.) They cannot only be survivable but also are poor for creating sudden incapacitation. The heart is a remarkable muscle with great resilience. The conclusion is obvious. A heart shot may not prevent an armed opponent from returning deadly fire. Even if the shot is effective enough to stop the heart, the brain may have enough oxygen to continue to function for as long as 15 seconds before unconsciousness. An opponent need not be on drugs to return fire. The output of adrenaline or epinephrine may be stepped up enough from excitement, fear or anger to keep the body responding. Frightening isn't it!

POWER CARTRIDGES

The .357 Magnum cartridge was introduced in January of 1935 by Smith & Wesson. It was widely acclaimed at the time as the most effective revolver cartridge ever made. Those that love the .45 ACP and similar large bore guns would argue that point. Even at that time in firearm history, a person armed with the .44 Special, .45 ACP or .44 Colt, to name a few, was able to defend himself very well.

THE 9mm SITUATION

Many police departments have changed to the 9mm semi-automatic handgun. From a ballistics point of view, the reasons for this change is not clearly understood. It is basically a light bullet moving at more velocity, which has some advantages and some disadvantages. Once again, form the point of view of ballistics, the disadvantages are far stronger. Of course, there are other reasons that have nothing to do with ballistics, such as the cartridges are relatively small and a large amount can be placed in a clip. This gives an abundant amount of shots before reloading. The 9 mm also has low recoil and a low sound level as compared to the bigger, heavier calibers that may (or may not) move slower.

One of the biggest problems created by the police change to 9 mm semi-automatic pistols, is that many officers today are not taught restraint. And this is on both a local and national level. Many people who received their firearms training in the past, both with revolvers and the Colt .45 model 1911, were taught the *double tap*. Put two rounds into the assailant and stop, or at least hesitate, and observe his response. If more strikes are required to do the job, then add a few more and check again on the reaction.

Now-days, it is common for an officer to empty a 13 round clip toward someone, generally spraying the area, and in some cases, reload and fire some more. This not only increases the danger to bystanders, but what does it say for the officer's training, shooting skill and response under stress? The double tap may be too little in many instances, but 13 to 26 rounds? If the reader is now wondering what this has to do with ballistics, it has plenty. This mentality started along with the change to the 9 mm cartridge and the semi-auto. The semi-auto, with proper training and proper design and manufacturing, is a fine weapon. If the 9 mm requires 13 hits to put someone down that is not drugged up, perhaps it is a poor choice. The FBI and other tests clearly show that from a ballistic point of view, there are better choices.

The 9mm and the RII charts can be best explained by a quote from the U. S. Department of Justice summary report. "Many of the 9mm cartridges place high because of their high velocity, and not necessarily because they expand."

Some experts believe a lot of the erroneous information about both the 9mm pistol and the .223 M16 rifle. Disregarding the weapons and their mechanics and dwelling strictly on the cartridge ballistics, both are not as good as what they replaced. (The .223 replaced the .30-'06 and the 9mm replaced various .45 ACPs and .357 magnums.)

CHAPTER 27

MISCELLANEOUS BALLISTIC INFORMATION

RAIN

While rain can make a hunt or a day on the range miserable, it has little or no effect on the projectiles. The statistical probability of a tiny bullet that is in the air for such an incredibly short period of time striking a rain-drop is thought by many experts to be small. If it contacted a drop or two or even 1,000, the trajectory or gyroscopic stability is not affected. At least not that testing by the U. S. military has discovered. Of course in testing, it is not possible to determine which bullets hit a rain-drop and which did not. (Unless shots are fired through a stream of water, but this will greatly exceed the proposed problem.) The chronograph and target show no difference.

Target shooters frequently do better in a light rain, but this is thought to be caused by the calmer wind and lower barometric pressure. The advantages of a light wind are obvious. Rain usually occurs during low barometric pressure readings. As we have discussed earlier in this book, the lower pressure means less air resistance because of less density. In other words, the air is thinner. Supposedly, a 1" reduction in barometric pressure will increase the bullets *effective* ballistic coefficient by over 3%. This is the reason the strikes will be higher on the target.

While the humidity will be higher during a rain, this has very little influence on ballistics or a trajectory.

TAPED MUZZLE

A piece of electric or masking tape stuck over a rifle muzzle while hunting is a good idea. It is both safe and practical. Chronograph tests have shown a drop of from zero to 3 or 4 f.p.s. This drop so small that it cannot be confirmed because the ammunition itself will vary by that much. No change in accuracy is noted.

In all probability, the air in the barrel that is pushed out ahead of the bullet will blow off or through the tape a microsecond before the bullet even touches it.

Why cover the end with tape? It is not unusual for a hunter to bump a barrel in such a way as to force in mud, a rock that can become wedged, or any of numerous other obstructions. It also is not rare for ice to form in the end of a barrel. Then the barrel is blown out and destroyed when a cartridge is fired. In rainy weather, the water can enter areas that may rust and be damaged even if the temperature is too warm for freezing.

But, the tape is like many things in life and can be overdone. One piece is good protection with no problems and one at a time is enough.

COLLIMATORS

Collimators are used to bore sight firearms; that is to align the bore and the sights to a target. With a bolt action rifle, the bolt can be removed and the target can be seen through the bore. The collimator permits something similar to be done on any gun only faster and better.

It is an optical system with a short tube with a lens at the end toward the shooter and a frosted piece at the other end. The lens focuses on a grid that while it is actually very close and tiny, it appears to be 2 feet to 6 feet square and 100 yards away. The unit is mounted above the muzzle end of the gun and held to the barrel by a stud, also called spud or arbor, which fits inside the bore. This aligns it properly and the sights are then adjusted parallel to the bore. By the use of trajectory tables for the cartridge used, the sights can be further adjusted to impact the bullet at the intended point.

Caution is required in several ways. The image is backwards (reversed) and the adjustment movements are also reversed. Up and down and left and right are all backwards. The second caution is that some units, even when new, are hard to focus and inaccurate.

The final adjustment with the trajectory tables requires knowledge of the height of the sights above the center line of the bore. The mid-range trajectory will be multiplied by 4 and this amount added to the sight height. Convert the answer to minutes of angle for the amount to raise the sight. At first this may seem wrong because the sight line will point below the bore and collimator. This is correct because it elevates the muzzle.

Test firing is still needed for a final finish adjustment to compensate for barrel jump and whip and all the other events discussed in internal ballistics. At short range, bore sighting without the sight elevation may be all that is needed. That is not to say that test firing should not be done properly. While it is true that in rare cases, no more adjustment will be needed, testing is always beneficial. Remember, not all units are accurate.

The reason the mid-range trajectory is multiplied by 4 is the drop is roughly 4 times the mid-range height. Drop is not given in all trajectory tables. A minute of angle is roughly 1 inch for each 100 yards. (Both of these subjects are covered in detail in other chapters. See the index.)

A collimator can also be used to keep a check on sight alignment and impact point. Damage to alignment from bumps and falls can be checked without a shot being fired if before the mishap, the relationship is noted between the collimator and the sights or scope. After a long trip to a hunting location, this method can find any damage caused by bumps and rough handling. Little changes may not show up, but the really troublesome problems will. It may prevent failure and save a hunting trip.

Changing telescope sights is another place where a good "bore sighter" can be handy.

This discussion is for basic knowledge only. Follow the factory directions with your collimator.

HEADSPACE

Headspace is the distance that a cartridge case can stretch out or move lengthwise after the breech has been closed and locked. It will be measured in a slightly different way according to the type of cartridge used. Another way of wording it would be the distance in the breech or chamber between the rear face and the front part where the cartridge is held or located. Its accuracy can be vital for both performance and safety.

Rimmed cartridges have a flange on the head of the case. With this type of cartridge, the headspace distance is from the breech block face to the rear chamber surface where the forward part of the rim is positioned.

Rimless bottle-necked cartridges have a shoulder where the case neck joins the body. This comes in contact with a stop (on the angled surface) in the chamber to prevent the case from going too deep. There is always a specific point on this surface. This type cannot be measured with ordinary equipment because it is the distance from the breech block face to the surface where the case rim is positioned. Special gauges are best for measuring this dimension.

Rimless straight case cartridges are stopped by the forward end of the case coming in contact with a square shoulder in the chamber. The headspace measurement is measured from the square shoulder to the breech block face. This type has a groove forward of the rim for the extractor.

Semi-rimmed cartridges with a rim that is bigger than the case body; headspace is from the forward edge of the rim to the chamber front. The stop is positioned ahead of the case mouth.

Belted cases, at least for headspace considerations, are considered similar to the rimmed case. It is sometimes called a modified rimmed design with a shallow but wide rim. It has an extractor groove because the rim is not suitable for grasping. The headspace is short because it is measured to the edge of the modified rim.

.22 Rimfire headspace is the short distance between the breech face and the shoulder that stops the rim. In some guns, it can be considered the difference between the bolt face and the end of the barrel.

The headspace, in each type or example, has one length that is correct. The permitted variation has a close tolerance. If it is too tight or small, the firearms breech action either will not close with the cartridge in place, or it will close too tightly.

If the headspace is too big, the bolt face will not properly support the case. The cartridge may move forward when the firing pin hits it and this may cause the blow to not be hard enough to ignite the primer. The case will stretch and it may rupture or split. In extreme situations of excessive headspace, the case may separate into two pieces. The primer pocket may enlarge enough to permit the primer to be blown out. The hot expanding gases may be released to the rear. (In other words, the chamber may no longer be sealed properly.) This can damage the gun or the shooter or both.

The lessons to be learned here are obvious. Head space should be checked on older guns. The measurement is critical and should be done only by a skilled gunsmith or someone with the knowledge and equipment. Any barrel change should be done only if the headspace is properly set. A change of rifle bolts also calls for a headspace check.

Careful examination of fired cases can frequently spot excessive headspace. Stretched cases or a case that has a crack running around it and is beginning to separate into two pieces is easy to spot. Some noticeable problems can be a symptom of excessive pressure while the head space is correct. Therefore, caution is urged.

PERSONAL COMPUTERS

One of the many miracles of modern times is the low cost personal computer. Programs can be purchased for almost any conceivable subject from A to Z: astronomy to zoology.

Many excellent programs are available for ballistics. Included are programs on trajectories, recoil data, rifle barrel twist, and on and on. The National Rifle Association even has a program that can be copied for personal use at no cost. (Copyrighted material for NRA members personal use only. The author highly recommends NRA membership. Contact them for information.)

This book is an excellent companion to computer programs. The computer can work problems and print out huge amounts of information in seconds that would take a math whiz a week. The problem arises when someone purchases a computer program, but they don't understand the principles behind ballistics. Many computer owners feed information into the computer as it requires, but do not really understand why that particular item is needed.

The formulas and knowledge contained in this book, along with a hand-held calculator, will enable most people to do *some* of what a computer could with several expensive programs. And a great amount of things that can be determined from this book are not in computer software.

For an accurate answer, there is one area of exterior ballistics where a computer is almost a necessity. Trajectories involving ballistic coefficients require either a computer or complete tables. These tables are too long and involved for inclusion in this book. (See the Index for more on this subject.)

Speed is the computers forte; also accuracy if it has been programmed correctly. For those that can afford the cost, home computers can be justified. You have this book, a calculator and some spare time? You'll do just fine with most subjects and what is more important, learn a lot in the process.

BALLISTICS OF LONG RANGE SHOOTING

The ultimate in precision shooting is the 1,000 yard range, and the accuracy obtained is nothing short of astounding. The best shooters can obtain groups of around 4 1/4 " to 4 3/4" with 10 shots at 1,000 yards. To truly understand this feat, measure off 1,000 yards (3,000 feet or slightly over 9/16 th. of a mile). Or, simply lean a small pocket size book against a tree and watch your car odometer as you drive 5 and 1/2 tenths of a mile. Stop and look back. Unless the tree is of a decent size, the average shooter probably couldn't hit the trunk in several shots. And without a good scope, most eyes will have trouble even seeing the book.

At 1,000 yards, all laws of ballistics appear to be exaggerated. Time seems to stand still as the bullet flies lazily - or so it seems - toward the target. The wind plays cruel tricks that a demented magician would be proud to brag about. Some shots wander off onto some far off trail and are never seen again. The very slightest variation in cartridge loading or the faintest beat of the heart can send the bullet jumping like a Texas rabbit. The same spirit of adventure that drives a person to climb a shear ice cliff in the Swiss Alps must also drive the 1,000 yard shooter.

NOTES ON LONG RANGE WORK

Handloading for long range shooting requires extreme care and dependable consistency.

Heavy bullets with high ballistic coefficients, such as spitzer boat-tails, receive praise for their excellent long range performance.

Some computer programs give minutes of angle (MOA) at these long distances to permit a scope to be set close enough to mark the target on the first shot. (Hopefully) These ranges can require huge MOA adjustments in elevation.

HUNTING: If a shooter can obtain bench-rest groups of 2", for example, at 100 yards, he will probably not be able to equal it in the field. Cross winds, elevation problems, range estimation, poor position; all of these and more will probably double it to 4" at 100 yards. This will be an 8" group at 200 yards, a 16" group at 400 yards and a 32" group at 600 yards. Even at 400 yards, this is too large to promise a hit in a vital area.

Study the trajectory charts for your cartridge. If long range shooting is expected, the gun should be sighted-in for a longer range. Many hunters have a gun sighted for 200 yards and think nothing of taking a 500 yard shot. They are afraid that if they change the zero to 350 yards, for example, the bullets trajectory will be very high at the shorter ranges of 100 or 200 yards.

They are right in that it will be higher, but for most cartridges suitable for long range work, it won't be too much high. A look at a chart will show that holding low a few **inches** will be easier than holding high a few **feet**. Note the sentence said a *suitable* cartridge for long range work. While many of us love the old .30-30, for example, it and others are not good long range cartridges, no matter how they are sighted in.

Always use a cartridge and gun that is suitable for the range involved. And what ever happened to stalking? Now, in legal terms and the media, stalking has taken on a whole new and sinister meaning. In the woods, it is still a skill that every hunter should be proud to possess. While a hunter can brag of the game he killed with one shot at 500 yards, he can be as boastful if he can stalk the range down to half or less.

Long range hunting shots should never be made in a strong wind. For example, with the .30-'06 Federal 150 grain bullet, a 10 m.p.h. 90 degree cross wind will blow the projectile off about 4.2" in 200 yards; no problem there. At 500 yards, it will be about 31.2". That is a major concern. Add to that situation the fact that a wind is usually gusty and can be much more than 10 m.p.h. And the best estimate of the wind between the shooter and the target will be just a guess.

VERTICAL SHOTS

Firing vertical shots is a subject people find interesting, although it has very limited practical value beyond the obvious safety considerations and scientific curiosity.

A bullet fired straight up will return to earth at a much lower speed than it left the gun. Going up it stays point first (mostly) with gyroscopic stability imparted by the rifling grooves in the barrel. The upward velocity drops at the beginning because of gravity (about 2% of the decrease) and drag (the 98% balance). Gen. Julian S. Hatcher stated that air resistance slowed a vertical shot 60 times as fast as gravity. The exact amount is not as important as the knowledge that gravity is a minor player on the upward flight. As the velocity decreases going up, the drag also decreases. The deceleration slows until the bullet momentarily stops. It then falls back toward earth.

Under perfect conditions, a bullet would fall base first with no pitching or upsetting movement. This exists mostly in concept, but testing by the U. S. Army, which will be explained later, showed a high percentage of bullets did return base first. When loaded reversed in the case so the bullet went up base first and returned nose first, it did not go near as high and returned faster. This is the result a thoughtful person would expect.

A bullet that is not fired straight up, in other words, at 70 or 80 degrees instead of 90 degrees, may arc over and keep its stability. The return to earth will be point first and much faster than a tumbling or base first bullet. Air resistance will be the cause for the different speed. The pull of gravity will be the same. (All else being equal.)

During the bullet's return fall to earth, it will accelerate until it reaches it's *terminal velocity*. That is its maximum speed no matter how far or long it falls. All objects have this limit. Skydivers take advantage of this fact and position their bodies so the drag will be more or less as they desire. Air compression, drag, the object's density and the air's density are factors in slowing the descent. Air density, as we mention several other places in this book, is less the higher above sea level we are, and also less the higher the temperature. It will also be less during periods of low barometric pressure as with poor weather and rain.

The fall will start because of gravity and as the velocity increases, drag increases and slows the acceleration. The air resistance will increase as the velocity increases. As stated in other parts of this book, the drag increases almost as the square of the velocity. If we double the velocity, the drag increases 3 times. (At subsonic speeds such as we have in a free fall.) When the aerodynamic drag equals the pull of gravity, terminal velocity is reached. The bullet will fall no faster. To be very simplistic, we could say that the weight and drag are in balance.

It is obvious that a bullet falling point first will be streamlined and have low air resistance and fall faster. (In the general area of 450 to 500 f.p.s.) Base first is slower. (About 300 to 350 f.p.s.) Tumbling is the slowest.

An interesting note on terminal velocity; an aircraft falling in either a normal or flat spin will not fall fast enough to damage the plane until impact with the ground. The air resistance will be strong and terminal velocity low. (Although still fatal on impact with the ground.) Nose down, only the best aircraft can survive a terminal velocity fall or dive. The terminal velocity will exceed the maximum safe structural velocity and the aircraft will come apart in the air before reaching the ground or pulling out of the dive. Even many high-speed fighter aircraft have this problem.

A bullet fired from an aircraft straight down toward the ground would be slowed to terminal velocity by drag if the altitude was high enough. Remember from earlier chapters that air resistance increases with velocity. Whatever the resistance at 1,000 f.p.s. it will be 3 times as much at 2,000 f.p.s.

Air density and kinematic viscosity, which are both explained elsewhere in this book, will have a slight effect on terminal velocity.

In popular usage, terminal velocity is sometimes used to indicate striking velocity. This is based on the definition of terminal that means end or final. In ballistics and firearms use, that is not correct. Although, to make it confusing, it is acceptable to call the science of how a bullet kills, terminal ballistics.

Gravity slows a bullet's velocity fired straight up but has no effect on its spin. Air drag slows a bullet's velocity fired straight up but has a small effect on spin. The result of these two actions is the bullet will still be spinning at the top and may still be spinning as it returns to earth. Although, a tumble is a frequent position at return.

DAMAGE-vertical shots

The danger from a falling bullet should not be ignored, although at 300 to 500 f.p.s. it may not be too deadly. The projectile weight is the main factor. A .22 Hornet .40 grain bullet falling at 375 f.p.s. will not do near as much damage as a 500 grain .45-70 Government bullet at the same velocity. Or if that does not get your attention, what about a 75mm shrapnel projectile that weighs 15.96 pounds and is tumbling down at terminal velocity. The speed would once again be slow, although higher than the earlier examples, but the weight alone would do a great deal of damage. Remember Newton's laws about energy in Chapter 3?

HEIGHT-vertical shots

The height obtained by any projectile can be computed, but it is a long process. A simple rule of thumb, which will be close enough for most purposes is, "The maximum height reached will be about two-thirds (2/3) the maximum horizontal range." Of course, as with all rules of thumb, it will be accurate for some cartridges and only fair for others.

For readers who don't take air resistance as seriously as they should, consider that in a vacuum, a bullet would take the same time to fall back to earth as it did to reach its maximum height. For the average bullet, this vacuum height would be in the area of 21 miles (almost 111,000 ft). The time would be about 85 seconds each way.

WIND DEFLECTION-vertical shots

Wind deflection on a bullet fired straight up can be computed using our formula for delay time which was explained in Chapter 20.

$$T_V = R/V$$

If we use the same .30-'06 bullet mentioned earlier (Federal 150 grain), we find that 9,000 ft. divided by 2,700 f.p.s. is 3.33. And 18 sec. minus 3.33 is 14.67 or 14.7 secs. An eight feet per sec. cross-wind would then blow or deflect the bullet 117.6 feet at the top. (14.7 * 8 = 117.6) (This is not technically correct, but it does show the basic idea.)

The descent was given as 31 seconds and for this we could, using the same stretched theory for demonstration, compute an additional drift of 248 feet and a total of 365.6 feet of drift by return to earth.

We must remember that the bullet is decelerating as it climbs and will slow to a point where it has no forward velocity. It will hang almost suspended for a brief moment and then the fall to earth will slowly start. Using the gravity figures we have counted on so strongly in other chapters, it is apparent that for 2 full seconds the bullet moves only about 16 feet up and 16 feet down. The wind is still blowing it sideways during this time. (If necessary, review Chapters 2 & 3)

SHOT PELLETS-vertical shots

Shot pellets falling back to earth follow the same laws, and accelerate to terminal velocity where weight and drag are in balance. The larger shot is heavier and falls faster. (Forget Italian physicists because this is not in a vacuum.)

This chart is reprinted from *The American Rifleman* of Dec. 1974 with their permission. They have been generous with several items for this book and it is much appreciated.

shot size	shot diameter in inches	ballistic coefficient	terminal velocity in f.p.s. Ingalls-Siacci
9	.08	.008	74
6	.11	.011	87
4	.13	.013	94
2	.15	.0154	103
BB	.18	.018	111
4 Buck	.24	.025	131
1 Buck	.30	.031	146
00 Buck	.33	.034	153

TESTING & RESEARCH-vertical shots

Man being the curious creature that he is, testing on vertical shots probably goes back to the earliest guns. Benjamin Robins, an English ballistician who is mentioned in other parts of this book, conducted experiments and wrote about them in a book published in 1761. The title was *Mathematical Tracts*.

A few interesting figures came from old tests made by U.S. Army Ordnance in 1919 and 1920. This research was conducted many years ago, but is still valid and sound today. New studies would probably gain little in information and be very costly. Testing was conducted from a platform surrounded by calm water and also from the center of an area circled by wet hard-packed sand. The locations were Miami and Daytona Beach, Florida.

A .30-'06 bullet with a 150 grain spitzer point was fired straight up at 2,700 f.p.s. It went up 9,000 feet in 18 seconds and returned to earth in 31 seconds for a total time of 49 seconds. (Actually 49.2 sec.) The velocity at impact was about 300 f.p.s. This was considered by the Army testing group to be the average. It is interesting to mention that Gen. Julian S. Hatcher was one of the officers involved in the test, although at that time, the promotion to General was some time in the future.

It was calculated that if the same was done in a **vacuum**, the altitude reached would be about 113,000 feet (about 21 1/2 miles). The trip would take 84 seconds each way for a total of 168 seconds. (Almost 3 minutes.) **Comparing the two, we can see that drag has a major effect.**

FLINCHING

A flinching or jerky movement by the shooter at the time the bullet is fired cannot cause a keyhole, even though some people believe so. It may cause the bullet to go astray because the bore may not be pointed properly, but it will still be stabilized and point forward.

EUROPEAN TABLES & CHARTS

European ballistic tables list two different pressures, which they call atmospheres. Physical atmosphere is in a column labeled with the abbreviation "atm." Multiplying the numbers in the column by 14.7 converts it to pounds per square inch (p.s.i.), which Americans are familiar with. Example: 2,500 atm. * 14.7 = 36,750 p.s.i.

A column labeled "at." is technical atmosphere (pressure of one kilogram per square centimeter). The conversion factor is 14.223. Example: the same 2,500 at. (Not atm.) * 14.223 = 35,558 p.s.i.

There is about a 3% difference between technical and physical atmospheres and both are used in the European tables.

LOT NUMBERS & CHANGES

Lot changes, whether in powder, cartridges, bullets, primers, or whatever, may produce different results. Even if a manufacturer has high standards with a tolerance variation as close as 5 %, the change will not be acceptable for bench rest or other super accurate work. It is common for people in this avocation to purchase large lots of powder so that little adjustment will be needed. Every time a change is made, the load should be developed from the starting point again. Most handloading manuals suggest this. Bullets and cartridges can be manually checked for conformity.

NATO PERFORMANCE STANDARDS

Through 1977-1979, the NATO armies held trials to find a new generation of ammunition. They selected the 5.56 mm cartridge with the

American 3.56 gm. M193 ball as the standard bullet. Later, the 3.95 gm. jacketed streamlined bullet was selected. (SS-109 by Fabrique Nationale.)

The following is a copy of some of the ballistic performance standards for the 5.56 mm NATO cartridge. It is reprinted because serious students of ballistics will find it interesting.

Terminal - Bullets shall completely perforate at 69.8° F. a 10 ga. (0.138" - 3.5 mm) thick steel plate of SAE 1010 or 1020 of Rockwell hardness B55 to B70 at 570 meters (623.4 yards) from the muzzle at 0° obliquity. (In simpler form, at a right angle of 90° to the line of fire.)

Chamber pressure average not to exceed 55,120 p.s.i.

Action time. The sum of primer ignition time, propellant burning time and the time taken by the bullet to reach the gas port, plus five standard deviations shall not exceed 3 milliseconds when fired at -65° F. (Note: that is **minus** 65° F.)

ABRIDGED BALLISTIC TABLE, 5.56 x 45 BALL SS 109

Range metres	Velocity m/sec.	Energy J	Drop mm	Elevation mils	Vertex mm
0	930	1,708	0	0	0
100	832	1,367	66	0.65	18
200	740	1,081	281	1.41	75
300	650	834	675	2.30	187
400	574	650	1,296	3.37	382
500	500	494	2,268	4.68	681

SOLUTION TO STEVE'S PROBLEM

The short story on page 6 has a simple solution.

The cold weather will increase density for more air resistance and drag and also slow the powder ignition, but it will be such a small amount we need not consider it. The drizzle will also make no difference in any way. The plastic electric tape over the muzzle is a good idea. It will keep out the rain and not change the trajectory. The bullet weight and type were not given, but the .30-'06 is a fine cartridge for a shot of this type and with the right bullet, it should prove to be satisfactory. Ammunition does not deteriorate as we might expect. The fact that the cartridges are 10 years old should not make any difference. So far in our study of the situation, Steve seems to have no problems.

Steve estimated the range at 300 yards and his rifle was sighted in for 300 yards. This sounds good for him. He guessed the high angle would be a longer shot and added a 4" allowance. Steve also thought the bullet would drop an extra amount because of the up hill angle. He raised his aim 5" to compensate for this extra drop. This was a total allowance of 9" and a bad mistake. In Chapter 17 it states that we should "aim as shooting for the horizontal yardage." His rifle was sighted in for the proper yardage and should have been corrected for the *lower* amount the bullet would hit because of the uphill shot. Steve made his adjustment the wrong way and his shot will be high by perhaps 12". (No bullet was specified, so accurate figures cannot be worked out.)

We have yet to discuss the wind that was blowing strong at Steve's location and not at the rest of the range. This will also be bad for Steve because it will start the bullet moving sideways and the bullet's lateral momentum will continue even after it moves into the area with little or no wind. By the time it reaches the game, it will have been blown off a large amount. He should have corrected into the wind. Not as much as his bullet required for a wind that covered the entire range, but it could not be disregarded altogether.

Altogether, Steve is bound to miss his first shot. If the game does not run off, he will also miss his second shot. He does not understand the ballistics that are involved, so each successive shot will have the same errors.

CHAPTER 28

GLOSSARY OF BALLISTIC TERMS

ACCELERATION OF GRAVITY: The changing rate of velocity of a free falling object caused by gravity. The pull of gravity is not the same at different locations. Lately, the standard of 32.1741 feet per second has been set by international agreement.

ACCURACY: The capacity of a weapon to strike the target in a small group. Precision. It diminishes with range and varies with the ammunition and firearm. It is affected by the shooters ability, weather, etc. It is how close a strike is to where it was intended to hit and measurements are to that one point only.

ACP: as in .45 ACP, .380 ACP, etc. Initials that stand for "Automatic Colt Pistol." Now, these and others that have ACP in their name are used world wide in many firearms other than Colt.

ACTION TIME: The time between the firing pin hitting the primer and the bullet exiting the muzzle. This time will not only change with different types of ammo, it will vary slightly from one shot to the next.

ACTION: The mechanism that makes a firearm work. The heart of the weapon. This may include the extractor, bolt, firing pin, breech, etc. Various types include the blow-back, blow-forward, bolt, box-lock, falling block, lever, pump, rising block, rolling block, automatic, semi-automatic, side lock, single shot, slide, top break, trap door, under-lever, etc. etc.

AIR DENSITY: The term for the air's thickness or compactness. Technically the ratio of mass to volume. Sea level standard is 29.921 in. Hg at 59 F. (0.075 lb. per cu. ft.) and drops fairly uniformly to the isothermal region at 35,332 ft. and 7.04 in. Hg .

AIR RESISTANCE: An aerodynamic force which retards the flight of a bullet. Caused primarily by pressure distribution and skin friction. It can be reduced by streamlined bodies with smooth contours. Along with gravity, it is one of the two primary forces acting on a projectile after it departs the muzzle.

AIR SPEED: Or more correctly, velocity, of a projectile through the body of air it is in. The air mass itself may be moving over the earths surface, and the reference is to the air mass only, not the earth it and the projectile are moving over.

ALLOY: A metal that is a mixture of either two or more metals or a metal and something else. The result may be an improvement in strength or some other benefit such as lower cost.

AMERICAN RIFLEMAN, The: A monthly publication of The National Rifle Association.

AMMUNITION: abbr. AMMO: Usually used to describe self contained cartridges of any type but can also be used for powder, bullets, primers and even grenades, mines, etc.

ANGLE OF DEPARTURE: Frequently given as the angle between the bore as the gun is aimed at the target and the bore the instant the bullet leaves the muzzle. It includes barrel jump. It is sometimes incorrectly based on the horizontal and is frequently confused with angle of elevation which has a different meaning.

ANGLE OF ELEVATION: The angle between the bore axis of the gun and the line of sight to the target, not to the horizon.

ANNEALING: A process of softening metal by controlled heating and cooling. It can also be done to alter mechanical or physical properties of the metal.

ARTILLERY: Large caliber guns that are too heavy to carry (except machine guns). They may be mobile, mounted or stationary.

ANTIQUE: A firearm, by Federal law, made in or before 1898. Also can describe a firearm which requires fixed ammunition made in or before 1898 which is no longer readily available. Various States and private groups have diverse definitions.

ASCENDING BRANCH: The portion of the trajectory from the muzzle to the highest part.

ASSAULT RIFLE OR WEAPON: Defined as a firearm of intermediate power that is capable of selective fire, i.e.. a change between full automatic and semi-automatic. This term is frequently used improperly by the media and anti-gun groups.

AUSTENITE: A term in metallurgy. A solid solution. Carbon or other alloy is a solute. Gamma iron as the solvent.

AUTOMATIC: A type of firearm that will feed and load cartridges, fire and then eject the case and continue as long as the trigger is depressed and cartridges are in the magazine. Many semi-automatic weapons are incorrectly called automatics.

AVERAGE: The number obtained by dividing the sum of several quantities by the number of quantities.

AXIS: The line, real or imaginary, about which a thing (mass) rotates or would rotate if it was free to do so.

AXIAL: A term for movement about an axis of rotation.

BALL: Word for the round ball used in muzzle-loading firearms. Also a colloquial term used to describe ammo with any shape that has a solid single bullet.

BALL CARTRIDGE: A military term for a regular metal-jacketed pointed bullet loaded to full velocity. Cartridges loaded with round balls, such as are used in shotguns for guard duty are not called ball cartridges but are called guard cartridges.

BALLISTIC COEFFICIENT: A ratio which denotes a projectiles ability to overcome air resistance and maintain velocity in comparison to a standard. The larger number is usually preferable.

BALLISTIC PENDULUM: An old time device for measuring projectile velocity. It is still used today to measure recoil.

BALLISTICS: The science of projectiles. **Interior ballistics** concerns the events until the projectile exits at the muzzle. **Exterior ballistics** covers events between the muzzle and the impact point. **Terminal ballistics** covers the event after impact. (Terminal is not the best word, but has became correct due to common usage.) **Forensic ballistics** is the legal study of all as applied to police investigation and court procedures.

BALLISTIC WAVE: Another term for the bow wave which precedes a projectile below Mach 1. It is caused by air compression.

BARREL: The tube which directs the projectile. The bore may be rifled or smooth and extends from the breech end to the muzzle.

BB: The round steel copper coated balls used in air guns. Sized at .176 to .177 inch. Also a term for the shot in shotgun shells that are sized at .181 inch. The size disparity creates errors and confusion.

BLACK POWDER: An earlier form of gunpowder. It was used from the 13th. century to the beginning of the 20th. It is still in use by muzzle-loaders.

BLOW BACK: A type of automatic or semi-automatic mechanism where the breech is held closed by springs. There is no positive lock. Also, the rearward movement of the gas that is used to operate the blow back mechanism.

BOAT-TAIL: A feature of some bullets where the base tapers down to a smaller diameter.

BOLT: Sometimes called breech-block, it is the sliding part of a breech-loader that shoves the cartridge into the chamber and holds it in place. It frequently includes the extractor and firing pin.

BOLT ACTION: A firearm that uses a bolt type of mechanism. Commonly rifles, but it is also used on shotguns and handguns.

BORE: The inside of a barrel and also the interior diameter usually measured across the lands. i.e. the smallest diameter. The chamber is not included.

BOUNDARY LAYER: A thin layer of air adjacent to a projectile's surface where the air flow is retarded. This expends energy and creates drag from a change in pressure distribution. This is in addition to the drag from skin friction.

BREECH: The area of a firearm where the cartridge is inserted. Sometimes also used for chamber and receiver. One book defines it as the entire firing mechanism, magazine, chamber, action, trigger, etc. Therefore, it would have to be considered a term with an imprecise meaning.

BREECH-LOADER: Applies to any firearm that accepts or loads the cartridges from the rear. The idea dates to 1811 and the Hall rifle.

BREECH PRESSURE: The pressure or rearward thrust caused by the case head against the breech or bolt. Not to be confused with the chamber pressure which will always be considerably higher.

BRINELL TEST: One of many scales and methods of testing the relative hardness of a material, usually metal. Also see Rockwell.

BUCKSHOT: Term for large shot pellets.

BULL BARREL: A heavy, large diameter barrel used mostly for benchrest and target shooting. They usually have no taper and a longer length than a hunting barrel. They can reduce or eliminate whip, vibration, etc.

BULLET: A projectile fired from a gun. It can be made of various materials and in different shapes. The word is frequently used incorrectly by the general public and the media when *cartridge* is what is meant.

BULLET DROP: The drop or descent of a bullet from the pull of gravity. It begins at the muzzle.

BURNING RATE: With gun powder, it is the speed that powder burns as compared to samples in controlled laboratory tests. A relative term.

CALIBER: The diameter usually measured across the lands inside the bore or the diameter of the projectile. Frequently, the two do not coincide as they should in standard practice.

CANNON: Artillery that can be either fixed or mobile. Usually larger in size than 37 mm but the 20mm aircraft gun is called a cannon.

CANT: The angle, lean or tilt of a gun that is not held vertical. It affects sighting and the bullet strike point.

CARBINE: A rifle with a short barrel usually less than 22 inches long. It was of military design and background for use by cavalrymen.

CARBON: An element in steel which determines its ability to be hardened. Also, as charcoal, it was a main ingredient in early gunpowder. (15%)

CARBURIZING: A method of adding carbon to soft metal. Involves heating and quenching.

CARTRIDGE: A complete self-contained round of ammunition including the case, powder, primer and projectile. Technically, shot-shells could be included, but they normally are not.

CASE: The container for the powder, projectile and primer. The largest and most visible part of a cartridge. It can be made of brass, copper, plastic, steel or paper, depending on its use.

CASE HARDENING: Hardening only the outer layer of a ferrous alloy while the interior remains softer. The depth can be from tissue paper thin up to .030 inch. Normally about .005 inch.

CENTER-FIRE: Term for a cartridge case with the primer located in a center pocket as opposed to rimfire.

CENTER OF GRAVITY: The point of a body through which the line of action of its weight passes. It is important in gyroscopic stability of bullets and the handling properties of guns.

CHAMBER: The part of a firearm that holds the cartridge at ignition. Located in the rear end of the barrel or in a revolver, multiple chambers are in the turning cylinder.

CHAMBER PRESSURE: The pressure in the chamber of the firearm created by the rapid burning of the powder. It is what makes the gun work. Usually it is measured at the peak or highest point which is soon after the bullet begins to move out of the case. Shot guns usually go to 12,500 p.s.i., handguns very wildly up to about 45,000 p.s.i. and rifles up to 55,000 p.s.i.

CHARGE: Amount of shot or powder in a cartridge, usually measured by weight. Also a term meaning to load a firearm.

CHOKE: A constriction in the muzzle end of a shot gun barrel that controls the shot dispersion. It comes in many amounts or degrees to deliver different patterns.

CHRONOGRAPH: An instrument that measures projectile velocity through time.

COEFFICIENT: In physics and ballistics, a number, constant for a given substance, used as a multiplier in measuring the change in a property under given conditions.

COEFFICIENT OF FORM: The shape of a bullet's forward end described by a mathematical index or number.

COLLIMATOR: An optical device used to bore sight a firearm.

COMBUSTION: Oxidation. If rapid enough, it creates heat and light.

COMPENSATION: A word with a meaning in ballistics that is different than any given by a standard dictionary. It is used in regard to the bullet leaving the barrel while the barrel is moving from recoil, jump, whip or another force. **Favorable compensation** is where faster velocity is low on the target in a small spread. **Unfavorable compensation** is where faster velocity is higher on the target in a larger spread.

CONVERSION FACTOR: Handy numbers used to convert one method of measurement to another. i.e. yards to meters.

CORDITE: A type of British powder extruded into strands with a cord like appearance.

CORROSION: Damage from rust or chemical action.

CUP: Short for *Copper Units of Pressure*. Used as a method of determining the internal pressure inside a firearm during use. A relative term based on the amount of flattening or crushing of a copper pellet using special laboratory equipment. CUP is generally used above the 10,000 mark. There is also a LUP gauge used below 10,000. (see LUP) Neither are the same as p.s.i. although there is a relationship.

CYANIDE HARDENING: A method of case hardening the outer shell of a mild steel object by dipping the red hot part in potassium cyanide, repeating as desired for added depth and hardness, and then quenching in water.

CYCLIC RATE: A term used to describe the maximum rounds per minute that a particular full-automatic firearm can operate. It is usually a theoretical amount which can not be reached due to heat and the few shots usually required to *get up to speed*.

CYLINDER: Used in revolvers to hold cartridges. It rotates from one chamber to another and can turn to either the left or right, depending on make and model.

DEFLECTION: Side movement of the projectile by wind, brush, etc. Deviation from the intended path. Sometimes called *drift*.

DENSITY: Mass per unit volume. Weight is a measure of mass so density may be either **absolute** (directly weighed) or **relative** (compared with the same volume of a standard under the same temperature and pressure). If compared to an equal volume of water, it is **specific gravity** and given in grams per cubic centimeter. For gases, it is expressed in terms of hydrogen or air.

DENSITY OF LOADING: See loading density.

DISPERSION: The maximum spread of the bullet holes on a target. It can be either horizontal or vertical.

DOUBLE ACTION: See revolver, double action.

DOUBLE BARREL: A gun with two barrels fastened together along their length. They can be over/under or side-by-side and shotgun, rifle or a combination of both.

DOUBLE BASE POWDER: Also known as nitroglycerin powder. It is called double because it has both nitroglycerin and nitrocellulose as ingredients. A single base powder has only the latter.

DRAG: Another term for air resistance.

DRAM: A unit of weight. Drams * .0039 = pounds.

DRIFT: Used to describe the side movement of a projectile caused by its rotation. Also used to indicate deflection by wind or brush, etc. Deflection is the preferred term for the later, although drift is in common use and therefore acceptable.

DUM DUM: In correct usage, a bullet designed and made at the British Arsenal in Dumdum, India (near Calcutta) in 1897-1898. (Mostly in .303 British caliber.) The bullet was designed for expansion. Today, the term is misused to describe any expanding bullet.

EFFECTIVE RANGE: The range at which a bullet can be expected to hit and kill a particular size game or score on a target. It is a theoretical distance that may or may not be correct and will vary with conditions.

ELASTIC LIMIT: An engineering term for the stress limit of a material that will deform it so that it will not return to its original shape. This point is usually given in pounds per square inch and it will be much lower than the materials tensile strength.

ELEVATION: The vertical adjustment of the sight and the barrel to control point of impact at a distance.

ENERGY: The power of force to do work. It can be *potential* or *kinetic* and is normally expressed in foot pounds.

EROSION: Gradual damage (to the bore) from wear caused by hot gas and friction.

EXPANSION RATIO: The ratio of the volume of the bore from the cartridge base to the muzzle, to the volume of the powder portion of the cartridge.

EXTREME SPREAD: A term used to evaluate hits. Measured on a target from the inside edge of a hole to the outside edge of another hole. The 2 holes being the most widely separated of the group. It may be the *group diameter* or it may not and so it may be incorrect to call it so. Sometimes it is called *maximum spread* or *group size*.

EXTREME VERTICAL OR HORIZONTAL DISPERSION: The longest vertical or horizontal distance between shots.

FIREARM: Technically, a firearm is a mechanical object that transfers energy through space using "gunpowder" as the propelling force. (That hardly does justice to a beautiful antique revolver or a modern big bore rifle with a top line scope.) More normally described as a weapon that propels a projectile by either compressed air or rapid burning of a propellant. Note: the legal definition will be long and involved and vary from one area to another.

FIRE-FORMING: Firing a case with a reduced load so it will expand and properly fit a chamber.

FLASH HOLE: The hole in the case that lets the primer flame pass through to ignite the powder. In muzzle-loaders, it is the hole that does the same between the bore and the pan.

FLASH POINT: The lowest temperature at which a flammable material will flash up but not sustain combustion. It will be somewhat lower than the ignition point.

FLIER: A bullet that strikes outside a group fired with one gun and ammunition lot.

FOOT POUND: Unit of measurement for kinetic energy. 1 foot pound is required to raise 1 pound up 1 foot.

FORCE: The power and its intensity that causes motion or a change or stoppage of motion of a body.

FORCING CONE: The tapered section at the front of a shotgun chamber that guides the shot into the bore. A *similar* area in a rifle is frequently called the throat or leade (lead). Note it is only similar, not exactly the same. The term is not used correctly for a rifle or handgun.

FORM FACTOR: A ballistic term indicating the effect air resistance has on a particular bullet shape as compared to a standard.

FREE BORE: When a cartridge is in the chamber, it is the area ahead of the bullet before the rifling. It can be an extension of the chamber or ahead of it. It will have straight walls with no taper. It is sometimes called a *long throat chamber,* although it may not have much length.

GAS OPERATED: A locking breech firearm, either automatic or semi-automatic, that utilizes some of the gas pressure to cycle the action.

GAUGE: A unit of bore size for a shotgun. Rarely used on other bores.

GRAIN: A unit of weight used to measure powder and bullets. 1 grain = 1 / 7000 of an avoirdupois pound. 1 oz. av. = 437.5 grain. It is incorrect to use *grain* to describe the powder pellets.

GRAVITY: The force that pulls object toward earth at the rate of acceleration of 32.1741 feet per second.

GREENHILL FORMULA: A formula developed by Sir Alfred Greenhill to determine the correct amount of rifling twist for proper stability.

GROOVES: The lengthwise helix rifling in the bore.

GROUP: Target hits fired at one setting with the same gun and ammunition lot. A cluster.

GUNPOWDER: A word used for black powder which is incorrectly used for all propellants.

GYROSCOPIC STABILITY: Lack of wobble and erratic behavior because of the spin from the rifling while adhering to complex laws of physical science.

HANDGUN: A firearm intended to be fired by one hand although most defensive combat techniques now use both hands. It may be a single shot, a revolver or a semi-automatic. Pistol is used synonymously although not all handguns are technically pistols.

HAND-LOADING: Another term for reloading where the case or hull is repeatedly used over again either to save money or to develop special loads.

HARDENING: Many processes are used for increasing the hardness of metal. All involve heating and cooling.

HEADSPACE: The distance in the breech or chamber between the rear face and the front part where the cartridge is held or located. Its accuracy can be vital for both performance and safety.

HELIX: The 3 dimensional curved path of a screw, rifling, or in some ways, a trajectory. Frequently called spiral, but helix is the proper term.

HIGH VELOCITY: A muzzle velocity between 3,500 and 5,000 feet per second as specified by the U.S. Army.

HOLDING OFF: Compensating for wind deflection by aiming to the side into the wind. No sight adjustment is made.

HOLD OVER: Compensating for range trajectory by holding high so the bullet will correctly drop to the target. No sight adjustment is made.

HOLLOW POINT: A bullet with a cavity at the nose to aid in expansion at impact and penetration.

HYPERVELOCITY: A muzzle velocity in excess of 5,000 feet per second as specified by the U.S. Army.

IGNITION: The term applied to the firing or burning of the powder. It is incorrect to consider the primer flash alone as ignition, although it will or may start the ignition. A certain high temperature is required, normally in the neighborhood of 550 degrees F.

IGNITION TIME: The short span of time between when the hammer or pin hits the primer and the gas pressure starts the bullet to moving. Normally about .0002 of a second. (2 ten-thousands)

INERTIA: The tendency to remain at rest or if moving, preserve the motion and direction unless affected by an outside force.

INGALLS' TABLES: Tables developed by Col. J. M. Ingalls for the computation of trajectory, remaining velocity, time of flight, etc.

INITIAL RECOIL: This is the small beginning recoil that happens before the bullet leaves the muzzle.

INTERIOR BALLISTICS: This is the area of ballistic science that deals with things that happen in the firearm while the bullet is still inside the bore. In the interior, so to speak.

INVERSE: Opposite or inverted order as a quantity is greater or less according as another is less or greater.

JACKET: The metal covering enclosing the bullet's core. Usually copper or steel over lead.

JOURNEE'S FORMULA: A formula that determines the approximate range for shotgun pellets by stating that it will be about 2,200 times the pellet diameter in inches.

KENTUCKY WINDAGE: Slang term for aiming to the side (hold-over) without sight adjustment to compensate for wind deflection.

KEYHOLE: The elongated keyhole shape in a target from an unstable, tumbling bullet.

KICK: A relative term for felt recoil. It is not the same as measured recoil.

KILLING POWER: An inexact and loose term on how effective a bullet is in killing for game or self-defense.

KINETIC ENERGY: The energy of a body resulting from its motion. As opposed to potential energy.

LANDS: The raised area of the bore in a rifled barrel that is between the grooves.

LEAD: (Pronounced leed.) The aiming forward (ahead) of a moving target or the distance involved to compensate for its speed.

LEADE: (Pronounced leed.) Also called *throat*. The area just ahead of the chamber that is cut on a taper by the chamber reamer. This aids the bullet in fitting into the bore. It is sometimes incorrectly called freebore which is straight instead of tapered. Note: The final *e* spelling is a colloquialism and used only in regard to ballistics and firearms. This spelling is not found in a standard dictionary.

LINE OF BORE: The imaginary straight line extending through and from the bore center line.

LINE OF DEPARTURE: Same as line of bore.

LINE OF SIGHT: A conceptual line through the sights or scope to the target.

LINE OF SITE: The imaginary straight line extending from the firearm to the target.

LOAD: This word has 3 meanings in firearm ballistics. **(1)** A single round of ammunition. **(2)** To insert ammunition into the magazine or chamber. **(3)** To reload or hand-load ammunition.

LOADING DENSITY: The ratio between the charge weight and the water weight that will fill the powder space in the case. It is frequently misunderstood by experts and thought to denote percentage of volume. **It does not.** If the case is filled with powder to the base of the bullet or it has an air space, the actual loading density may be identical.

LOCK TIME: The time it takes the hammer or firing pin to fall and detonate the primer. A short time is best because it reduces the chance of gun movement away from the point of aim. Modern firearms can run as short as .0020 of a second. Muzzle-loaders can be up to 5 times as long.

LUP: Short for Lead Units of Pressure. Used as a method of determining the internal pressure inside a firearm during use. A relative term based on the amount of flattening or crushing of a lead pellet using special laboratory equipment. LUP is generally used below the 10,000 mark. There is also a

CUP gauge used above 10,000. (see CUP) Neither are the same as p.s.i. although there is a relationship.

MAGAZINE: A container for storing ammunition for feeding into a firearm. It may be detachable or a fixed part of the firearm. Also, a military term for a storage area for ammunition and powder.

MAGNUM: It normally means a cartridge or shot-shell that is loaded heavier or larger than normal. Usually they give higher velocities or have more shot. The term is of British derivation. It can be meaningless as far as actual velocities, killing power or pressure is concerned. That is because in some examples, other cartridges may have more velocity, accuracy and killing power in a standard cartridge.

MASS: The quantity of a body as measured in its relation to inertia. Obtained by dividing its weight by the acceleration due to gravity. One of the three fundamental quantities, along with length and time, in mechanical measurement.

MAXIMUM MEAN RADIUS: A term used in military ammunition specifications to give the grouping ability of the ammo. After a shot group has been fired, the center is determined and the distance is measured to each hit. This distance is totaled and then divided by the number of shots in the group. Also described as the average of the radial distances from the center of impact.

MAXIMUM RANGE: The range at which a cartridge or shot-shell is effective. The figures are not to be used as suitable for hunting.

MEAN RADIUS: A way of measuring group size based on the average distance from the center of the group.

MEAN VERTICAL OR HORIZONTAL DEVIATION: This is the average distance from the center of impact of a target group.

MEPLAT: The blunt forward end of a bullet measured by its diameter.

MIDRANGE TRAJECTORY: The highest vertical point in the curved trajectory above the line of sight. Measured in inches.

MIL: The U.S. Army's method of angular measurement. MOA is the civilian. MIL is 1 / 6400 of a circle. This is equal to 1 / 1000 of the range and chosen for its convenience. Technically, it would be 1 / 6283 of a circle.

MINUTE OF ANGLE: M.O.A. is used to give group size and sight adjustment. Roughly 1" at 100 yards and 1 / 60 th. of 1 degree. More precise figures are included in Chapter 18.

MOMENT: Force in rotary motion. Similar to torque. They are the twisting or rotating force applied to an object. Moment is the product of a specified force or mass and its distance from its axis or plane.

MOMENTUM: The impetus of a moving object. It is equal to the product of its mass and its velocity.

MUSHROOM: A term used to describe a bullets' shape after expanding on impact because of the resemblance to the fungi with an umbrella-like top. Not all bullets react in this way.

MUZZLE: The open forward end of the barrel.

MUZZLE BLAST: The push of hot gas escaping the bore as the projectile exits. It has a big effect on kick, recoil and in some cases, trajectory.

MUZZLE ENERGY: Bullet energy as it exits the muzzle. Usually given in foot pounds.

MUZZLE JUMP: A sudden rise of the barrel following ignition. Usually affects trajectory because it will happen before the projectile exits the barrel. It may change impact point if ammo is varied. The main cause is the gun design which places center line of the bore above the mass of the firearm.

MUZZLE VELOCITY: The projectile's speed leaving the bore.

NATIONAL RIFLE ASSOCIATION, The: A non-profit organization formed in 1871 to represent the interests of gun owners and sportsman. It is involved in safety education, marksmanship training for both police officers and civilians, and many other worth-while projects. (It is invaluable in this historic period where many well meaning groups incorrectly believe that preventing honest people from possessing firearms will cause the criminals from using theirs. The lower elements of society will still have and use weapons. After all, laws are only followed by the lawful.)

NITROCELLULOSE: The base for modern powder made of cotton or a similar cellulose product, and impregnated with sulfuric and nitric acids.

OBTURATE: To close a gun breech so the gas can not escape except out the bore. This is usually accomplished by the cartridge case itself.

OGIVE: The curved area near the front of a projectile extending from the tip or meplat to the main cylindrical portion (bearing surface) and the radius involved. Incorrectly used to describe all portions ahead of the bearing surface.

OVER BORE CAPACITY: A quality of a cartridge that has too much volume in relation to the bore. This does not permit the complete and efficient burning of the charge. To a lesser degree, barrel length, bullet weight and caliber are also involved.

PARABELLUM: This word is normally used in connection with the Lugar pistol and its 9 mm cartridge. It means "prepare for war". Supposedly, it comes from a Latin expression of around the 4th. century, *si vis pacem, para bellum.* ("If you want peace, prepare for war.") Many are following these words when they carry a concealed weapon, even if they do not think of it in this manner.

PARABOLIC CURVE: The normal way to describe a projectile's curved trajectory. This is incorrect because the second half of the curve does not match the first half because drag slows the bullet. The highest point is at about 55% of the curve. Correctly, the trajectory is an unsymmetrical line called a ballistic curve.

PATTERN: The distribution of shot pellets normally tested at 40 yards.

PELLET: The round spherical shot used as projectiles in shot-shells. Also skirted or waisted pellets as used in spring or air guns.

PI: (π) Used in math. formulas to represent the ratio of the circumference of a circle to its diameter. For most use, 3.1416 is more than adequate for the infinite number.

PIEZO-ELECTRIC CRYSTAL: A mineral, usually quartz, that generates a small electric charge (electromotive force) when acted upon by pressure. It is used in ballistic laboratories to measure chamber pressure in a Piezo quartz crystal gauge.

PISTOL: This word is used in place of handgun to cover semi-auto, revolvers, etc. Some books limit its use to guns with fixed chambers, i.e. semi-autos but not revolvers. While this is technically correct, common usage now makes it synonymous with handgun.

PITCH: This word has three meanings in connection with firearms. The only ballistics definition concerns the pitch of rifling. It is the angle of the rifling helix to the bore axis. The *twist* angle.

POINT BLANK RANGE: An expression with 3 definitions. **(1)** In colloquial use (slang), it is used to indicate a very close range with no precise limits. This is common in pulp fiction and TV detective shows when they are written by writers with no knowledge of firearms. This is incorrect usage. **(2)** The first range that requires no elevation change either up or down to hit point of aim. The point where the projectile crosses the line extending through the sights because of the angular difference with the bore line. This first point will be just a few yards range even when the 2nd. primary point is 100 yards or 300 yards. (See sketches in Chapter 17.) This usage is seldom used but is correct. **(3)** Common correct usage is to describe the maximum distance that will result in an accurate hit in the vital area with the sights set on the target correctly. If the diameter of the vital area is 8", then the sights would be adjusted for the range that would not let the bullet pass more than 4" above or below the center of the circle.

POTASSIUM NITRATE: A chemical frequently called saltpeter. One of the three main ingredients in early black powder. (75%)

POWDER: The black or modern smokeless propellant that is rapidly burned to create the gas that pushes the projectile from the gun. Modern propellants are no longer powders, but the term persists.

POWDER EFFICIENCY: Thermal efficiency. The propellants capacity to produce heat energy for its weight. Also used to describe the velocity and energy obtained from a given chamber pressure and/or given amount of propellant.

POWER: In math., the multiplication of a quantity by itself as 4 is the second power of 2 (2^2). Also force, strength, etc. with the energy of linear or rotational motion.

PRECISION: In target shooting, it is how well shots are grouped and the measure is to each other no matter how close or far from a target center.

PRESSURE: As used in firearm ballistics, it is the thrust of the gas (force) created by igniting the powder charge. This thrust is against the interior of the case, the chamber, base of the bullet, etc. Safe pressure can vary from 5,000 p.s.i. for a black powder gun and 15,000 p.s.i. for a .38 Special revolver to 54.000 p.s.i. for a .270 Winchester rifle and 55,000 p.s.i. in a .22-250 Remington rifle. All are averages and the .22-250 is not the highest, but darn close.

PRIMER: Usually a metal cup holding an explosive mixture that, by a sudden blow, detonates and ignites the powder charge. Lead styphuate and previously, metallic fulminate are used. All U.S. made primers are now non-corrosive.

PROJECTILE: Any object such as a bullet, shot pellets, shell, etc. that is projected, hurled or shot from a gun. Technically, a bullet, for example, does not become a projectile until it exits the barrel and takes flight.

PROOF: An intentional over-load to test a firearm for strength. The *proof-load* is usually about 25 % over normal.

PROPELLANT: noun. The powder charge that is ignited to propel or push the projectile. Smokeless powder is the most commonplace, but black powder, CO_2 or compressed air would also be correct usage with a firearm that used that method.

PROPELLENT: adjective. Note the spelling with an *e* instead of an *a*.

PYROCELLULOSE: A product in smokeless powder that has a lower nitration than guncotton.

QUENCH: A term used in heat-treating. It is an immersion of the metal in oil, water, air, etc. to cool it rapidly after heating. The result, if done with care and to specifications, will either harden, stress relieve, etc., i.e.. change the properties to a desired state.

RANGE: Distance to an intended target. Also a place where shooting is safe and organized.

RATIO: An expression of relationship or proportion between two numbers or similar things.

RECOIL: The backward push of a firearm in measured thrust. Incorrectly called kick which is felt by the shooter, but is not the same as the recoil.

RECOIL OPERATED: A locking breech firearm, either semi-automatic or automatic that utilizes some of the recoil energy to cycle the action.

RELOADING: Also called hand-loading, it is the recharging of a used case by installing a new primer, powder and bullet.

REVOLVER: A gun that has a cylinder that holds the cartridges and rotates to align each, one at a time, with the barrel. Usually a handgun, but a few long barrel guns, even shotguns, have been made using this principle.

REVOLVER, DOUBLE ACTION: A revolver that can be fired by pulling the trigger without *thumbing back* the hammer. The hammer, whether exposed or concealed, is cocked by the internal mechanism. The trigger does double duty. It cocks the action and releases the hammer.

REVOLVER, SINGLE ACTION: A revolver that can only be fired by cocking the hammer back manually before the trigger is pulled. The trigger does just the single job of releasing the hammer.

RICOCHET: A word with a French back-ground that means to rebound or skip. Bullets that hit an object at an oblique angle can glance away in a new direction.

RIFLING: The name used for the helix grooves in the bore of a firearm. They create the spin to the projectile which is needed for gyroscopic stability and accuracy. Rifling is not used in just rifles but also in handguns, cannons and shotgun barrels intended for slugs.

RIMFIRE: A type of cartridge where the primer is located around the rim instead of in a center pocket. Today, most are .22 caliber. In the last half of the 1800's, many other sizes were popular. They died out, partly because they are not practical to reload.

ROCKWELL: One of the many scales or methods of determining the relative hardness of an object, usually a metal such as steel.

SABOT: Pronounced *say-bo* or *sab-o*. A word of French origin for a carrier that fits around a projectile which is smaller than the bore. The sabot falls away when exiting the barrel. Originally designed for military use for bullets with extreme points on the rear. Now popular for shotgun slugs.

SATURDAY NIGHT SPECIAL: Originally the term was applied to cheap handguns that were little more than junk. Now it is used incorrectly to describe any handgun, regardless of condition or value. The term is slang and not accepted by knowledgeable people. Anti-gun groups and occasionally the media use it for all handguns, even the most prized and expensive. Therefore, the term is meaningless. See *Suicide Special*.

SEATING DEPTH: The depth or distance the bullet is placed in the open forward end of the case.

SECTIONAL DENSITY: With bullets, it is a ratio of its mass in pounds to the square of its diameter in inches (cross section). SD = bullet weight in pounds divided by the bullets diameter in inches squared.

SELECTIVE FIRE: A fully automatic firearm where the shooter *selects* either full automatic or semi-automatic.

SEMI-AUTOMATIC: A firearm that is self loading but will fire only one time when the trigger is pulled. Holding back the trigger will not fire another cartridge. The trigger would have to be released and pulled again.

SHOT: The spherical pellets used in shotgun shells and some special cartridges for rifles and handguns. They come in many sizes for different needs. Usually lead, steel or bismuth composition, but other materials are occasionally used.

SHOTGUN: Normally thought of as a smooth-bore shoulder gun for shooting a number of pellets at small game or clay targets. Some modern shotguns have rifled barrels to impart spin and gyroscopic stability to slugs.

SIGHTING IN: The adjusting of the sights for elevation and windage so the projectile will strike a pre-intended point at a pre-intended range.

SINGLE ACTION: See revolver, single action.

SLUG: Occasionally it is used as a synonym for bullet but it is more commonly used for large shotgun projectiles. In ballistics and science, it is called *the engineer's unit of mass*. 1 slug = 1 lb. per (ft, per sec.^{2})

SMOKELESS POWDER: Usually considered modern powder, although it was discovered in 1832 and perfected in 1884. Interestingly, it is not smokeless or a powder. It is made in both single base and double base types.

SODIUM NITRATE: It was frequently an ingredient in black powder (75%) in place of potassium nitrate.

SPECIFIC GRAVITY: The term used to describe the ratio of the mass of a solid or a liquid to the mass of an equal volume of water at a standard temperature. With a gas it is usually expressed in terms of hydrogen or air.

SPEED OF ROTATION: Used in connection with rifling and bullet stability and expressed in revolutions per second instead of the revolutions per minute which is common with slower turning objects.

SPEED OF SOUND: A major point in the study of airflow. It varies from 573.8 Knots per hour at -69.7 degrees F. to 661.7 Knots at 59.0 degrees F. The best figure for bullets is 1,120 feet per second but it will change slightly with temperature and temperature *only*. (See pages 147-149.)

SPIRAL: A term incorrectly used to describe the rifling grooves in a barrel. A spiral has a constantly decreasing or increasing radius in a single plane. See Helix.

SPITZER: A bullet or bullet point that has a tapered end with a sharp tip

SQUIB: Squib load. A cartridge loaded to less than normal powder charge. It will produce less pressure and velocity.

STANDARD DEVIATION: A mathematical measure of variation from the average that gives emphasis to the extreme variations. It is common in velocity studies.

STOPPING POWER: An inexact term for the capacity of a firearm, cartridge and/or bullet to stop a drugged-up felon or dangerous hunted game. Stopping is harder to do than killing. Death may come later from blood loss from a wound that failed to disable.

STRIKING POWER: Energy available at point of impact. Depending on the cartridge and range, it may be considerably less than muzzle energy.

STRIKING VELOCITY: Velocity at point of impact. Depending on the cartridge and range, it may be considerably less than muzzle velocity.

SUICIDE SPECIAL: Term originally used to describe cheap pocket guns made in the late 1800's. The term became popular in the 1940's and recently has been used incorrectly for all handguns. See *Saturday Night Special*.

SULPHUR: A chemical that comprised about 10% of black powder.

TEMPERATURE EFFECT: The change in a projectile's performance caused by a change in temperature. It can be the atmospheric temperature or the propellent temperature.

TENSILE STRENGTH: A figure given in pounds per square inch that is used to compare one material to another. It is the resistance to deformation at the fracture point and is obtained by dividing the maximum load during the test by the original cross-sectional area.

TERMINAL VELOCITY: See *Velocity, Terminal*.

THROAT: see leade.

TRIGONOMETRY: Mathematics that deals with the ratios between the sides of triangles, the relations between these ratios, and the application of these facts, in finding the unknown sides or angles of any triangle.

TORQUE: A force that produces a rotating or twisting motion. In firearms it will twist the gun in the opposite direction as the rifling. It can be very noticeable in big magnum handguns and light rifles with high velocity cartridges. All firearms will have it in some amount, but it may not be noticeable. Torque may be called *moment*.

TRAJECTORY: The curved path of a projectile in flight. It can be measured by *drop* or *midrange height*.

TRANSVERSE: Across or crosswise as with an axis.

TUMBLING: The disorderly end over end roll of an unstable bullet.

TWIST: The rifling grooves helix (incorrectly called a spiral) is expressed in a twist rate of turns per inch to either the right or left. A 1:10 twist would complete one turn in 10 inches.

VELOCITY: A rate of motion or speed in a particular direction in relation to time. Projectile velocity is expressed in feet per second (f.p.s.) and can be measured or calculated at any place from the muzzle to impact.

VELOCITY, ESTIMATED: A figure based on experience and comparison.

VELOCITY, INSTRUMENTAL: A figure based on measurements by a chronograph.

VELOCITY, MUZZLE: A figure taken at the muzzle. Usually taken from a short distance to avoid damage to the equipment from the blast of gas and then converted to zero range.

VELOCITY, STRIKING: A projectile's velocity at impact.

VELOCITY, TERMINAL: Incorrectly used to indicate striking or down range velocity. Correctly, it is the velocity of a falling object after the velocity has stabilized where the air drag and pull of gravity have cancelled each other out. Therefore, the falling velocity will remain the same unless acted upon by another force or changed in some way.

VIBRATION: A rapid rhythmic motion back and forth; a quiver. The barrel of a firearm, especially rifles, will go through a small motion sometimes said to be a vibration. Whether vibration is the proper word is debated, but the motion is real. Other motions as jump and whip, etc. are separate and different although all occur at almost the same instant in time.

WADCUTTER: A bullet with a flat, blunt end that is designed to punch a clean hole in a target. It is good for its purpose but suitable for only short range because it is poor from a ballistic standpoint.

WAVE: A series of advancing impulses through the air or a liquid. The shock waves from a transonic or supersonic bullet. Also, the wave of air ahead of the projectile and the gas behind the bullet at hyper-velocity.

WHIP: A sudden, quick movement, usually associated with a rifle barrel as the bullet exits the muzzle.

WIND DEFLECTION: See deflection.

WINDAGE: A word with four meanings. Both the horizontal (lateral) movement of a sight and another term for wind deflection. Also in cannon and muzzle-loaders, the extra clearance between the smaller projectile and the larger bore and the gas that will escape through the gap.

WOBBLE: Sometimes incorrectly spelled *wabble*. The movement of a bullet that is unstable because of damage or a flaw in manufacture. It can also be caused by poor gyroscopic stability where it will show up at extreme long range. The latter is rare because of the distance and time required.

YAW: To swing the axis of a bullet from the line of flight. Usually given as an angle and unintentional.

INDEX

Aberdeen Proving Ground, 70
Acceleration, 4, 17, 60, 139, 347
Action time, 53, 347
Accuracy, 180, 226, 347
Airguns, 290-291
Air resistance, 2, 146-149, 151-153, 159-173, 203-204, 260, 276-277, 340-341, 347
Alcohol, 22, 27
American National Standards Institute (A.N.S.I.), 73
American Rifleman, The, 66, 68, 78, 108, 115, 127, 154, 173, 177, 270, 273, 278, 284, 304, 316, 343, 348
Anvil, 19
Armor-piercing bullets, 155
Askin, Col. Charles, 108
Atomic weight, 25-26, 28
Autofrettage stress relief, 75
Automatic, Semi-auto, 62-63
Ayoob, Massad, 318

Bacon, Rodger, 20
Ballistic coefficients, 159-173, 233, 276-277, 349
Ballistic efficiency, 40
Ballistic pendulum, *see pendulum, ballistic*
Ballistrate (ballistrite), 22, 29-30
Ball powder, 30
Barometric pressure, 167, 344
Barrel, diameter, 116
    effect on velocity, 157
    European length, 116
    flex, handgun, 288-289
    length, 57, 70, 110, 112-115, 118, 273
    life, 39, 116-117, 177
    manufacturing, 76, 78, 109, 137
    obstruction, 36, 57
    octagon shape, 114
    recoil, 64, 66-67,
    straightening, 109-110
    time, 56-58
    tip problems, 166, 190
    tracer, 185
    varmints, best for, 184-185
    weight, 116, 289
    whip, flex, etc. 4, 50-58, 213, 288-289
Barrier material, 325-326
Bearing surface, 183, 189
Bell, Alexander Graham, 207
Berden, Hiram, 19
Bernoulli, 2
Blackpowder, 20-21, 60
Blueing, 118
Bomb, ballistic, 24
Bore diameter, 121-122, 130, 349
Boulengë, Paul Emil le, 144
Boundary layer, 148, 260, 349
Boxer, Edward, 19
Boyle's law, 204
Bracketing, 211
British Small Arms Committee, 99
Browning, 57, 63, 252
Brush deflection, 5, 106-107
Buckshot, *see shotgun*
Bulk density, *see density, bulk*
Bulk powder, 25
Bullets, armor-piercing, military, 186
    boat-tail base, 183, 186-188, 349
    cast, 136

conical, 185
design and performance, 92, 94-95, 122-124, 182-195, 248, 311
design experiments, 192-195
drop, 350
dropped at muzzle, 154-155
expansion, 191
hollow points, (*see Hollow point*)
jacket, 183, 192
lead, 123, 169, 182, 184
mass, 331
nomenclature, 183
shape, 92, 94-95, 159-173, 321, 330-331
spin, 306-307
tip problems, 166, 190
tracer, 185
varmints, best for, 184-185
weight, 8, 116, 289
Burning, degressive, 23
Burning, neutral, 23
Burning, progressive, 23
Burning rate, 23, 350

Calculus, 11
Caliber, 114, 350
Cannelure, 183
Cannon, 70, 136, 292-295
Canting, 219-220, 350
Carbohydrate, 26, 29
Carbon, 20, 25-26, 73-74, 350
Carbon dioxide, 21, 28, 290-291
Carbon monoxide, 21, 28
Carnot, 31
Cartridge, center fire, 174-175
    details, 174-181, 350
    rim-fire, 174-175
    shape, 55
    volume, 177-178

Case, 3-4, 157, 179, 180-181, 351
    reading for pressure signs, 45-47
Case hardening of steel, 75, 352
Celluloid, 25
Cellulose, 26
Chamberline, Col. Frank, 308
Chaney, Officer Steve, 319-321
Chapman, John, 117
Charcoal, 20
Choke, Ithaca & Charles Daly, 279
    longevity, 253
    pattern testing, 254-257
    percent in circle at 40 yards, 253-254
    percent of reduction, 253
    rifle barrels, 117-118
Chronograph, 3, 144-145, 273, 351
Clark University, 44
Clausius, 31
Coefficient of drag, 153
Coefficient of form, 161-162, 351
Collimators, 336-337
Combustion, 27
Compensation (barrel movement), 53, 351
Computers, 338-339, 351
Concealed weapons, 318, 333
Concentric bore, 117, 181
Conical shape, 189
Conversion tables, 8-9
Cooking-off by temperature, 37-38
Copper crusher pressure test (CUP), 42-44, 352
Cordite, 29-30, 352
Coriolis, Gaspard, & effect, 99
Corrosion, 80
Coulomb, 44
Cross sectional area, 305-306
Cycloidal movement, 96
Cylindro-concoidal bullet shape, 189

Cylindro-ogival bullet shape, 189

Damascus, 73
Davis, William C. Jr., 66, 68, 70, 115, 154, 173, 278
Decomposition, 27-28
Deflection, 96, 105-108, 155, 170, 217, (*also see wind deflection*)
Delay time, 241, 243-246, 342
Density, air, 152-153, 202-203, 236, 347
    bulk, 178-179
    loading, 178-179, 352
    projectile, 242
Department of Justice, 322, 326-329, 334
Detonation, 24
Deviation, vertical & horizontal, 226
Diphenylamine, 27
Dispersion, vertical & horizontal, 226
    at range, 229-230
Double-base powder, 29-30, 352
Drag, 146-152, 352
Dram equivalents, 31
Drift, 96-104
Drug induced terminal ballistic problems, 318, 321
DuPont, 30, 161, 165

Earth's rotation effect, 99-101
Elastic properties, 2-4, 54-55, 79
Elevation angle, 202, 209, 231-235, 244, 353
Energy, 13, 16, 27, 35, 41-42, 49, 132-133, 162, 260, 302-304, 311
Equation of state, 33
Erosion, 80
Ether, 22, 27
Evaluation of target groups, 225-230

Expansion, 301, 308-310, 315-316
    ratio, 41, 112-113, 353
Exterior ballistics explained, 1
Extreme spread, 226, 353

Federal Aviation Administration handbook, 86
Federal Bureau of Investigation tests, 325-326
Federal Cartridge Corp. 118
Flash tube cartridge design, 175-176
Flechettes, 193
Force, 12, 15, 354
Forcing cone, 275-276, 354
Form factor, 354 (*see coefficient of form*)
Forsyth, Alexander, 18
Francotte, 65
Freebore, 127-130, 354
Freemantle, 99
Friction, 41
Fulminate of mercury, 18

Gain twist, 126-127
Galilei, Galileo, 2, 13, 142
Gas expansion, 3-4, 20, 60-61
Gauss, Carl F., 7
Geepound, 15
Gelatin, ballistic tests, 309-310
Geneva Convention, 329-330
Glock handgun, 80
Gravity, 13-15, 96, ,103, 294, 354
Grains, powder, 21
Greenhill, Sir Alfred & formula for rotation, 133-135, 354
Groove, definition, 120
    depth, 123-124
    depth, decreasing, 127
    measuring diameter, 130-131
    number of, 122-123
Gyroscopic drift, 5, 96-99

stability, 81-104, 354

Hague, 330
Handgun, 53, 280-289
    barrel length / velocity, 284-285
    long range shooting, 286
    recoil, 60, 68
    rifling, 285-286
    sights, 281-283
    trajectory, 206, 280-284
Hardening of steel, 75, 355
Halford, Sir Henry, 99
Hatcher, Gen. J. S., 98, 215, 304, 308, 321-323, 340, 344
Headspace, 4, 337-338
Heat treatment of steel, 75-77
Heel, 183
Helix, 120, 355
Helix angle, 120-121
Helmholtz, 31
Hercules powder, 29
Holland & Holland Ltd., 274
Hollow point bullets, 103-104, 327-330, 355
Hooke, Robert & Hooke's law, 54
Hoxie Ammunition Co., 193
Humidity, 167-168, 205
Hydrogen, 25, 27-28, 32
Hydrogen sulfide, 21

Igniters, 19
Ignition time, 56, 355
IMR powders, 30
Ingalls, Col. J. M. & tables, 164-165, 193, 355
Interior ballistics definition, 1

Jet effect, 187
Journee's formula, 236, 264-265, 355

Keith, Elmer, 122, 312, 316

Kelvin, 31
Keyhole, 87, 135, 307-308
Kick, *see recoil*
Kinematic viscosity, 202-203, 301-302
Knox Gelatin Co. 310
Kotter, Augustinius, 311
Krupp, 164

Lag time, *see delay time*
Lands, definition, 120, 356
Lateral jump, 102
Lead pressure test, 42-44, 356
Lead styphnate, 18
Loading density, 356, *also see density loading*
Lock time, 56, 356
Long range, 312-314, 339-340
Lowry, E. D., 166

Malter Arms Co., 193
Manganese, 74
Marlin Multi-Groove rifling, 123
Mass, 15-17, 62, 104, 141, 157, 357
Maximum mean radius, 227
Mayer, 31
Mayevski, Col., 164
Meplat, 183, 189, 357
Micrometer, how to use tip, 46
Miles per hour, 8
MILS, 224-225, 292-293, 357
Minute of Angle (MOA), 223-230, 245, 249, 357
MOL (MOLE), 31-32
Molecules, 3-4, 26, 28, 31
Molybdenum, 74
Momentum, 17, 35, 141, 304-305, 321, 357
Mushroom, *see bullet expansion*
Muzzle blast, 60, 358
    energy, 41, 358

flash, 61
loader, 36, 39, 291-292

National Rifle Association, The, 38, 78, 108, 119, 127, 154, 166, 173, 230, 273, 284, 338, 358
NATO standards, 344-345
Newton, Sir Isaac, 2, 12, 59, 69, 82, 141, 164
Nitric acid, 27
Nitrocellulose, 3, 22, 26, 30, 358
Nitrogen, 21, 25, 28
Nitrogen dioxide, 26
Nitroglycerine, 22, 26, 30, 358
Nobel, Alfred, 22, 29
Normalizing of steel, 76
Norma-Precision, 37, 149
Notation, scientific, 9, 13
Numerical integration, 171
Nutation, 5, 88-89

Ogive, 106, 162, 164, 188, 358
Olsen, Fred, 30
Operation sequence of firearms, 2-5
Organic compound, 26
Oxygen, 25-26, 28, 31-32, 35

Parasitic drag, 151
Pattern testing, 255-257, 267
Pellets, *see Shotgun*
Pendulum, ballistic, 2-3, 140-142, 349
Penetration, 106, 300-301, 308-310
Perfect gas law, 33
Plastic in gun construction, 80
Point-blank range, 220-221, 359
Pope, Harry, 126
Potassium chlorate, 18
Potassium nitrate, 20-21, 359
Poudre B, 22
Powder, black, 60, 349

chemistry, 25
compressing, 180
double base, 29
position in case, 37
progressive, 29, 48
single base, 29
velocity effect, 157
Power factor, 230
Precession, 83-89
Precision, 226
Pressure, 3-4, 21-22, 24-25, 31-32, 34, 40, 47, 58, 70, 126, 341, 360
average, 49
gradient, 41
limit, 50
peak, 47
Piezo-electric testing, 44-45, 359
testing methods, 42-46
Primer, 3, 18-19, 46, 157
Propellant, 3, 20-39, 37
Psychological factors, 332-333
Pyrodex, 30

Radical, 26
Rain interference, 6, 335, 345
Random, 293
Range finding, 217-219
Range, maximum effective, 231-236, 268-270, 353
Rankin temperature, 32-33
Recoil, 4, 21, 54, 59-72, 271, 316, 360
Reduced loads, 36-38
Relative Incapacitation Index (RII), 322-324
Relative quickness, 24
Remington, 63-64, 67
Revolver, 286-288, 361
Revolving disk for velocity, 142-144

Ricochet, 105-106, 361
Rifling, antique, 124
   factory specifications, 126
   handgun, 285-286
   twist, 3, 104, 119-138, 361
   unusual types, 124-126
Rimfire, 18, 361
Robins, Benjamin, 2, 140, 343
Rocket Ball bullet design, 192
Rockwell hardness test, 76-77, 361
Rodman, Col. Thomas, 23, 42
Rotation speed, 131-133
Rule of thumb, 39, 61, 154, 259, 342

Sabot, 192, 270-271, 279, 361
Saltpeter, 20
Schönbein, Christian, 22
Sectional density, 136, 159-161, 311, 361
Selenium, 74
Semi-auto pistols, 287-288, 362
   velocity loss, 153-154, 362
Shock waves, 61, 146-150
Shotgun, 66-67, 71, 250-279, 362
   buckshot, 265-268
   chamber length, 271-272
   chart of gauges, 278
   double barrel problems, 64, 257
   forcing cone, 275-276
   range, 236
   recoil, 64, 66-67, 71
   slugs, 268-270, 279
   targeting, 255-256
   tracers, 268
   Wannsee target, 256-257
Shotgun pellets, 331-332, 343
   amount in a charge, 279
   bismuth, 261, 264
   deformation, 259-260
   lead vs. steel, 261-263
   maximum range, 264-265

   size, 258-259
Sight line definition, 197, 199, 208, 281
Sights, 209-211, 249, 362,
   handgun, 281-283
   short range sight in, 211-212, 220-221
Silicon, 73
Site line definition, 197
Slugs, 15, 362 *(also see Shotgun....)*
Smith & Wesson cartridge patent, 174-175, 192-193
Sobrero, Ascanio, 22
Society of Automotive Engineers, (S.A.E.), 73
Sodium nitrate, 362
Softening of steel, 76
Sonic boom, 147-149
Specific gravity, 135, 362
Speed of sound, 146-149, 186, 246, 362
Spheres, ballistics of, 169, 276-277, 301
Spin drift, *(see gyroscopic drift)*
Spitzer bullet shape, 189
Sporting Arms and Manufacturers Institute (SAAMI), 48
Spreader loads, 260-261
Springfield, 44, 62, 73, 76, 122, 222
Stability, 27, 104
Stainless steel, 78-79
Standard deviation, 227-230, 363
Steel classifications, 73-74
Stellite, 79
Stress relieving of steel, 75
Sulphur, 20-21, 73-74, 363
Sulphur dioxide, 32
Swamped barrel, 274-275
Symmetry, 191

Taper heel, 186

Tartaglia, Nicholas, 2
Taped muzzle, 6, 335-336, 345
Temperature, 3-4, 22-23, 27, 31-33, 38, 41, 55, 146, 167, 204, 213, 274
Tempering of steel, 75
Terminal ballistics, 1, 296-334
    expansion, 301
    medical aspects, 298-300
    penetration, 300-301
    weapon choice, 297-298
Testing, ballistic coefficients, 166-168
    expansion, penetration, etc. 308-310
    sampling size required, 227-228
Thermodynamics, 20, 31, 33
Titanium, 79-80
Torque, 17, 61-62, 363
Trajectory, 5, 100, 196-222, 279, 336
    cannon, 293-294
    described, 198-200, 363
    drop, 197, 199, 201-202, 223-224
    effect of time, 198
    elevation angle, 202, 268
    handgun, 280-284
    inverted, 200
    mid-range height, 207-208, 268
    on a slope up or down, 213-217
    vacuum, 154, 222, 231-233
Trigonometry, 10
Tubb, G. David, 119
Tungsten, 74
Twist (*also see rifling*), 363
    measuring, 130

Unbalance, Static & dynamic, 89-93
Uniform motion, 140

Union Metallic Cartridge Co. 193
University College London, Ballistics Research Laboratory, 251, 273

Vanadium, 74
Vaseline, 30
Velocity, 13, 17, 35, 48, 113-115, 118, 139-158, 190-191, 242-246, 330, 363
    estimated, 139, 364
    high, 156-158
    instrumental, 139, 364
    muzzle, 145-146, 260, 364
    radial, 83
    remaining, 139
    semi-auto, 153-154
    shot pellets, 272-274
    summit, 139
    terminal, 139, 364
Vertical drift, 101-102
    shooting, 340-342
Vibration, barrel, 51
Vielle, Paul, 22, 27, 29
Volume, bore area, 49
    cartridge, 177-178
    gas expansion, for, 34

Wannsee target, 256-257
War Department, 74
Water, 28
Wave expansion, 147-149, 364
Webster, Dr. Arthur G., 44
Weight, 8, 14-16, 83, 190-191, 242
Whelen, Col. Townsend, 72
White, H. P., Laboratory, 36
Winchester (and divisions), 29-30, 51, 63, 68, 105, 120, 166, 252, 265, 270, 279
Windage, 237, 364
Wind angle, 239-241

Wind deflection,155, 237-249
    on slope shots, 217
    shot pellets, 272
    vertical shots, 342
Wind displacement formulas, 244-245
Wind proximity to shooter or target, 246-248
Wind velocity estimating, 239
Wobble, 102, 364

Yaw, 5, 97-98, 169, 198, 307-308, 364
    explanation, 89-90

ABOUT THE AUTHOR

Robert A. Rinker has a unique and diversified background that is perfectly suited to author a book on firearm ballistics. His love affair with firearms started at the age of six when his father, a military officer and newspaper man, taught him the manual of arms and gun safety. He started reading and studying about firearms and ballistics while still a youth. He has owned and fired most types of firearms, both antique and modern.

He also acquired from his father a love for the printed word. He has written many non-fiction articles, mostly on aviation and firearms. His fiction writing has been well received, including a 400 page novel, *Slate Creek*.

While attending school, Mr. Rinker was employed as an armed police officer at a major airport. He also worked as a tool and die apprentice during the day while attending college at night. When his foreman found out about his knowledge of firearms, some of his early machine-shop work involved repair and restoration on antique guns from the supervisor's private collection.

Educated as an industrial engineer, he owned and operated his own tool-die and machine shop where some of the work was on firearms, including big guns for the U. S. Navy. This work included design and research on modern ship-board weapons. He has also been trained and employed as a commercial-instrument-multi-engine aircraft pilot and flight instructor. He taught aeronautics to Naval Aviators during the Vietnam War.

He is one of many people who's life has been saved by having a weapon handy during a bad situation. No shots were fired because the gun's presence was enough to convince the criminals to back off.

He has degrees in both Industrial Engineering and Aviation Science.